T5-DGH-268

HANDBOOK OF INTEGRATED CIRCUITS: FOR ENGINEERS AND TECHNICIANS

HANDBOOK OF INTEGRATED CIRCUITS: FOR ENGINEERS AND TECHNICIANS

JOHN D. LENK

Reston Publishing Company
A Prentice-Hall Company
Reston, Virginia

Library of Congress Cataloging in Publication Data

Lenk, John D.
Handbook of integrated circuits.

Includes index.
1. Integrated circuits. I. Title.
TK7874.L37   621.381'73   78-7260
ISBN O-8359-2744-X

© 1978 by Reston Publishing Company, Inc.
A Prentice-Hall Company
Reston, Virginia 22090

All rights reserved. No part of this book may be
reproduced in any way, or by any means, without
permission in writing from the publisher.

10 9 8 7 6 5 4 3 2 1

Printed in the United States of America

To
Irene,
Karen,
Mark,
Brandon,
Justin,
and Mr. Lamb.

CONTENTS

CHAPTER 1   INTRODUCTION TO INTEGRATED CIRCUITS, 1

1.1.   Packaging integrated circuits, **1**
1.2.   Internal construction of integrated circuits, **3**
1.3.   Differences between IC and discrete component circuits, **13**
1.4.   Supply voltages for integrated circuits, **16**
1.5.   Temperature considerations for integrated circuits, **16**
1.6.   Basic integrated circuit types, **17**

CHAPTER 2   PRACTICAL CONSIDERATIONS FOR INTEGRATED CIRCUITS, 21

2.1.   Selecting the IC package, **22**
2.2.   Mounting and connecting integrated circuits, **23**
2.3.   Working with integrated circuits, **29**
2.4.   Layout of integrated circuits, **39**
2.5.   Power dissipation problems in integrated circuits, **43**
2.6.   Effects of temperature extremes on integrated circuits, **50**
2.7.   Power supplies for integrated circuits, **51**

CHAPTER 3   THE BASIC IC OP-AMP, 61

3.1.   Op-amp circuits, **62**
3.2.   Frequency response (bandwidth) and gain, **73**
3.3.   Op-amp characteristics, **91**

CHAPTER 4 OPERATIONAL TRANSCONDUCTANCE AMPLIFIERS (VARIABLE OP-AMPS), 107

4.1. Circuit description of typical OTA, **108**
4.2. Definition of OTA terms, **112**
4.3. Effects of control bias on OTA characteristics, **113**
4.4. Basic design considerations for OTA units, **114**
4.5. Typical OTA applications, **120**

CHAPTER 5 LINEAR APPLICATIONS FOR OP-AMPS, 151

5.1. Basic op-amp system design, **152**
5.2. Zero offset suppression, **156**
5.3. Voltage follower (source follower), **158**
5.4. Unity gain with fast response, **160**
5.5. Active filter circuits using op-amps, **163**
5.6. Audio circuits using op-amps, **171**
5.7. Other op-amp linear applications, **173**

CHAPTER 6 NONLINEAR APPLICATIONS FOR OP-AMPS, 175

6.1. Peak detectors, **175**
6.2. Oscillators using op-amps, **178**
6.3. Temperature sensors using op-amps, **182**
6.4. Linear staircase and ramp generators, **184**
6.5. Other op-amp nonlinear applications, **187**

CHAPTER 7 LINEAR IC PACKAGES AND ARRAYS, 189

7.1. IC dual differential comparator, **189**
7.2. IC differential amplifier, **202**
7.3. IC high frequency (video) amplifiers, **205**
7.4. IC audio amplifiers, **222**
7.5. IC IF amplifiers for AM/FM and FM radios, **228**
7.6. IC linear four-quadrant multipliers, **249**
7.7. IC voltage regulators, **266**
7.8. IC balanced modulators for communications circuits, **292**
7.9. IC arrays, **299**
7.10. IC phase-locked loops, **305**

CHAPTER 8   DIGITAL IC BASICS, 319

8.1.   Logic forms, **319**
8.2.   Selecting logic integrated circuits, **351**
8.3.   Interpreting logic IC datasheets, **354**
8.4.   Basic interfacing circuit, **357**
8.5.   MOS interface, **358**
8.6.   ECL and HTL level translators, **381**
8.7.   RTL interface, **385**
8.8.   Summary of interfacing problems, **388**

CHAPTER 9   SOLVING DESIGN PROBLEMS WITH INTEGRATED CIRCUITS, 389

9.1.   DC motor control with pulse width modulation, **389**
9.2.   Induction motor variable speed control, **397**
9.3.   Digitally-controlled power supplies, **403**
9.4.   Recovering recorded digital information with integrated circuits, **415**
9.5.   Power control using an IC zero-voltage switch, **430**
9.6.   Digital voltmeter using an IC dual ramp system, **442**
9.7.   Industrial clock/timer with back-up power supply operation, **451**
9.8.   Battery-powered frequency counters, **462**

PREFACE

Handbook of Integrated Circuits is an outgrowth of the author's popular *Manual for Integrated Circuit Users*. This book is written for integrated circuit (IC) users, rather than for designers of ICs. The book is, therefore, written on the basis of using existing, commercial ICs to solve design and application problems. Typical users include design specialists who want to integrate electronic units and systems, or technicians who must service equipment containing ICs. Other groups that can make good use of this approach to ICs are the experimenters and hobbyists.

The approach found in this book serves a two-fold purpose: (1) to acquaint the readers with ICs, in general, so that the users can select commercial units to meet their particular circuit requirements, and (2) to show readers the many other uses for existing ICs, not found on the manufacturer's datasheet.

This book assumes that the reader is already familiar with basic electronics, including solid-state, but may or may not have a knowledge of ICs. For this reason, Chapter 1 provides an introduction to ICs. Such topics as the how and why of IC fabrication techniques, a comparison of IC to discrete component circuits, a description of the basic physical types, and circuit types, found in commercial ICs, as well as some basic design considerations are covered.

With basics out of the way, Chapter 2 discusses practical considerations for ICs. No matter what IC is used, it must be mounted (both in experimental form for design, and in final production form), leads must be soldered and unsoldered, power must be applied, and heat sinks may be required. These subjects are described in detail.

As the reader may already know, there are two basic IC types: linear and digital. Linear ICs are discussed in Chapters 3 through 7. Digital ICs are discussed in Chapter 8. Considerable emphasis is placed on interpreting manufacturer's datasheets. For example, in Chapter 3, a typical IC datasheet is analyzed, characteristic by characteristic.

Chapter 3 covers the basic IC op-amp, whereas Chapter 4 describes the operational transconductance amplifier (OTA) or so-called "variable op-amp." Chapters 5 and 6 describe linear and non-linear applications, respectively, for the op-amps covered in Chapters 3 and 4.

Chapter 7 is devoted to Linear IC packages and arrays. Such ICs include diode, transistor, and amplifier arrays, as well as IC regulators, comparators, differential, video, IF, RF, audio and DC amplifiers, balanced modulators, and phase-locked loops (PLL).

Chapter 8 describes digital IC basics, concentrating on what type of digital ICs are available, logic forms, their relative merits, selecting logic ICs, interpreting logic IC datasheets, and basic interfacing problems common to all logic ICs.

Chapter 9 describes how digital and linear ICs can be combined to solve a variety of design and application problems. Highlights of Chapter 9 include motor control systems, recovery of recorded digital data, power control techniques, digitally-controlled power supplies, clock/timers for industrial use, digital voltmeters, and frequency counters.

This book concentrates on guidelines for the selection of ICs, and the related external components, on a trial-value basis, assuming a specified design goal and a given set of conditions. The book concentrates on simple, practical approaches to IC use, not on IC analysis. Theory is kept to a minimum. Design tradeoffs between desired performance and available characteristics are discussed from a simplified, practical standpoint.

Since the book does not require advanced math or theoretical study, it is ideal for the experimenter and the engineering designer. On the other hand, the book is suited to schools where the basic teaching approach is circuit analysis, and a great desire exists for practical design.

Many professionals have contributed their talent and knowledge to the preparation of this book. The author gratefully acknowledges that the tremendous effort to make this book such a comprehensive work is impossible for one person, and he wishes to thank all who have contributed directly and indirectly.

The author wishes to give special thanks to the following: Don Aldridge, Richard Brunner, Vern Gregory, Karl Huehne, Loren Kinsey, Terry Kiteley, Thomas Mazur, Al Mouton, Ed Renschler, Rod Russell, Neil Wellenstein, Brent Welling, Henry Wurzburg, Dave Zinder and, especially, Lothar Stern of Motorola Semiconductor products, Inc.; David Morgan and Goetz Steudel of RCA Corporation, Solid State Division; Texas

Instruments Incorporated; George Bruce of Spartanburg Technical College; Richard Castellucis of Southern Technical Institute; Robert Coughlin of Wentworth Institute and Wentworth College of Technology; Robert Dreyer; Vito Fiore of Rock Valley College; John Kelley of Capitol Radio Engineering Institute; and Joseph A. Labok of Los Angeles Valley College.

John D. Lenk

1

INTRODUCTION TO INTEGRATED CIRCUITS

Typically, an integrated circuit, or IC, consists of transistors, resistors, and diodes etched into a semiconductor material. The material is usually silicon, and is sold or used in the form of a "chip." Since all of the components are fabricated on the same chip, construction of an IC is called "monolithic." All of the devices are interconnected (by techniques similar to those used in printed circuit boards) to perform a definite function or operation. Thus, the IC concept is one of a complete (or nearly complete) circuit, rather than a group of related semiconductor devices.

To make the IC package an operable unit, it must be connected to a power source, an input and an output. In most cases, the output must also be connected to external components such as capacitors and coils, since it is not practical to combine these relatively large parts on the very tiny semiconductor chip.

1.1. PACKAGING INTEGRATED CIRCUITS

In theory, an IC semiconductor chip could be connected directly to the power source or input. However, this is not practical because of the very small size of the chip. IC chips are almost always microminiature. Instead of

1

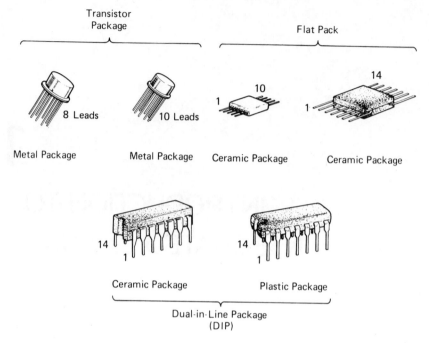

Fig. 1-1. Typical IC packages

a direct connection, the chip is mounted in a suitable package and connected to the external circuit through leads on the package.

There are three basic IC packages: the *transistor package*, the *flat-pack*, and the *dual-in-line package* (or DIP). Some typical examples of these are shown in Fig. 1-1.

In the transistor package, the chip is mounted inside a type of transistor case, such as a TO-5 style case. Instead of the usual three leads found on a discrete transistor case (emitter, collector, and base), there are 8, 10, 12, or more leads to accommodate the various power sources and input/output connections required in a complete circuit.

In the flat-pack, the chip is encapsulated in a rectangular case with terminal leads extending through the sides and ends.

In the dual-in-line package, the chip is encapsulated in a rectangular case longer than the flat-pack. In general, the DIP has replaced the flat-pack for most applications. The DIP is used for digital applications, whereas the transistor package is used with linear ICs. This is not always true, however, and there are no hard and fast rules governing IC packages.

Although there has been some attempt at standardization for IC terminal connections, the various manufacturers still use their own systems. It is therefore necessary to consult the data sheet for the particular IC when making connections to and from external circuits.

1.2. INTERNAL CONSTRUCTION OF INTEGRATED CIRCUITS

There are many methods for fabricating the semiconductor chip of an IC, and new methods are being developed constantly. Because of the many methods, and because we are primarily interested in designing with existing IC units, we will not discuss all of the methods here. Instead, we shall describe two of the most popular techniques: the bipolar (or two-junction) IC and the MOS (metal oxide semiconductor) IC. Both techniques are similar to the techniques used in fabrication of corresponding discrete transistors.

1.2.1. Bipolar (two-junction) IC construction

Figure 1-2 shows the basic internal construction and symbol of an *npn* silicon planar transistor. As shown, the starting material for the single transistor is a uniform crystal of *n*-type silicon. *P*-type silicon can be used for the starting material of the discrete transistor, and will result in a *pnp* transistor. However, most bipolar ICs are *npn*.

The collector of the transistor is formed by the *substrate* of *n*-type silicon, which has been *passivated* (or coated) with an oxide layer. A circular trench is etched out of the oxide. The trench is filled with a *p*-type crystal by a diffusion process requiring precise time and temperature control. The

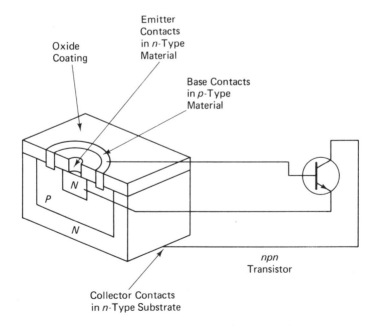

Fig. 1-2. Basic internal construction of an *npn* planar transistor

p-type material forms the transistor base element. Another disc-shaped area is etched at the center and filled (by diffusion) with an *n*-type crystal that forms the transistor emitter. The result is a discrete *npn* transistor. Metalized contacts are attached to the three elements. The oxide layer prevents shorts between the metalized contacts and protects the emitter base and collector base from contamination.

Note that the construction shown in Fig. 1-2 is for a single transistor. In an IC, the same basic process is used to fabricate many electrically isolated transistors on a single silicon substrate. Dozens or hundreds of transistors are formed on the chip of a typical IC. Figures 1-3 through 1-5 show basic process for fabrication of two transistors on an IC chip.

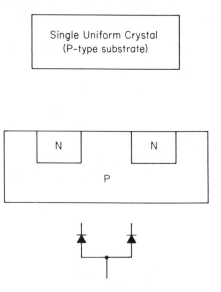

Fig. 1-3. Diffusion of *n*-type areas into *p*-type substrate to produce two diodes with common anodes and isolated cathodes

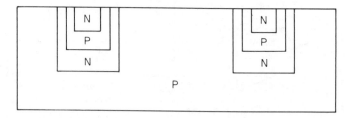

Fig. 1-4. Diffusion of *p*-type and *n*-type materials into a *p*-type substrate to form two transistors

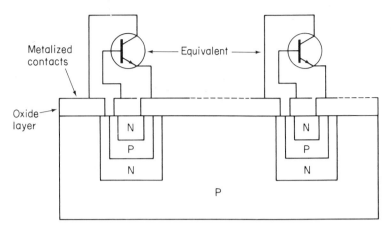

Fig. 1-5. Addition of metalized contacts to transistor elements formed in *p*-type semiconductor chip

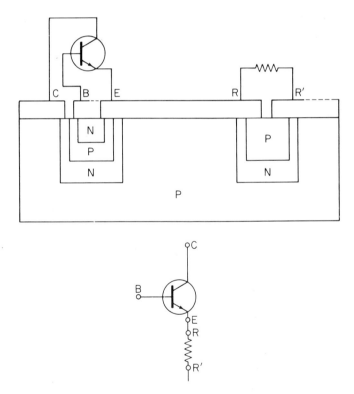

Fig. 1-6. Connection of contacts to *p*-type region to form integrated transistor and resistor

As shown in Fig. 1-3, the first step is to diffuse two regions of similar crystallike material into a substrate of dissimilar material. Here, two *n*-type regions are diffused into a *p*-type substrate. This is the starting point for the formation of two *npn* transistors on the chip. However, without further processing, the fabrication shown in Fig. 1-3 would result in two diodes with common anodes, but isolated cathodes.

Transistors are formed by diffusion of additional *n*-type and *p*-type regions, as shown in Fig. 1-4. Note that each transistor shown in Fig. 1-4 is similar to the single transistor of Fig. 1-2. The silicon substrate is then coated with an insulating oxide layer, and the oxide is opened (etched) selectively to permit metalized contacts and interconnections between elements (and between transistors) as required. With contact arrangement shown in Fig. 1-5, two separate and electrically isolated *npn* transistors are formed in a *p*-type substrate.

When resistors are required in the IC, the *n*-type emitter diffusion and two contacts are made to a *p*-type region (formed concurrently with the transistor base diffusion), as shown in Fig. 1-6. Here an *npn* transistor with a resistance connected to the emitter is integrated in a *p*-type substrate.

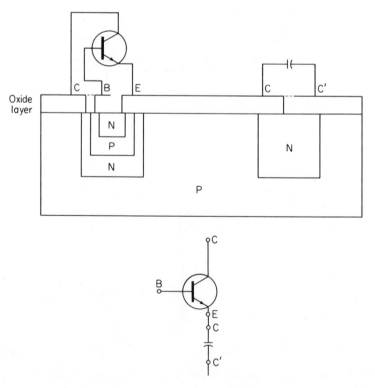

Fig. 1-7. Use of oxide as a dielectric to form integrated capacitor

When capacitors are required in the IC, the oxide itself is used as a dielectric, as shown in Fig. 1-7. Here, an *npn* transistor with a capacitance connected to the emitter is integrated in a *p*-type substrate.

Figure 1-8 shows a very simple IC with a combination of the three types of elements on a single chip. Figure 1-9 shows the physical arrangement of a typical IC semiconductor chip. The IC shown is a complete voltage regulator circuit containing approximately 24 transistors, 18 resistors, and 10 diodes.

1.2.2. MOS IC construction

The MOS principle is at present used mostly to produce some form of FET (field-effect transistor). In discrete form, the most common device is the MOSFET (also known as the IGFET, or insulated gate field-effect transistor). The most frequent use of MOS in ICs is the *complementary* MOS (generally known as CMOS, COS/MOS, or a similar term, depending on

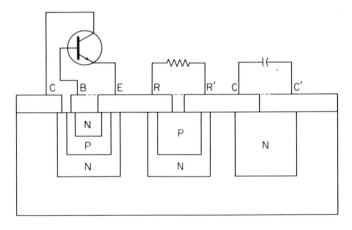

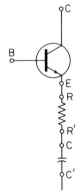

Fig. 1-8. Physical arrangement of typical IC semiconductor chip

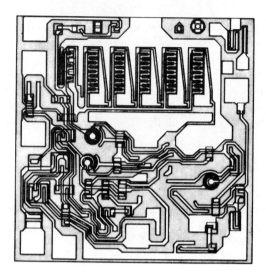

Fig. 1-9. Physical arrangement of typical IC semiconductor chip

the manufacturer). The CMOS fabrication technique involves using an
n-channel MOSFET and a p-channel MOSFET on a single chip. Some
MOS logic devices (particularly logic memories and control units) are
formed using only p-channel (PMOS) or n-channel (NMOS) fabrication
techniques.

Figure 1-10 shows the development of a n-channel MOSFET. The
basic MOSFET is essentially a bar of doped silicon, or a similar substrate
material, that acts like a resistor. The terminal into which current is injected
is called the *source*. The source terminal is similar in function to the cathode
of a vacuum tube. The opposite terminal is called the *drain* terminal and
can be likened to a vacuum tube plate. However, in a MOSFET the
polarity of the voltage applied to the drain and source can be interchanged.

In Fig. 1-10a, the substrate is a high-resistance p-type material. Two
separate low-resistance n-type regions (source and drain) are diffused into
the substrate, as shown in Fig. 1-10b. The surface of the structure is covered
with an insulating oxide layer, illustrated in Fig. 1-10c. Holes are cut into
the oxide layer, allowing metallic contact to the source and drain. The *gate*
metal area is overlayed on the oxide, covering the entire channel region.
Similar metal contacts are made to the drain and source, as shown in Fig.
1-10d. The contact to the metal area covering the channel is the gate
terminal. There is no physical penetration of the metal through the oxide
into the substrate; since the drain and source are isolated by the substrate
bulk, any drain-to-source current that occurs in the absence of gate voltage
is very low.

The metal area of the gate, in conjunction with the insulating oxide
layer and the semiconductor channel, forms a capacitor. The metal area is

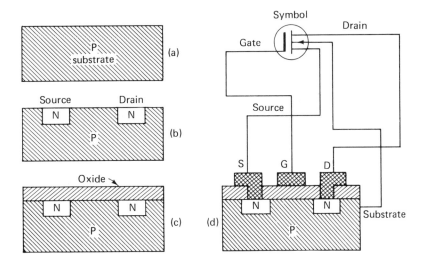

Fig. 1-10. Typical n-channel MOSFET construction and symbol

the top plate, and the substrate (bulk) material is the bottom plate of the capacitor.

Figure 1-11 shows operation of the MOSFET. Positive charges at the metal side of the metal-oxide capacitor induce a corresponding negative charge at the semiconductor side. As the positive charge at the gate is increased, the negative charge induced in the semiconductor increases until the region beneath the oxide becomes an n-type semiconductor region, and current can flow between the source and drain through the induced channel. Drain current flow is enhanced by the gate voltage and can be controlled or modulated by it. Channel resistance is directly related to the gate voltage.

Note that it is possible to make a MOSFET with a p-channel by reversing all of the material types. Likewise, it is possible to form both n-channel and p-channel MOSFETs on the same substrate. This results in the complementary COS/MOS or CMOS types used in digital circuits. Figure 1-12 shows the basic physical construction and corresponding diagrams, for both p- and n-channels of a CMOS IC. The device illustrated is representative of the CMOS IC devices manufactured by Motorola.

The p-channel device consists of a lightly n-doped silicon substrate with heavily doped p-type diffusions into this substrate. Between the drain and source is the gate-oxide region, which serves as the insulation between the metal gate and the substrate. The basic operation of the p-channel device involves placing the metal gate at a negative potential with respect to the substrate. The induced electric field causes an inversion of the n-type substrate into a p-type region. This inversion occurs only between the drain and source diffusions. The inverted area of the substrate is called the *channel*. The carriers in a p-channel are holes.

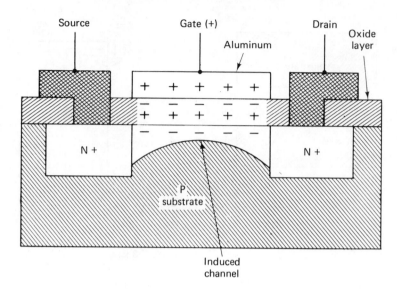

Fig. 1-11. Operation of typical MOSFET showing channel enhancement

The *n*-channel device consists of a *p*-doped silicon substrate with *n*-doped drain-and-source diffusions. When the metal gate is placed at a positive potential with respect to the substrate, an electron-dominated channel between the two diffusions is created in the *p*-type substrate, resulting in the flow of current between the drain and source. The magnitude of current flow is controlled primarily by the gate-to-substrate potential difference, or bias.

The complementary MOSFET is always biased in such a manner that the drain-to-substrate junction and source-to-substrate junction are reverse biased. Thus, the substrate is always the most positive voltage on the *p*-channel device, and is always the most negative voltage on the *n*-channel device.

Physically, the drain diffusion is identical to the source diffusion. The two diffusions, however, are usually distinguished when the device is used in a circuit. The diffusion at the *least potential difference* with respect to its substrate is called the source.

Basic MOS complementary inverter. By combining *p*- and *n*-channel devices on a single IC chip, it is possible to form a MOS complementary inverter. The basic inverter circuit is shown in Fig. 1-13. Here, the *p*-channel source is connected to the supply voltage (+V) with the *n*-channel source connected to ground. The gates of both channels are tied together and represent the input. The output is taken from both drains (also tied together).

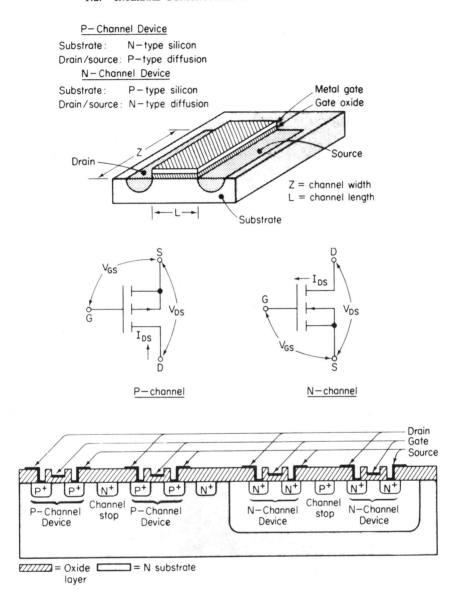

Fig. 1-12. Basic physical construction and diagrams for *p*- and *n*-channel CMOS IC

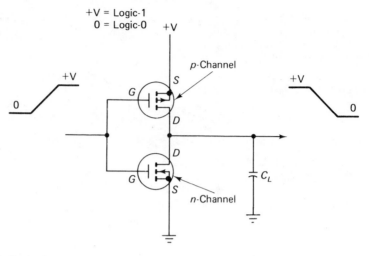

Fig. 1-13. Basic MOS complementary inverter (CMOS inverter)

As discussed, the complementary inverter circuit is used extensively in digital logic ICs. In the circuit of Fig. 1-13, the logic levels for the inverter are $+V$ for a logic 1 and ground for a logic 0. With a 1 input ($+V$), the p-channel section of the substrate has a zero gate voltage and is essentially cut off. The p-channel conducts very little drain current (typically a few picoamperes of leakage current). The n-channel section of the IC chip is forward biased and its drain voltage (with only a few picoamperes of leakage drain current allowed to flow) is near ground or at a logic 1. The load capacitance C_L represents the output load, plus any stray circuit capacitance.

With a 0 input (ground), the n-channel element is cut off and permits only a small amount of leakage drain current to flow. The p-channel element is forward biased, thus making the p-channel drain at some voltage near $+V$. Capacitor C_L is charged to approximately $+V$.

No matter which logic signal is applied at the input (or appears at the output) power dissipation is extremely low. This is because both stable states (1 and 0) are conducting only a few picoamperes of leakage current (since both channels are in series, and one channel is always cut off). Power is dissipated only during switching, an ideal situation for logic circuits.

In addition to the lower power dissipation, another advantage of MOS devices for digital logic circuits is that no coupling elements are required (the gate acts as a coupling capacitor). Since no capacitors are needed, it is relatively simple to fabricate MOSFET logic elements in IC form (fabrication of capacitors is often a major stumbling block for IC design). OR, NOR, AND, and NAND gates, with either positive or negative logic, can be implemented with MOSFETs. Thus, almost any logic circuit combination can be produced using the basic MOS complementary inverter. MOSFET logic can also be used over a wide range of power-supply voltages.

1.3.   DIFFERENCES BETWEEN IC AND DISCRETE COMPONENT CIRCUITS

Although the basic circuits used in ICs are similar to those of discrete transistors, there are certain differences. For example, inductances are not found as part of an IC. It is not practical to form a useful inductance on a material that contains transistors and resistors. Likewise, large-value capacitors (about 100 pF and above) are not usually formed as part of an IC. When a large-value capacitor, or an inductance of any type is a necessary part of a circuit, these components are part of the external circuit.

   Integrated circuits often use direct-coupled circuits to eliminate capacitors. This makes it possible to operate at low frequencies (down to direct current in most cases). Figure 1-14 shows how a transistor (Q_3) is used to eliminate the need for a capacitor between Q_1–Q_2 and Q_4–Q_5. By eliminating the capacitor, the frequency range of the circuit is extended at both the low and high ends.

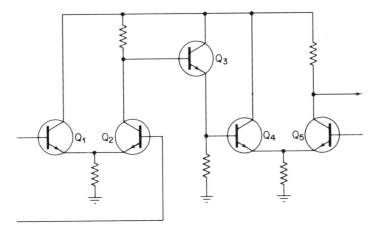

Fig. 1-14. Transistor Q_3 substitutes for capacitor in a typical IC

   Transistors are often used in place of resistors in IC packages. Usually such a transistor is a FET since the basic FET acts somewhat like a resistor, as discussed in Sec. 1.2.2. Figure 1-15 shows how a FET can be substituted for a resistor in an IC. In this circuit, the FET gate is returned to one side of the supply. With such an arrangement, the FET takes up less space than a

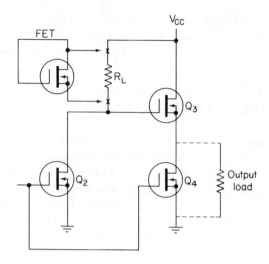

Fig. 1-15. How a FET can be substituted for a resistor in an IC

corresponding resistor and provides a much higher power dissipation capability. Integrated transistors can also be connected to form diodes.

Although IC transistors are essentially the same as discrete transistors (except for some added capacitance produced across the substrate and transistor junctions), integrated resistors are significantly different from discrete versions. Discrete resistors are normally made in standard forms, and different values are obtained by variations in the resistivity of the material. In ICs the resistivity of the material cannot be varied. Thus, the value of the resistor depends primarily on its physical shape. An IC resistance value R is determined by the product of its diffusion-determined sheet or chip resistance R_S, and the ratio of its length L to its width W (that is, $R = R_S \times L / W$). As a result, small-value resistors are short and squat, whereas large-value resistors are long and narrow.

The value of an integrated capacitor C is equal to the product of its area A and the ratio of dielectric constant E to the thickness D of the oxide layer (that is, $C = A / (E / D)$). Because D is kept constant, capacitor values vary directly with area.

As a point of reference, a 1000 Ω IC resistor occupies about twice as much area as a bipolar IC transistor, whereas a 10 pF capacitor occupies three times the area of a transistor. MOS ICs require only one-third of the process steps needed for bipolar IC. The most significant feature of MOS ICs is the large number of semiconductor circuit elements that can be put on a small chip. The size relationship of MOS and bipolar ICs is shown in Fig. 1-16. This high circuit density available in MOS means *large-scale integration*

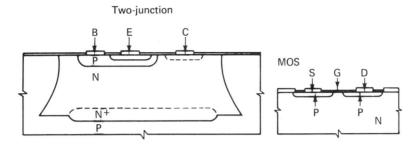

Fig. 1-16. Comparison of two-junction and MOSFET size

(LSI), instead of *medium-scale integration* (MSI), or *small-scale integration* (SSI) found in bipolar ICs. For example, it is possible to put 5,000 devices on a silicon chip only 150×150 mils square. Each transistor in a typical MOS/LSI array requires as little as 1 square mil of chip area, a great reduction over the bipolar IC transistor which requires about 50 square mils.

MOS/LSI has other advantages over bipolar MSI or SSI. These include lower cost per circuit function, fewer subsystems to test, fewer parts to assemble and inspect, increased circuit complexity per package, lower power drain per function, and a choice of standard or custom products.

The last advantage is of particular importance to the designer. Most major MOS device manufacturers offer a complete line of standard logic ICs. Typically, the line will include gates, switches, registers, dividers, counters, generators, memories, coders, decoders, and general purpose logic. The great majority of design problems can be solved with these standard, off-the-shelf devices.

In addition to the standard logic ICs, many MOS manufacturers offer a custom production service. That is, they will produce complete logic devices from the customer's logic drawings. In effect, the customer draws the desired logic, defines the inputs and outputs, and describes the test procedures. This information is then sent to various MOS manufacturers for bids on completed hardware. Some MOS suppliers also provide software for both standard and custom devices. Likewise, manufacturers will sometimes produce MOS logic elements from customer's software.

From this description, it would appear that the customer need only tell the manufacturer what is wanted, check the bids, and then wait (indefinitely) for the finished hardware. However, it is necessary that the customer or designer have a working knowledge of MOS logic devices in order to make an intelligent comparison of the manufacturer's services. For example, the designer should know the switching characteristics of MOS devices, and the basics of complementary logic.

1.4. SUPPLY VOLTAGES FOR INTEGRATED CIRCUITS

Typical IC supply voltages are generally less than 35 V. However, future development of IC fabrication and packaging methods may increase this figure. The limiting factor in maximum IC supply voltages is usually the collector-to-emitter breakdown voltage of the transistors.

Typically, supply voltages are applied to ICs in the same polarity required for *npn* transistor stages. That is, the collector voltage is positive, with the emitter at ground, zero voltage, or at some negative value. If the polarity is reversed, the normally reverse-biased collector-to-substrate insulation junction will conduct very heavily and cause a portion of the metalized contact pattern to be destroyed. For this reason, a protective diode is often used in the dc supply voltage line.

Any length of wiring or ribbon on a printed-circuit (PC) board has inductance that can develop significant voltages in response to high-frequency or fast-rise currents. Such voltages are added to the dc supply voltage of an IC mounted on the board. In addition, the supply-lead metalization of the IC may develop high-frequency signals as a result of stray internal feedback. Adequate compensation for both effects can be achieved by bypassing the supply leads of high-gain units with small external capacitors.

The practical considerations for connecting power supply leads to ICs are discussed fully in Chapter 2.

1.5. TEMPERATURE CONSIDERATIONS FOR INTEGRATED CIRCUITS

The basic temperature considerations for ICs are essentially the same as those for discrete component circuits. The heat dissipated by the circuit components must be transferred to the outside of the package without the temperature at any point on the circuit chip becoming excessive.

In an IC, the heat is dissipated on top of the silicon chip. The heat sources, therefore, are highly localized, with the exact distribution determined by internal circuit layout. Because the silicon chip mounted on a metal header is a good heat conductor, the heat rapidly diffuses throughout the chip, and the entire chip may be considered to be essentially at the same temperature. It is more meaningful to examine the dissipation capability of the overall chip than to determine the limits of each of the various regions on the chip. The dissipation capability of a monolithic silicon circuit chip is determined primarily by the encapsulating material, the chip mount, the terminating leads, and the volume and area of the IC package.

Generally, heat transfer in an IC is conducted through the silicon chip and through the case. The effects of internal free convection and radiation, and lead conduction are small and may be neglected. The value of thermal resistance from chip to case is dependent upon the chip dimension, the package configuration, and the location of the selected case reference point. The effect of temperature on an IC transistor is different from that on an IC resistor. The effect for resistors is generally detrimental, whereas the effect for transistors is generally beneficial.

Integrated transistors on the same circuit chip have a number of advantages over discrete units as a result of their proximity. In fabrication, adjacent transistors receive almost identical processing and thus are closely matched in characteristics. Because of the close spacing, minimum temperature differences occur between components, and this close match is maintained over a wide operating range. In addition, ICs can contain many more transistors per given area than discrete components. (Typically, six bipolar IC transistors, or 300 MOS IC transistors, occupy the same area as one discrete transistor.)

Integrated resistors have a relatively large value variation with temperature. Typically, IC resistors will vary ±25%, and can vary as much as −50% to +100%. This temperature dependence makes it difficult to achieve close tolerances on absolute values of resistors. However, the ratios of IC resistors can be closely controlled during the fabrication process. As a result, it is desirable that IC design be made dependent on ratios rather than on absolute values of resistors.

The practical considerations for power dissipation and thermal design problems of ICs are discussed in Chapter 2.

1.6.   BASIC INTEGRATED CIRCUIT TYPES

There are two basic types of ICs: linear and digital.

1.6.1.   Linear integrated circuits

Linear ICs are the IC equivalents of basic transistor circuits. Examples are amplifiers, oscillators, mixers, frequency multipliers, modulators, limiters, voltage regulators. Although linear ICs are complete functioning circuits, they often require additional external components (in addition to a power supply) for satisfactory operation. Typical examples of such external components are: a resistor to convert a linear amplifier into an operational amplifier, a resistor–capacitor combination to provide frequency compensation, and a coil–capacitor combination to form a filter (for bandpass or bandrejection).

There are three general groups of linear ICs: the operational amplifier (or op-amp) and operational transconductance amplifier (OTA), the special purpose IC (such as a voltage regulator), and the linear component arrays. Generally, linear ICs use bipolar fabrication techniques. However, MOS is used, and there has been a trend to combine bipolar and MOS fabrication for certain linear ICs. An example is the RCA BiMOS op-amp which has MOS input, bipolar intermediate stages, and MOS output.

Op-amp and OTA ICs.   The most common form of linear IC is the operational amplifier. Such amplifiers are high-gain, direct-coupled circuits where gain and frequency response are controlled by external feedback networks. With these networks, the op-amp can be used to produce a broad range of intricate transfer functions and thus may be adapted for use in many widely differing applications. Although the op-amp was originally designed to perform various mathematical functions (differentiation, integration, analog comparisons, and summation), the op-amp may also be used for many other applications. For example, the same op-amp, by modification of the feedback network, may be used to provide the broad, flat frequency-gain response of video amplifiers, or the peaked responses of various types of shaping amplifiers. This capability makes the op-amp the most versatile configuration used for linear ICs.

The OTA is often called a variable op-amp since its operating characteristics can be controlled by an external bias. Like a transistor, the OTA permits the user to vary not just voltage but also power, bandwidth, slew rate, input current, and output current. Unlike a conventional op-amp where the design characteristics are set during manufacture, the OTA can be programmed and/or the signal modulated to select the optimum gain, speed, bandwidth, power, and so on needed for a specific design.

The major design characteristics of the op-amp and OTA are described in Chapters 3 and 4, respectively. Chapters 5 and 6 cover typical design applications for both the op-amp and OTA.

Special purpose linear ICs.   These ICs are usually designed to replace several stages of discrete-component circuits. Typical examples include voltage regulators, modulators, IF strips in AM or FM radio receivers, sound circuits (IF amplifier-limiters, discriminator and audio voltage amplifiers) in TV receivers, remote amplifiers and similar specialized multistage circuits. With few exceptions, special purpose ICs can provide multiple circuit functions at performance levels equal to or greater than those of their discrete-component counterparts.

Linear component arrays. Such arrays consist of groups of unconnected active devices fabricated on a single chip, and mounted in a common IC package. Typical arrays include groups of transistors, diodes, Darlington pairs, constant-current differential amplifiers, and combinations of devices such as a Darlington pair, Zener diode, programmable unijunction transistor, bipolar transistor, and SCR on a single chip.

The components of an array that are fabricated simultaneously in the same way on a silicon chip have nearly identical characteristics. The characteristics of the various components track each other with temperature variations because of the proximity of the components and the good thermal conductivity of silicon. Thus, IC arrays are especially suited for applications in which closely matched device or circuit characteristics are required and in which a number of active devices must be interconnected with external components, such as tuned circuits, large-value or variable resistors, and large bypass or filter capacitors. For example, diode arrays are particularly useful in the design of bridge rectifiers, balanced mixers or modulators, gating circuits, and other configurations that require identical diodes.

Transistor arrays make available closely matched devices that may be used in a variety of circuit applications (for example, push–pull amplifiers, differential amplifiers, multivibrators, and dual-channel circuits). The individual transistors in the array may also be used in circuit stages that are located in differential signal channels, or in cascade or cascode circuits. Arrays of individual circuit stages are very useful in design of equipment that has two or more identical channels, such as stereo amplifiers, or they may be interconnected by the use of external coupling elements to form cascade circuits.

Chapter 7 describes the major design characteristics for a number of special purpose ICs, as well as for IC component arrays. Chapter 7 also describes some typical applications for special purpose ICs and component arrays.

1.6.2. Digital integrated circuits

Digital ICs are the integrated circuit equivalents of basic transistor logic circuits. Like their discrete component counterparts, digital ICs are used in computers and other digital electronics to form such circuits as gates, counters, registers, coders, decoders, dividers, memories, etc. A digital IC is a complete functioning logic network, usually requiring nothing more than an input, output, and power source. However, the interconnection of digital IC packages to form logic systems often requires considerable analysis in design.

Digital circuits are generally repetitive and concerned with only two levels of voltage or current. Thus, digital ICs do not require accurate control of transitional-region characteristics (transconductance linearity, for example). As a result, digital ICs are standardized into basic designs and are produced in large quantities as low-cost, off-the-shelf devices.

Both bipolar and MOS fabrication techniques are used for digital ICs, with the trend toward MOS. The major design considerations for digital ICs are discussed in Chapter 8. A cross section of digital and linear IC design applications is given in Chapter 9.

2

PRACTICAL CONSIDERATIONS FOR INTEGRATED CIRCUITS

It may appear that there are few practical considerations for ICs. If you are to believe the data found in some engineering literature, an IC requires only that the power be applied for the package to be ready for immediate use. While this is essentially true, there are a number of points to be considered in selecting and using ICs. For example, each of the three basic package types has certain advantages and disadvantages. These should be considered when selecting ICs to meet specific design requirements. Once the package type has been chosen, there are various alternate methods of mounting the IC to be considered.

For the technician who must work with ICs, removal and replacement of the packages (unsoldering leads, bending leads to fit existing mounting patterns, etc.) can present problems. For the designer who must work with ICs in the experimental stage, there are problems of oscillation, circuit "ringing" and noise. As in the case with any solid-state circuit, there are always temperature and power-supply limitations to be considered.

All of these problems have some effect on ICs, and the problems are discussed in this chapter as a starting point for the reader who is interested in the overall practical side of ICs. Considerations that apply to a specific IC type or circuit are discussed in the related chapters.

2.1. SELECTING THE IC PACKAGE

As discussed in Chapter 1, there are three basic types of IC packages, TO-5 (metal can) style, flat-pack, and dual-in-line (DIP). There is a further division in that the dual-in-line packages are available in both ceramic and plastic. In some cases, the designer has no choice of package style since the particular IC is available in only one style. For example, where the IC operates at high power, and dissipates considerable heat, the metal can is required, because it permits the use of heat-sinks and/or possible mounting directly on a metal chassis. In other cases, the choice of package style is wide open. The following notes are included to help the designer make this choice.

If cost is a major factor, or if a large volume of ICs is required, the DIP is generally the best choice. DIPs are ideally suited for mounting on PC boards, since there is more spacing between the leads (typically 0.1 inch) than with other package types. During production, DIPs can be inserted manually or automatically into mounting holes on PC boards, and soldered by various mass-production techniques.

The choice of ceramic versus plastic is again a matter of cost versus reliability. In general, a ceramic IC will provide a better hermetic seal (to protect the silicon chip), but at a higher cost than plastic. There are exceptions to this rule of course. For example, a ceramic IC with poor plating on the leads could fail faster than a plastic IC with good plating.

The real *weak spot* in any IC package is at the point where *the leads enter the case or body.* Usually, there is a glass or plastic seal at these points. The seals can be broken, thus exposing the chip and unplated metal inside the package if the leads are bent or twisted during production or repair. Moisture and other undesirable elements can then enter the IC package. While this may not cause immediate failure, it will almost certainly cause ultimate failure. If nothing else, the exposed bare metal will corrode, and affect the IC performance.

One method used to offset the effects of possible lead-seal damage during production is to bake the entire PC board after all the ICs have been mounted and soldered. This technique applies only to plastic ICs. Typically, the bake cycle is about one hour at 125 °C (or at lower temperatures for longer periods of time, depending on the heat characteristics of the IC and of other components mounted on the same board). The effect of the prolonged bake is to remove any moisture and to improve the seal. Of course, the technique will be of no value to an IC with a seriously damaged lead-seal.

If reliability is the major factor, the ceramic flat-pack is generally the best choice. Flat-packs have an excellent history of reliability. Likewise, flat-packs are smaller and lighter than DIPs, all other factors being equal.

Ceramic flat-packs are usually the choice for any airborne application, except where high power is involved. (High power requires a metal can.)

As in the case of any solid-state device, it is always good practice to check the manufacturer's test data regarding failure rates of package types in special environments. For example, if the IC must be operated in the presence of thermal shock (extreme high temperatures followed by immediate extreme low temperatures), then check failure rates (due to thermal shock) of different package types. Often, the manufacturer can provide considerable data regarding failure rates, reaction to environments, etc., that will prove very helpful in selecting IC package types.

2.2. MOUNTING AND CONNECTING INTEGRATED CIRCUITS

Once the package type has been selected, the IC must be mounted and connected to other components. The selection of a particular method for mounting and connecting ICs depends on the type of IC package, the equipment available for mounting and interconnection, the connection method used (soldering, welding, crimping, etc.) the size, shape, and weight of overall equipment package, the degree of reliability, and ease of replacement, and on the never-to-be-forgotten cost factor. The following sections summarize mounting and connection methods for the three basic IC package types.

2.2.1. Socket mounting and connection

During the experimental stage of design, any of the IC packages can be mounted in commercially available sockets. This eliminates soldering and unsoldering the leads during design and test. Such sockets are generally made of Teflon or some similar material, and are usually designed for mounting on a PC board. However, some IC sockets are designed for metal chassis mounting. In other cases, the IC can be soldered to the socket that is in the form of a plug-in PC card. The card is then plugged into or out of the circuit during design and testing. Sockets manufactured by Azimuth Electronics, Barnes Development Co., Jefferson Products, Inc., Heath, and Sealectro Corp. are typical of the temporary mounts required during design. In some cases, the same mounts can be used on a permanent basis if required.

2.2.2. Ceramic flat-pack mounting and connection

Figure 2-1 shows five methods for making solder connections to flat-packs. Note there is a notch in one end of the package. This is a reference point to

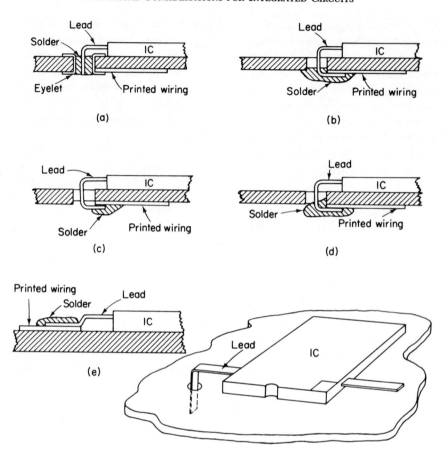

Fig. 2-1. Typical soldering techniques for flat-pack ICs

identify the lead numbering. Usually, but not always, this reference notch is nearest to lead No. 1. Always consult the manufacturer's data regarding IC lead numbering. Although there has been some attempt at standardization, most manufacturers prefer their own lead identification system.

In the *straight-through method* (Fig. 2-1a), the leads are bent downward at a 90° angle and are inserted in the circuit board holes. When assembled during manufacture, all leads are connected simultaneously by dip soldering or wave soldering. During repair, the leads are soldered one at a time. A disadvantage of the straight-through method is that the IC package must be held firmly in position during the soldering operation.

The *clinched-lead, full-pad soldering method* (Fig. 2-1b) requires an additional operation (clinching the lead), but has the advantage that the IC does not have to be held in position during soldering.

The *clinched-lead, offset-pad* (Fig. 2-1c) and the *clinched-lead, half-pad* (Fig. 2-1d) methods are variations of the clinched-lead, full-pad methods since the hole is not filled with solder.

In the *surface-connection method* (Fig. 2-1e) the connections are made on the package side of the board. This method is often used when ICs must be mounted on both sides of the board. No holes are required in the board. The *reflow solder technique* is often used when surface-connection ICs are assembled during production. With reflow soldering, both the surface contacts and the leads are tinned and covered with solder. Then the leads are set on the surface contacts, and heat is applied to all contacts (or all contacts on one side). The heat causes the solder to reflow and make a good connection between leads and surface contacts.

The mounting patterns shown in Fig. 2-1 use *in-line lead and pad arrangements*. Although such arrangements simplify lead forming, they result in very close spacing between leads (typically 32 mils) and require the use of high-precision production techniques in both board manufacture and assembly of ICs on the board, particularly when the leads must be inserted through holes in the PC board. Another disadvantage of the in-line arrangement is the limited space available for routing circuit conductors between adjacent solder pads.

Some of these disadvantages can be overcome by the use of *staggered lead arrangements*, shown in Fig. 2-2c. In these staggered arrangements, the lead holes and terminal pads for adjacent leads on the same edge of a flat package are offset by some convenient distance from the in-line axis. Although a staggered lead arrangement requires somewhat more PC board area per IC than the in-line arrangement, staggered leads provide several advantages: (1) tolerances are far less critical; (2) larger terminal pads can be used; (3) even with larger pads, more space is available for routing circuit conductors between adjacent terminal connections; and (4) larger lead holes can be used to simplify lead insertion.

In a staggered lead arrangement, a good compromise between loss of available PC board area and gain in the number of conductors that can be routed between adjacent circuit terminal pads can be achieved by the use of an offset between adjacent pads. Also, the staggered lead arrangement provides great flexibility in circuit wiring configuration. For example, if offset lead holes 30 mils in diameter are used with solder pads 80 mils in diameter, then at least one 8 mil wide printed conductor can be routed between adjacent solder pads, assuming the dimensions of Fig. 2-2c. If 60 mil diameter pads are used, up to two 8 mil wide printed conductors can be routed between adjacent pads.

For some applications, it may be necessary or desirable to use *welded connections* rather than soldered connections. Figure 2-3 shows three methods that may be used for making such welded connections. In general, the

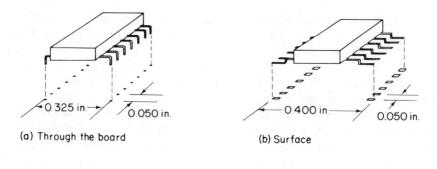

(a) Through the board   (b) Surface

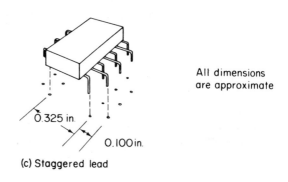

All dimensions
are approximate

(c) Staggered lead

Fig. 2-2. Typical mounting patterns for flat-pack ICs

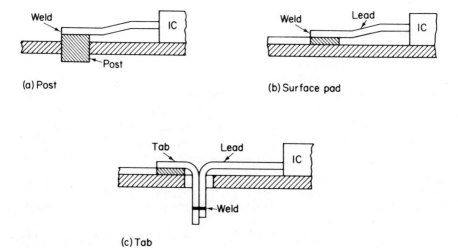

(a) Post   (b) Surface pad

(c) Tab

Fig. 2-3. Typical welding techniques for flat-pack ICs

mechanical space considerations described for solder connections apply equally well to welded connections.

Although for some applications welding may provide more reliable connections than soldering, welding has a disadvantage in that with conventional welding equipment only one connection can be made at a time.

The "tab" method of welding shown in Fig. 2-3c is good from a maintenance standpoint because the IC can be easily removed if replacement becomes necessary. The tab method used "cross-wire" resistance welding in which the weld is made at the ends of the IC lead and the terminal tab. The IC can thus be removed simply by clipping the leads just above the weld point. However, the replacement process is not as simple. The IC lead must be soldered to the tab.

2.2.3.   TO-5 (metal can) style package mounting and connection

The most direct method for mounting TO-5 style packages is shown in Fig. 2-4a. Here, the leads are simply inserted in the proper plated-through holes in the PC board, and connection is completed by dip- or wave-soldering.

Although the method of Fig. 2-4a requires minimum handling of the IC (trimming of terminal leads to approximate lengths may be necessary), it

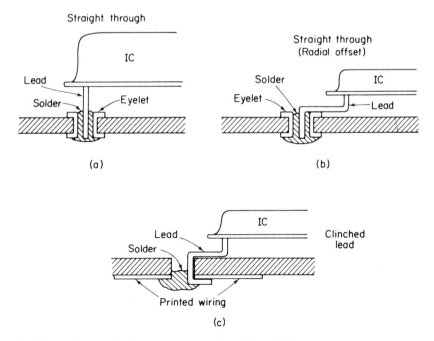

Fig. 2-4. Typical soldering techniques for TO-5 style ICs

does require extremely precise drilling and "through-plating" of the lead holes and preparation of the solder pads. The method also has disadvantages, namely that automatic insertion of the device leads in such limited space can present problems, and that the device must hold in position during the soldering operations.

The methods shown in Fig. 2-4a and 2-4b make it possible to achieve effective "wicking" of the solder around the lead. Figure 2-4c shows a method in which the holes are not plated through and the leads are clinched before the soldering operation is performed. Because the electrical connection depends upon the solder being on the pad and not in the hole, the lead-hole diameter can be made larger, and therefore this method permits easier lead insertion (possibly with an automatic tool during production). Also, the clinching of the leads helps to hold the IC in position during soldering.

Figure 2-5 shows the lead hole arrangement for straight-through mounting of the 10-lead TO-5 style package. This arrangement provides approximately 11 mils clearance between adjacent pads, the smallest clearance possible without danger of shorting. This separation is insufficient to accommodate a printed conductor of conventional size (8 mils wide with 8 mils spacing).

Figure 2-5 also shows a radially offset lead-circle pattern for a 12-lead TO-5 style package in which the leads are formed to increase the effective lead-circle diameter (530 mils). This configuration permits the use of 80 mil diameter pads with sufficient spacing between pads to accommodate three 8 mil wide printed conductors. The radially offset pattern also permits clinched-lead mounting and represents a good compromise from the standpoint of mounting arrangement, reliability, ease of maintenance, and cost.

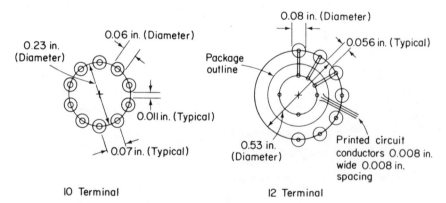

Fig. 2-5. Typical mounting patterns for TO-5 style ICs

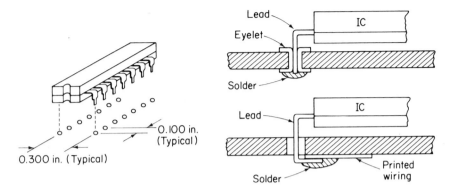

Fig. 2-6. Typical mounting pattern and soldering techniques for dual-in-line (DIP) ICs

2.2.4. DIP package mounting and connection

Figure 2-6 shows the mounting arrangement used for DIP ICs. Because the package configurations are very similar, the mounting arrangement and terminal-sorting techniques used for DIPs are much the same as those used in the in-line method (Fig. 2-2a) for the flat-pack ICs. The DIP IC is longer, however, and the soldering operation is made simpler because of the greater spacing provided between lead terminals.

The terminals of the DIP may be soldered to a PC board by any of the through-the-board techniques (shown in Figs. 2-1a through 2-1d) used for the in-line method of mounting ceramic flat-packs. The DIP terminal leads are larger than those of the flat-pack. The larger-size terminals are more rigid and more easily inserted in the mounting holes (either in the PC board or in the IC sockets).

Another significant feature of the DIP is the sharp step increase in width of the terminals near the package. This step forms a shoulder upon which the package rests when mounted on the board. Thus, the package is not mounted flush against the board, and it is possible to run printed circuit wiring directly under the package. Also, convection cooling of the package is increased, and the IC can be more easily removed if replacement is required.

2.3. WORKING WITH INTEGRATED CIRCUITS

It is assumed that the reader is familiar with common handtools used in electronics, such as diagonal wire cutters, long needle-nose pliers, soldering tools (both pencil type and soldering guns), insulated probes, nut drivers and wrenches. These same tools are used in IC work. However, certain additional tools and techniques are also required (or will make life much easier if used).

The main problem in working with ICs, besides the small size, is *heat*. Typically, ICs are made of silicon, which cannot withstand high temperatures for any length of time. A temperature of 200 °C is about tops for any IC. The leads cannot be subjected to continuous and excessive heat when soldering and unsoldering leads.

Practically all ICs are miniature. This demands the use of small soldering tools, pliers with fine points, and very delicate handling in general. The problems of fine wire leads and seals where the leads enter the packages are always present when working with ICs.

The following paragraphs summarize the major problems found in working with ICs during experimental design, production, and repair.

2.3.1. Lead bending

In any method of mounting ICs that involves bending or forming of the leads, it is extremely important that the leads be supported and clamped *between the bend and the seal*, and that bending be done with extreme care to avoid damage to lead plating. Long-nose pliers can be used to hold the lead, as shown in Fig. 2-7. In no case would the radius of the bend be less than the diameter of the lead. In the case of rectangular leads, the radius of the bend should be greater than the lead thickness. It is also extremely important that the ends of the bent leads be perfectly straight and parallel to assure proper insertion through the holes in the PC board.

2.3.2. Working with PC boards

ICs are often mounted on PC or etched circuit boards. Likewise, ICs may be mixed with other components (resistors, transistors, coils, etc.) on the same PC board. In turn, the board may be bolted to a main chassis or plugged into a connector on the chassis. In either case, the component leads must pass through eyelets (sometimes known as *soldering cups*, or simply *holes*) in the board. The eyelets make contact with the printed or etched wiring. The component leads must be soldered to the eyelets as shown in Fig. 2-8.

When properly executed, removal and replacement of components in a PC board is not impossible, although it may appear that way in the case of ICs. The following techniques are recommended to replace components on etched or printed boards. The techniques will be satisfactory for most practical applications. Of course, you must modify the procedure as necessary to fit the particular equipment

Basic PC rework procedure.   It is best to remove the board from the equipment, unless the back of the board is accessible. Unless specified

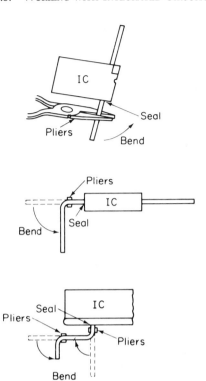

Fig. 2-7. Lead bending techniques for ICs

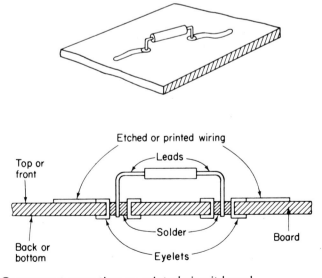

Fig. 2-8. Component mounting on printed circuit board

otherwise in the service literature, use electronic grade 60/40 solder and a 15 W pencil soldering tool. A higher-wattage soldering tool, if applied for too long a period, can ruin the bond between the etched wiring and the insulating base material by charring the glass epoxy laminate (or whatever is used for the base). However, a 40 to 50 W soldering tool can be used if the *touch-and-wipe* technique is followed (as described later in this chapter). The author prefers a chisel tip for the soldering tool, about one-sixteenth inch wide. However, this is a matter of personal preference.

If the component is to be removed and replaced with a new part, cut the leads near the body of the component as shown in Fig. 2-9. This will free the leads for individual desoldering. Grip the lead with long-nose pliers; apply the tip

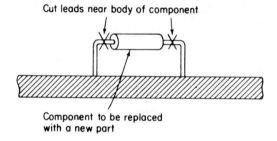

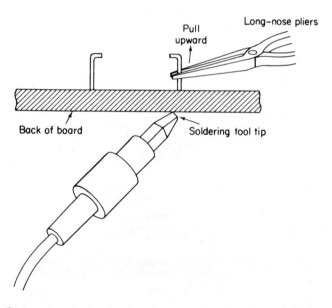

Fig. 2-9. Basic printed circuit rework procedure (when component is to be replaced)

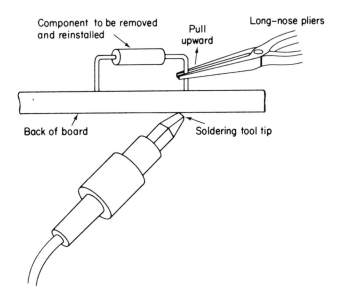

Fig. 2-10. Basic printed circuit rework procedure (when component is to be removed and reinstalled)

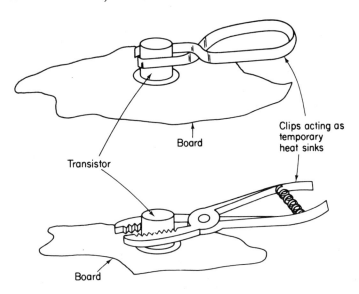

Fig. 2-11. Using clips as transistor heat sinks during soldering and desoldering

of the soldering tool to the connection at the back of the board; and then pull gently to remove the lead.

If the component is to be removed for test and possibly reinstalled, do not cut the leads. Instead, grip the lead from the front with long-nose pliers and apply the soldering tool to the connection at the back of the board. Lift the lead straight out as shown in Fig. 2-10.

Note that the pliers will provide a heat sink for the component being removed. This will keep the component from heating if it is necessary to apply the soldering tool for a long time (which should be avoided). If an IC or transistor with a metal case is to be desoldered, an alligator clip attached to the case (shown in Fig. 2-11) will provide a temporary heat sink. Soldering heat sinks are also available commercially.

When the lead comes out of the board, the lead hould leave a clean hole. If not, the hole should be cleaned by reheating the solder and placing a sharp object such as a toothpick or enameled wire into the hole to clean out the old solder. Some technicians prefer to blow out the hole with compressed air. This is *never* recommended when the board is in or near the equipment, since the solder spray could short other circuits or parts in the equipment.

Removal and replacement of ICs on PC boards.   It may appear that an IC cannot be removed from a PC board or socket without destroying or seriously damaging the IC. In any package style, all leads must be desoldered before the IC can be removed. This brings up obvious problems. You cannot use the standard desoldering procedure discussed in foregoing sections (pulling each lead with long-nose pliers while applying heat to the connection). That is, such a procedure is not possible in some cases (due to arrangement of the leads) and is highly impractical in most cases.

There are two practical solutions to the problem. First, you can use a *desoldering tool tip* that will contact all of the lead connections simultaneously. Such tips are shown in Fig. 2-12. There are commercial versions of IC desoldering tips, or you can make your own by cutting or grinding a conventional tip to fit the particular need.

As an alternative procedure for removing ICs, you can use the *solder gobbler*, which was developed originally as a desoldering tool for ICs. There are many versions of solder gobblers (also known as *vacuum desolderers*). The tool shown in Fig. 2-13 is hand-operated and is used mostly for repair work. There are vacuum-pump-operated models available, used mostly for assembly-line work.

As shown in Fig. 2-13, the solder gobbler tool consists of a soldering tool, collector tip, and bulb. In use, the bulb is squeezed, the collector tip is placed on the solder area, and when the solder is molten, the bulb is released, drawing the solder into the collector tip. Then the solder can be forced from the tip by squeezing the bulb again. The tool can also be used for soldering.

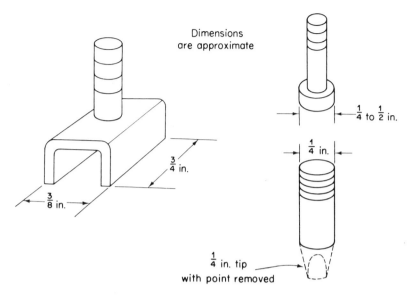

Fig. 2-12. Typical desoldering tool tips for IC removal

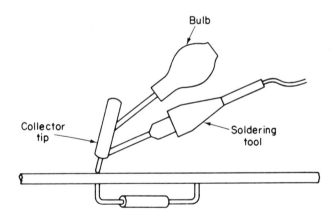

Fig. 2-13. Typical hand-operated desoldering tool (solder gobbler)

The main advantage of the special desoldering tip is speed. The main disadvantage is the excess heat that might damage the IC semiconductor material. When using a special tip that covers all contacts simultaneously, be ready to remove the IC immediately (that is, just as soon as the solder is molten). It is not necessary to use the desoldering tips for soldering the IC back in place. This produces considerable unnecessary heat.

The main advantage of the solder gobbler is the absence of excess heat. Also, the solder will be removed from the hole (usually). Of course, it takes

more time to remove an IC with a solder gobbler than it does to remove on
with a desoldering tip.

When installing a replacement IC, you must exactly follow the mounting pattern (including lead bending, if any) of the original IC. We say
exactly since there is rarely enough space in IC equipment to do otherwise.

Touch-and-wipe soldering of ICs.   As an alternate to desoldering tips
or solder gobblers for ICs, the touch-and-wipe method can be used. With
this method, a hair-bristle soldering brush, that has its bristles shortened to
about one-quarter inch in length, and a conventional soldering tool of 50 W
or less is used.

The tip of the tool is touched to the area where the IC lead comes
through the board, with the brush ready to wipe the melted solder away.
The solder is removed by a series of touch-and-wipe operations. The tool is
touched to the board only long enough to melt the solder and is removed as
the brush is wiped across the joint. After a few touch-and-wipe operations,
the IC lead should be free of the solder pad or eyelet on the board.

Miscellaneous soldering techniques.   ICs are often used with discrete
components. The following notes apply to such components.

In the soldering of large metal terminals (switch terminals, potentiometer terminals, etc.) found in solid-state equipment, ordinary 60/40 elec-
tronic-type solder is satisfactory. However, a larger soldering tool is required
for such terminals. The 15 W soldering tool recommended for ICs, and
printed or etched circuits are usually too small for the large terminals. A 40
to 50 W soldering tool will do most jobs in solid-state electronics. A soldering
gun is also useful.

When soldering large terminals, use good electronic soldering practices; that is, apply only enough heat to make the solder flow freely and
apply only enough solder to form a good electrical connection. Too much
solder may impair operation of the circuit. An excess of solder can also cover
a cold solder joint.

Clip off any excess wire that may extend past the solder connection. If
necessary, clean the solder connection with nonacid flux-remover solvent.

Many ICs are used with solid-state components mounted on strips.
Usually, the strips are made of a ceramic material. Strip mounting is used
instead of (or in addition to) PC mounting, especially on equipment where
parts are not of the plug-in type.

Typical strip mounting is shown in Fig. 2-14. Often, the notches in
these strips are lined with a silver alloy. This should be verified by reference
to the service literature. Application of excessive heat or repeated use of
ordinary 60/40 tin/lead solder can break the silver-to-ceramic bond. Occa-

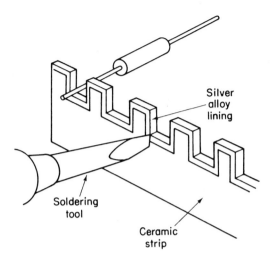

Fig. 2-14. Strip soldering techniques

sional use of ordinary solder is permissible, but for general repair work, solder containing about 3% silver should be used.

When removing or installing a part mounted on ceramic terminal strips:

1. Use a 50 W soldering tool. The tip should be tinned with silver-bearing solder.
2. Apply heat by touching one corner of the soldering tool tip to the base of the notch as shown in Fig. 2-14. Do not force the tip of the tool into the terminal notch. This may chip or break the ceramic strip.
3. Use the minimum amount of heat required to make the solder flow freely.
4. Apply only enough solder to form a good bond. Do not attempt to fill the notch with solder.

If it is necessary to hold a bare wire in place while soldering (a bare wire is sometimes used as a bus when several ICs are mounted on the same

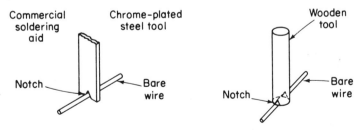

Fig. 2-15. Typical wire-holding tools

board), a handy tool for this purpose can be made by cutting a notch into one end of a wooden tool as shown in Fig. 2-15. There are commerical versions of such wire holding tools. Usually, the commercial soldering aids are made of chrome-plated steel. (The solder will not stick to the chrome plating.)

2.3.3. Handling and protecting MOS ICs

Damage caused by static discharge can be a problem with MOS ICs. Electrostatic discharges can occur when a MOS IC is picked up by its case and the handler's body capacitance is discharged to ground through the series capacitances of the device. This requires proper handling, particularly when the MOS device is out of the circuit. In circuit, a MOS IC is just as rugged as any other IC of similar construction.

MOS ICs are often shipped with the leads all shorted together to prevent damage in shipping and handling (there will be no static discharge between leads). Usually, a shorting spring, or similar device, is used for shipping. The spring *should not be removed until after* the IC is soldered into the circuit. An alternate method for shipping or storing MOS ICs is to apply a conductive foam between the leads. Polystyrene insulating "snow" is not recommended for shipment or storage of MOS ICs. Such snow can acquire high static charges that could discharge through the IC.

When removing or installing a MOS IC, first turn the power off. If the MOS IC is to be moved, your body should be at the same potential as the unit from which the IC is removed and installed. This can be done by placing one hand on the chassis before removing the MOS IC.

MOS IC *protection methods.* Because of the static discharge problem, manufacturers provide some form of protection for a number of the MOS ICs. Generally, this protection takes the form of a diode incorporated as part of the substrate material. A diode (or diodes) can be fabricated as part of the monolith chip. The protection scheme used by Motorola in their complementary MOS ICs is shown in Fig. 2-16.

The diode is designed to break down at a lower voltage than the MOS IC junctions. Typically, the diode will break down at about 30 V, whereas the junctions breakdown at 100 V. The diode can break down without damage, provided the current is kept low (as is usually the case where there is a static discharge).

The *single diode* scheme protection by clamping positive levels to V_{DD}. Negative protection is provided by the 30 V reverse breakdown. The *resistor diode protection* method adds some delay, but provides protection by clamping positive and negative potentials to V_{DD} and V_{SS}, respectively. The resistor is included to provide additional circuit isolation.

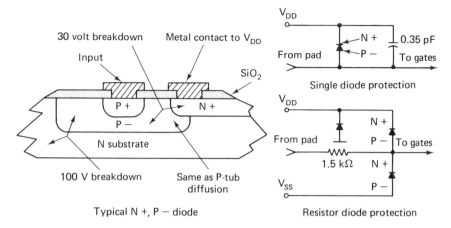

Fig. 2-16. Gate protection for Motorola complementary MOS devices

2.4. LAYOUT OF INTEGRATED CIRCUITS

Because the layout of ICs is such a broad field, no attempt is made here to cover the entire subject. Instead, we shall summarize the problems that apply to all types of ICs.

From a practical standpoint, each type of IC has its own problems. For example, most linear ICs have high gain and are thus subject to oscillation if undesired feedback is not controlled by good circuit layout. However, because of the differential input used by most linear ICs (refer to Chapters 3 and 4), the pick up of signal noise is usually not a major problem. On the other hand, digital ICs rarely oscillate because of low gain, but are subject to noise signals. Proper circuit layout can minimize the generation and pick up of such noise. The following paragraphs describe those circuit layout problems that IC users must face at one time or another.

It is assumed that the reader is already familiar with good design practices applicable to all electronic equipment. It should be noted that most ICs are mounted on PC cards or boards (there are exceptions, of course). All of the problems that apply to discrete components on PC boards also apply to ICs mounted in the same way.

2.4.1. Layout of digital ICs

All logic circuits are subject to noise. Any circuit, discrete or IC, will produce erroneous results if the noise level is high enough. Thus, it is recommended that noise and grounding problems be considered from the very beginning of layout design.

Wherever dc distribution lines run an appreciable distance from the supply to a logic chassis (or a PC board), both lines (positive and negative) should be bypassed to ground with a capacitor, at the point where the wires enter the chassis.

The values for power-line bypass capacitors should be on the order of 1 to 10 μF. If the logic circuits operate at higher speeds (above about 10 MHz), add a 0.01 μF capacitor in parallel with each 1–10 μF capacitor. Keep in mind that even though the system may operate at low speeds, there are harmonics generated at higher speeds. The high-frequency signals may produce noise on the power lines and interconnecting wiring. A 0.01 μF capacitor should be able to bypass any harmonics present in most logic systems.

If the digital ICs are particularly sensitive to noise, as is the case with the TTL logic form (refer to Chapters 8 and 9), extra bypass capacitors (in addition to those at the power and ground entry points) can be used effectively. The additional power- and ground-line capacitors can be mounted at any convenient point on the board or chassis, provided that there is no more than a 7 inch space between any IC and a capacitor (as measured along the power or ground line). Use at least one additional capacitor for each 12 IC packages and possibly as many as one capacitor for each 6 ICs.

The dc lines and ground return lines should be large enough to minimize noise pickup and dc voltage drop. Unless otherwise recommended by the IC manufacturer, use AWG No. 20 or larger wire for all digital IC power and ground lines.

In general, all leads should be kept as short as possible, both to reduce noise pickup and to minimize the propagation (delay) time down the wire. Typically, present-day logic circuits operate at speeds high enough so that the propagation time down a long wire or cable can be comparable to the delay time through a logic element. This propagation time should be kept in mind during the layout design.

Do not exceed 10 inches of line for each nanosecond of fall time for the fastest logic pulses involved. For example, if the clock pulses (usually the fastest in the system) have a fall time of 3 nS, no logic signal line (either PC or conventional wire) should exceed 30 inches.

The problem of noise can be minimized if ground planes are used. (That is, if the circuit board has solid metal sides.) Such ground planes surround the active elements on the board with a noise shield. Any logic system that operates at speeds above approximately 30 MHz should have ground planes. If it is not practical to use boards with built-in ground planes, run a wire around the outside edge of the board. Connect both ends of the wire to a common or "equipment" ground.

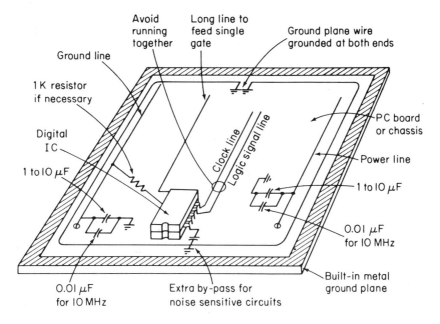

Fig. 2-17. General rules for digital IC layout

Do not run any logic signal line to a clock line for more than about 7 inches, because of the possibility of cross-talk in either direction.

If a logic line must be run a long distance, design the circuits so that the long line feeds a single gate (or other logic element) rather than several gates. External loads to be driven must be kept within the current and voltage limits specified on the IC datasheet.

Some digital IC manufacturers specify that a resistor (typically 1 kΩ) be connected between the gate input and the power supply (or ground, depending upon the type of logic), where long lines are involved. Always check the IC data sheet for such notes. The general rules for digital IC layout are given in Fig. 2-17.

2.4.2. Layout of linear ICs

The main problem with layout of linear ICs is undesired oscillation from feedback. Virtually all of today's IC op-amps (Chapters 3 through 6) are capable of producing high gain at high frequencies. Since the ICs are physically small, the input and output terminals are close, creating the ideal conditions for undesired feedback. To make matters worse, most linear ICs are capable of passing frequencies higher than those specified on the data sheet.

For example, an op-amp to be used in the audio range (say up to 20 kHz with a power gain of 20 dB) could possibly pass a 10 MHz signal with some slight gain. This higher-frequency signal could be a harmonic of signals in the normal operating range and, with sufficient gain, could feed back to the input and produce undesired oscillation.

In laying out any linear IC, particularly in the experimental stage, always consider the circuit as being radio-frequency (RF), even though the IC is not supposed to be capable of RF operation, and the circuit is not normally used with RF.

As is discussed in Sec. 2.7, all linear IC power-supply terminals should be bypassed to ground. This provides a path for any RF. The layout should be such that the capacitors are as near to the IC terminals as possible. Do not mount the capacitors at the power-supply end of the line.

Note that the lead between the IC terminal and the capacitors has some inductance. This inductance can combine with the capacitor to form a resonant circuit. If the circuit resonates at some frequency (including funda-

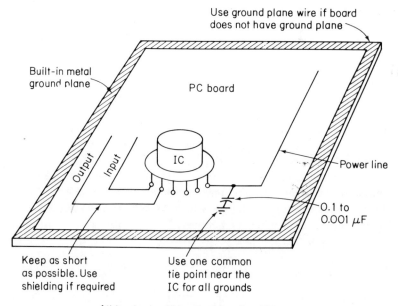

Fig. 2-18. General rules for linear IC layout

mental, harmonic, or subharmonic), the signal could be passed to the IC to produce oscillation.

Keep IC input and output leads as short as practical. Use shielded leads wherever practical. Use one common tie point near the IC for all grounds. Resonant circuits can also be formed by poor grounding or by ground loops in general.

As a general rule, ICs mounted on PC boards (particularly with ground planes) tend to oscillate less than when conventional wiring is used. For that reason, an IC may oscillate in the experimental stage, but not when mounted in final layout form. Once all of the leads have been connected to an IC and power is applied, monitor all IC terminals for oscillation with an oscilloscope before signals are applied. Except in a few rare cases, there should be no evidence of RF (or any alternating current) signals at any IC terminal under no-signal conditions. Of course, there can be power-line hum, noise, etc., that is not the fault of the IC or the layout. The general rules for linear IC layout are given in Fig. 2-18.

2-5. POWER DISSIPATION PROBLEMS IN INTEGRATED CIRCUITS

As discussed in Sec. 1.5, the basic rules for ICs regarding power dissipation and thermal considerations are essentially the same as those for discrete transistor circuits.

The maximum allowable power dissipation (usually specified as P_d or P_D or "maximum device dissipation" on IC datasheets) is a function of the maximum storage temperature T_S, the maximum ambient temperature T_A, and the thermal resistance from the semiconductor chip to case. The basic relationship is:

$$P_D = \frac{T_S - T_A}{\text{thermal resistance}}$$

All IC data sheets do not necessarily list all of these parameters. It is quite common to list only the maximum power dissipation for a given ambient temperature and then show a "derating" factor in terms of maximum power decrease for a given increase in temperature. For example, a typical IC might show a maximum power dissipation of 110 mW at 25 °C, with a derating factor of 1 mW/°C. If such an IC is operated at 100 °C, the maximum power dissipation is: $100 - 25$ or 75 °C increase; $110 - 75 = 35$ mW.

In the absence of specific datasheet information, the following typical temperature characteristics can be applied to the basic IC package types. No IC should have a temperature in excess of 200 °C.

Ceramic flat pack:

> thermal resistance = 140 °C per watt
> maximum storage temperature = 175 °C
> maximum ambient temperature = 125 °C

TO-5 style (metal can) package:

> thermal resistance = 140 °C per watt
> maximum storage temperature = 200 °C
> maximum ambient temperature = 125 °C

Dual-in-line (ceramic) package:

> thermal resistance = 70 °C per watt
> maximum storage temperature = 175 °C
> maximum ambient temperature = 125 °C

Dual-in-line (plastic) package:

> thermal resistance = 150 °C per watt
> maximum storage temperature = 85 °C
> maximum ambient temperature = 75 °C

2.5.1. Working with power ICs

Most present-day ICs require low power (typically less than 1 W) and can be operated without heat sinks. However, there are some power ICs that must be used with a heat sink (either an external heat sink or by direct contact with a metal chassis). Power ICs generally use some form of metal can package.

   The datasheets for these power ICs usually list sufficient information to select the proper heat sink. Also, the datasheets or other literature often make recommendations as to mounting for the power IC. Always follow the IC manufacturer's recommendations. In the absence of such data and to make the reader more familiar with the terms used, the following sections summarize considerations for power ICs.

2.5.2. Maximum power dissipation

From a designer's standpoint, an IC is a complete, predesigned, functioning circuit that cannot be altered in regard to power dissipation. That is, if the power supply voltages, input signals, output loads, and ambient temperature are at their recommended levels, the power dissipation will be well within the capabilities of the IC. With the possible exception of the data required to select or design heat sinks, the designer need only follow the datasheet recommendations. Of course, if the IC must be operated at a temperature higher than the rated ambient, the power dissipation must be derated as previously described in Sec. 2.5.

2.5.3.   Thermal resistance

ICs designed for power applications usually have some form of thermal resistance specified to indicate the power dissipation capability, rather than a simple maximum device dissipation. Thermal resistance can be defined as the increase in temperature of a semiconductor material (transistor junctions), with regard to some reference, divided by the power dissipated, or °C/W.

Power IC datasheets often specify thermal resistance at a given temperature. For each increase in temperature from this specified value, there will be a change in the temperature-dependent characteristics of the IC. Since there is a change in temperature with changes in power dissipation, the semiconductor chip (or transistor junction) temperature also changes, resulting in a characteristic change. Thus, IC characteristics can change with ambient temperature changes and with changes produced by variations in power dissipation. As will be discussed in Sec. 2.5.4, most ICs have circuits to offset the effects of temperature.

In power ICs, thermal resistance is normally measured from the semiconductor chip to the case. On those ICs where the cases are bolted directly to the mounting surfaces with a threaded bolt or stud, the thermal resistance is measured from the chip to the mounting stud or flange.

2.5.4.   Thermal runaway

The main problem in operating any semiconductor device (transistor or IC) at or near its maximum power dissipation limits is a condition known as *thermal runaway*. Heat is generated when current passes through a transistor junction. If all of this heat is not dissipated by the case (an impossibility), the junction temperature will increase. This, in turn, causes more current to flow through the junction, even though the voltage and other circuit values, remain the same. In turn, this causes the junction temperature to increase even further, with a corresponding increase in current flow. If the heat is not dissipated by some means, the transistor (or IC) will burn out and be destroyed.

Most ICs are designed to prevent thermal runaway. The usual arrangement is to place a diode in the reverse bias circuit for one or more transistors in the IC. The diode is fabricated on the same semiconductor chip as the transistors and thus has the same temperature characteristics. The circuit is arranged so that the reverse bias is increased (by a decrease in diode resistance) when there is an increase in temperature. When temperature increases because of an increase in transistor current (or vice versa), the diode resistance changes and increases the reverse bias. This offsets the initial change in current caused by temperature changes.

Many different temperature compensation circuits have been developed by IC manufacturers. However, the IC user need not be concerned as long as the data sheet limits are observed, since the circuit is already designed into the IC.

2.5.5. Operating ICs without heat sinks

If an IC is not mounted on a heat sink (as is typical for about 90 percent of the ICs in use today), the thermal resistance from case-to-ambient is so large in relation to that from junction-to-case (or mount) that the total thermal resistance from junction-to-ambient air is primarily the result of the case-to-ambient term.

The thermal resistance for a TO-5 (metal can) style case is 140 °C. Assuming that the ambient temperature is 25 °C and that the maximum temperature for any IC is 200 °C, the maximum permissible temperature rise is 175 °C $(200 - 25 = 175)$. Thus, the maximum theoretical power dissipation for any TO-5 style IC is:

$$\text{Maximum power dissipation} = \frac{\text{maximum permissible temperature increase}}{\text{thermal resistance}}$$

or

$$= \frac{175}{140} = 1.25 \text{ W.}$$

In practice, this would be an absolute maximum figure. Rarely, if ever, are ICs operated above about 1 W without heat sinks.

2.5.6. Operating ICs with heat sinks

After about 1 W (or less) it becomes impractical to increase the size of the case to make the case-to-ambient thermal resistance term comparable to the junction-to-case term. For this reason, most power ICs are designed for use with an external heat sink. Sometimes the chassis or mounting area serves as the heat sink. In other cases, a heat sink is attached to the case. Either way, the primary purpose of the heat sink is to increase the effective heat-dissipation area of the case and provide a low heat-resistance path from case to ambient.

To properly design (or select) a heat sink for a given application, the thermal resistance of both the IC and heat sink must be known. For this reason, power IC datasheets usually specify the junction-to-ambient thermal resistance, which must be combined with the heat-sink thermal resistance to find the total power dissipation capability. Note that some power IC datasheets specify a maximum case temperature rather than thermal resistance. As is discussed in Sec. 2.5.7, maximum case temperature can be combined with heat-sink thermal resistance to find maximum power dissipation.

Heat-sink ratings.   Commercial fin-type heat sinks can be used with TO-5 style ICs. Such heat sinks are especially useful when the ICs are mounted in Teflon sockets that provide no thermal conduction to the chassis or PC board. Commerical heat sinks are rated by the manufacturer in terms of thermal resistance, usually in terms of °C/W. When heat sinks involve the use of washers, the °C/W factor usually includes the thermal resistance between the IC case and sink. Without a washer, only the sink-to-ambient thermal resistance is given. In either case, the thermal resistance factor represents temperature increase (in °C) divided by wattage dissipated. For example, if the heat sink temperature rises from 25 to 100 °C (a 75 °C increase) when 25 W are dissipated, the thermal resistance is $75/25 = 3$. This can be listed on the datasheet as a thermal resistance of 3, or simply as 3 °C/W.

All other factors being equal, the heat sink with the lowest thermal resistance (°C/W) is best. That is, a heat sink with 1 °C/W is better than a 3 °C/W heat sink. Of course, the heat sink must fit the IC case and the space around the IC. Except for these factors, selection of a suitable heat sink should be no particular problem.

Calculating heat-sink capabilities.   The thermal resistance of a heat sink can be calculated if the following factors are known: material, mounting provisions, exact dimensions, shape, thickness, surface finish, and color. Even if all of these factors are known, the thermal resistance calculations are still only approximate. As a very approximate rule-of-thumb:

$$\text{Heat sink}\quad \text{thermal resistance (in °C/W)} = \sqrt{\frac{1500}{\text{area}}}$$

where the area (total area exposed to the air) is in square inches, material is one-eighth inch aluminum, and the shape is that of a flat disc.

From a practical design standpoint, it is better to accept the manufacturer's specification for a heat sink. The heat-sink thermal resistance actually consists of two series elements: the thermal resistance from the case to the heat sink that results from conduction (case-to-sink) and the thermal resistance from the heat sink to the ambient air, caused by convection and radiation.

Practical heat-sink considerations.   To operate an IC at its full power capabilities, there should be no temperature difference between the case and ambient air. This occurs only when the thermal resistance of the heat sink is zero, and the only thermal resistance is that between the junction and case. It is not possible to manufacture a heat sink with zero resistance. However, the greater the ratio of junction-to-case versus case-to-ambient, the more closely the maximum power limit can be approached.

When ICs are to be mounted on heat sinks, some form of electrical insulation is usually required between the case and heat sink. As is discussed in Sec. 2.7, many IC cases are not at electrical ground; instead, they can be connected to some other point (above or below ground) in the internal circuit.

Because good electrical insulators usually are also good thermal insulators, it is difficult to provide electrical insulation without introducing some thermal resistance between case and heat sink. The best materials for this application are mica, beryllium oxide (Beryllia), and anodized aluminum, with typical °C/W ratings of 0.4, 0.25, and 0.35 respectively.

The use of a zinc-oxide-filled silicon compound (such as Dow Corning #340 or Wakefield #1200) between the washer and chassis, together with a

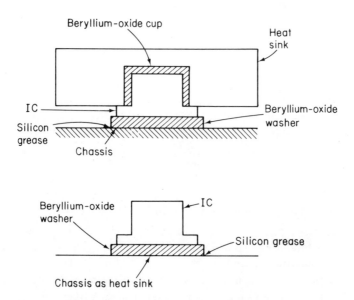

Fig. 2-19. Typical mounting arrangements for IC heat sinks

moderate amount of pressure from the top of the IC helps to decrease thermal resistance.

If the IC is mounted with a fin-type heat sink, an insulated cap (such as Beryllia) should be used between the case and heat sink. Figure 2-19 shows both methods of mounting ICs and heat sinks.

When a washer is added between the IC case and heat sink, a certain amount of capacitance is introduced. In general, this capacitance will have no effect on operation of ICs unless the frequency is about 100 MHz. Rarely, if ever, do power ICs operate above the audio range. Thus, few problems should be found.

Forced-air cooling. When there are a large number of ICs operated in a confined area, it may be necessary to use forced air (from a fan, blower, etc.) to keep the ambient air within an acceptable tolerance. Under these conditions, the ICs can be treated in the same way as are transistors, for the purposes of calculating the required amount of forced air.

2.5.7. Calculating power dissipation for ICs with heat sinks

The maximum dissipation capability of an IC used with a heat sink is dependent upon three factors. (1) The sum of the series thermal junctions (semiconductor chip) to ambient air, (2) the maximum junction temperature, and (3) the ambient temperature. The following are some examples of how power dissipation can be calculated.

Assume that it is desired to find the maximum power dissipation of an IC/heat-sink combination. The following conditions are specified: a maximum junction temperature of 200 °C (typical for any silicon semiconductor) a junction-to-case thermal resistance of 2 °C/W for the IC, and a heat sink with a thermal resistance of 3 °C/W and ambient temperature of 25 °C.

Note that the thermal resistance of the heat sink includes any thermal resistance produced by the washer between the IC case and heat sink. (If this factor is not known, add a factor of 0.5 °C/W thermal resistance for any washers between case and heat sink.)

First, find the total junction-to-ambient thermal resistance: junction-to-case (2) + sink-to-ambient (3) = junction-to-ambient (5). Next, find the maximum permitted power dissipation:

$$\frac{\text{maximum junction temperature }(200) - \text{ambient }(25)}{\text{junction-to-ambient }(5)} = \frac{200 - 25}{5} = 35 \text{ W (maximum)}$$

If maximum case temperature is specified instead of maximum junction temperature, the calculations for maximum power dissipation are as follows: (Assume that a maximum case temperature of 130 °C is specified instead of a 200 °C maximum junction temperature.)

First, subtract the ambient temperature from the maximum permitted case temperature.

$$130° - 25° = 105 \text{ °C}$$

Then, divide the case temperature by the heat-sink thermal resistance:

$$\frac{105}{3} = 35 \text{ W maximum power}$$

2.6. EFFECTS OF TEMPERATURE EXTREMES ON INTEGRATED CIRCUITS

The effects of temperature extremes (either high or low) will vary with the type of circuit involved, case style, and fabrication techniques of the manufacturer. Thus, no fixed rules can be made. However, the following general rules can be applied to most ICs.

In some instances, the IC will fail to operate at temperature extremes, but will return to normal when the operating temperature is returned to the normal range. In other cases, the IC will fail to operate properly once it has been subjected to a temperature extreme. In effect, the IC is destroyed once it is operated or stored at an extreme temperature. The effects of high-temperature extremes are generally worse than those of low temperatures. This is primarily because high temperatures can cause thermal runaway (Sec. 2.5.4).

In general, high temperatures cause the IC characteristics to change. An increased operating temperature also produces increased leakage currents, increased sensitivity to noise, increased unbalance in balanced circuits, increased "switching spikes" or transient voltages for transistors in digital ICs, and an increase in the ever-present possibility of burnout.

If the power-supply voltages, input signals, output loads, and ambient temperatures specified on the datasheet are observed, there should be no danger of temporary failure (or total destruction) for any IC. However, as a final check, multiply the rated thermal resistance by the maximum device dissipation, and then add the actual ambient temperature. If the result is less than 200, the IC should be safe.

For example, the typical thermal resistance of a DIP plastic IC is 150 (refer to Sec. 2.5). Assume that an IC has a maximum device dissipation of 600 mW, and that the ambient temperature is 50 °C. Under these conditions:

$$150 \times 0.6 = 90; \ 90 + 50 = 140$$

Since 140 (which is the junction temperature of the transistors within the IC under these conditions) is less than 200, the IC should be safe. In practice, the ambient temperature could probably go up to about 100 or 110 °C without permanent damage, but this would be approaching the danger point.

When an IC is operated at its low temperature extreme, the IC is likely to underperform. Usually, the IC will not be destroyed by extremely low temperature (except where the low temperature is prolonged). However, the IC will simply not perform as specified. For example, at low temperatures gain and power output will be different for op-amps and other linear ICs; operating speed will be reduced for digital ICs; and the drive or output load capabilities of digital ICs will be reduced.

In the absence of any specific information from the manufacturer, the following rule can be applied to derating an IC when operated at low temperatures. Derate the characteristic by 1 percent for each degree below the rated low operating temperature limit. Make certain to use the low *operating* temperature limit, not the low storage temperature limit.

For example, if the rated operating low temperature limit is -5 °C and the IC operates at -25 °C, derate all performance characteristics by 20 percent ($25-5=20$). In the case of digital ICs, allow for a 20 percent reduction in drive or output load capability as well as circuit speed. In no event should the IC be operated below the rated storage temperature. As a general rule, the low storage temperature limit is 10 to 20 °C below the operating limit. Always consult the datasheet for operating and storage temperature limits.

2.7. POWER SUPPLIES FOR INTEGRATED CIRCUITS

As a general rule, a linear IC requires connection to both a positive and negative power supply. This is because most linear ICs use one or more differential amplifiers in their circuits (refer to Chapter 3). Digital ICs generally require only one power supply. However, there are exceptions to both rules.

When two power supplies are required for a linear IC, the supplies are usually equal or symmetrical (such as $+6$ V and -6 V; $+12$ V and -12 V). This is the case with the IC of Fig. 2-20, that normally operates with $+12$ V and -12 V. A few linear ICs use unsymmetrical power supplies, and there are linear ICs that require only a single supply. However, such cases are the exception. And, in some cases, it is possible to operate a linear IC that normally requires two supplies from a single supply by means of special circuits (external to the IC). Such circuits are discussed in Sec. 2.7.6.

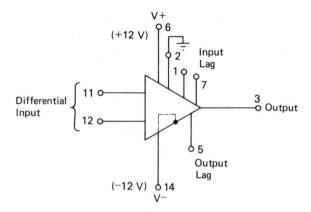

Fig. 2-20. Typical linear IC op-amp operating with symmetrical 12 V power supplies

2.7.1. Labeling of IC power supplies

Unlike most discrete transistor circuits in which it is usual to label one power supply lead positive and the other negative without specifying which (if either) is common to ground, it is necessary that all IC power supply voltages be referenced to a common or ground (that may or may not be physical or equipment ground).

As in the case of discrete transistors, manufacturers do not agree on power supply labeling for ICs. For example, the circuit of Fig. 2-20 uses $V+$ to indicate the positive voltage and $V-$ to indicate the negative voltage. Another manufacturer might use the symbols V_{EE} and V_{CC} to represent negative and positive, respectively. As a result, the IC datasheet must be studied carefully before applying any power source.

2.7.2. Typical IC power-supply connections

Figure 2-21 shows typical power-supply connections for both linear and digital ICs. The digital IC shown requires only one 5.2 V supply, with the positive connected to V_{CC} and the negative connected to V_{EE}. Keep in mind that either V_{CC} or V_{EE} could be at physical (equipment) ground. For example, if V_{CC} is at the physical ground, V_{EE} will still be negative, but will be 5.2 V below ground. Likewise, if V_{EE} is at physical ground, V_{CC} is positive and is 5.2 V above ground.

The protective diodes shown are recommended for any power-supply circuit in which the leads could be accidentally reversed. The diodes permit current flow only in the appropriate direction. The linear IC of Fig. 2-21 requires two power sources (of 12 V each) with the positive lead of one and

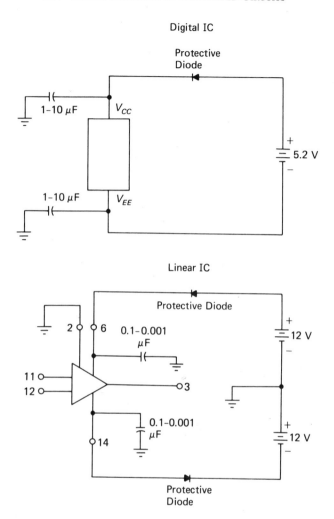

Fig. 2-21. Typical power-supply connections for digital and linear ICs

the negative lead of the other tied to ground or common.

The two capacitors shown in Fig. 2-21 provide for decoupling the power supply (signal bypass). Usually, disc ceramic capacitors are used. The capacitors should always be connected as close to the IC terminals as practical, not at the power-supply terminals. A guideline for linear IC power-supply decoupling capacitors is to use values between 0.1 and 0.001 μF.

In addition to the capacitors shown in Fig. 2-21, the ICs (both linear and digital) may require additional capacitors on the power lines. (Refer to Sec. 2.4.1.)

2.7.3. Grounding metal IC cases

The metal case of the linear IC shown in Fig. 2-21 is connected to terminal 2 *and to no other point in the internal circuit.* Thus, terminal 2 can and should be connected to equipment ground, as well as to the common or ground of the two power supplies.

The metal case of some ICs, both digital and linear, may be connected to a point in the internal circuit. If so, the case will be at the same voltage as the point of contact. For example, the case might be connected to pin 14 of the IC shown in Figs. 2-20 and 2-21. If so, the case would be below ground (or hot) by 12 V. If the case is mounted directly on a metal chassis that is at ground, the IC and power supply will be damaged. Of course, not all ICs have metal cases; likewise, not all metal cases are connected to the internal circuits. However, this point must be considered *before* using a particular IC.

2.7.4. Calculating current required for linear ICs

The datasheets for linear ICs usually specify a nominal operating voltage (and possibly a maximum operating voltage), as well as a total device dissipation. These figures can be used to calculate the current required for a particular IC. Use simple dc Ohm's law and divide the power by the voltage to find the current. However, certain points must be considered.

First, use the actual voltage applied to the IC. The actual voltage should be equal to the nominal operating voltage, but in no event higher than the maximum voltage. Second, use the total device dissipation. The datasheet may also list other power dissipations, such as device dissipation, which is defined as the dc power dissipated by the IC itself (with output at zero and no load). The other dissipation figures will always be smaller than the total power dissipation.

2.7.5. Calculating power required for digital ICs

Digital ICs usually operate with pulses. Thus, current is maximum in either of two states (0 or 1, high or low, etc.), but not in both states. Most digital IC datasheets list the current drain for the maximum (or worst-case) signal condition. As an example, I_{PDL} indicates the current drain when the logic signals (pulses) are low. If only I_{PDL} is listed for a digital IC, it can be reasonably assumed that the low state produces maximum current drain. I_{PDH} indicates the high-state current drain and, with no other listing on the datasheet, should be the maximum current drain. If both I_{PDL} and I_{PDH} are listed, it is obvious that the higher of the two indicates the maximum current drain state.

When current drains for both states are listed, some manufacturers recommend that the current drains be averaged to calculate power. For example, if the I_{PDL} is 10 mA and the I_{PDH} is 50 mA, add the two currents together and divide by 2 for an average current of 30 mA ($10 + 50 = 60$; $60/2 = 30$). The current requirements for digital ICs are also affected by the operating speed of the logic circuits and the type of loads into which the IC must operate.

For example, a digital IC requires more current as the operating speed is increased. Generally, the datasheet will list a nominal operating speed and a maximum operating speed, together with the current drain at the nominal speed figure. Of course, the IC should never be operated beyond the maximum limit. When operating between the nominal and maximum speeds, the additional current can be approximated by adding 0.5 to 1 mA for each 1 MHz of speed increase. For example, assume that the nominal speed is 15 MHz with a maximum speed of 25 MHz, the nominal or average current drain is 30 mA, and the IC is operated at the maximum speed. Then: $25 - 15 = 10$ MHz; $10 \times 0.5 = 5$ mA; $10 \times 1 = 10$ mA; $30 + 5 = 35$; $30 + 10 = 40$, and the IC requires between 35 and 40 mA.

Capacitive loads generally cause a digital IC to draw more current than pure resistive loads. However, it is not practical to juggle the direct effect of a capacitive load on an IC, so no general rules are given. The problem of loads and drives for digital ICs are discussed more thoroughly in Chapters 8 and 9.

2.7.6. Power supply tolerances for ICs

Typically, digital IC power supplies must be kept within ± 5 to $\pm 10\%$, whereas linear ICs will generally operate satisfactorily with $\pm 20\%$ power sources. These tolerances apply to actual operating voltage, not to maximum voltage limits. The currents (or power consumed) will vary proportionately.

Power-supply ripple and regulation are both important. Generally, solid-state power supplies with filtering and full feedback regulation are recommended. As in the case of discrete transistors, ripple (and any other power supply noise) must be kept to a minimum for noise-sensitive circuits (such as TTL digital ICs and high-gain linear ICs). Ideally, ripple (and all other noise) should be 1% or less.

When two power supplies are required, such as with a linear IC op-amp, there is a *voltage offset* problem. For example, if the V+ supply is 20% high, and the V− supply is 20% low, there will be an unbalance and a voltage offset condition, even though the IC op-amp circuits are normal. The problems of unbalance and voltage offset are discussed further in Chapters 3 through 6.

The effects of operating ICs beyond voltage tolerance are essentially the same as those experienced when the IC is operated at temperature extremes. That is, a high power-supply voltage causes the IC to overperform whereas low voltages result in underperformance. A low voltage does not usually result in damage to the IC, as is the case when the IC is operated beyond the maximum rated voltage.

2.7.7. Operating ICs from a single power supply

A linear IC op-amp (and any linear IC using differential amplifiers) is generally designed to operate from symmetrical positive and negative power-supply voltages. This results in a high common-mode rejection capability, as well as good low-frequency operation (typically a few Hz down to dc). (Common-mode rejection is discussed in Chapters 3 through 6.) If the loss of very low frequency operation can be tolerated, it is possible to operate linear IC op-amps from a single power supply, even though they were designed for dual supplies. Except for the low frequency loss, the other IC operating characteristics should be unaffected.

The following notes describe a technique that can be used with most IC op-amps to permit operation from a single power supply with a minimum of design compromise. The same maximum device ratings that appear on the datasheet are applicable to the IC when operating from a single polarity power supply, and must be observed for normal operation. Likewise, all of the considerations discussed thus far in this chapter apply to single-supply operation.

For example, power-supply decoupling capacitors are still required. The importance of decoupling capacitors, whether with single-supply or dual-supply operation, cannot be overemphasized. Today's IC op-amps are high-gain, high-frequency devices. Stray signals coupled back through the power supply can create instability problems. The decoupling capacitors should be placed as close as possible physically to the IC to minimize the effects of inductance in the power-supply leads. Circuit interconnections should be laid out in such a manner that the lead lengths are short enough to minimize pickup.

Split Zener.   The technique described here is generally referred to as the split Zener method. The main concern in setting up for single-supply operation is to *maintain the relative voltage· levels*. With an IC designed for dual-supply operation, there are three reference levels: $+V$, 0, and $-V$. For example, if the datasheet calls for plus and minus 10 V supplies, the three reference levels are: $+10$ V, 0 V, and -10 V.

For single-supply operation, these same reference levels can be maintained by using $++V$, $+V$, and ground (that is, $+20$ V, $+10$ V and 0 V), where $++V$ represents a voltage level double that of $+V$. This is

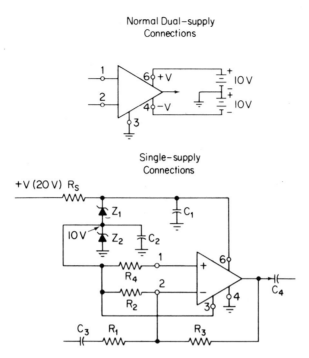

Fig. 2-22. Connections for single power supply operation (with ground reference)

illustrated in Fig. 2-22 where the IC is connected in the split Zener mode. Note that there is no change in the *relative voltage levels* even though the various IC terminals are at different voltage levels (with reference to ground). Terminal 4 (normally connected to the -10 V supply) is at ground. Terminal 3 (normally at ground or common) is set at one-half the total Zener voltage ($+10$ V). Terminal 6 (normally connected to the $+10$ V supply) is set at the full Zener voltage ($+20$ V).

Differential input. With single supply, the differential input terminals (1 and 2), that are normally at ground in a dual-supply system, must also be raised up one-half the Zener voltage (to $+10$ V). Under these circumstances, the output terminal (5) is also at one-half the Zener voltage, plus or minus an offset voltage error owed to input offset voltage, input offset current, and impedance unbalance. (Refer to Chapter 3 for information on these op-amp characteristics.)

Offset problems. To minimize offset errors caused by unequal voltage drops caused by the input bias currents across unequal resistances, it is recommended that the value of the input offset resistance R_4 be equal to the parallel combination of R_2 and R_3. This is in keeping with the standard op-amp practice as discussed in Chapter 3.

As with any op-amp, the deviation between absolute Zener level will also contribute to an error in the output voltage level. Typically, this is on the order of 50 to 100 μV per volt of deviation of Zener level. Except in rare cases, this deviation should be of little concern.

Ground reference. Note that the IC of Fig. 2-22 has a ground reference terminal (terminal 3). Not all IC op-amps have such terminals. Some ICs have only $+$V and $-$V terminals or leads even though the two levels are referenced to a common ground. That is, there is no physical ground terminal or lead on the IC, only $+$V and $-$V terminals. Figure 2-23 shows the split Zener connections for single-supply operation with such ICs. Here, the input terminals (A and B) are set at one-half the total Zener supply voltage; the $-$V terminal is set at ground; and the $+$V terminal is at the full Zener voltage ($+20$ V).

Negative supply. Figure 2-22 and 2-23 both show connection to positive power supplies. Negative power supplies can also be used. With a negative supply, the $+$V terminal is connected to ground, the $-$V terminal

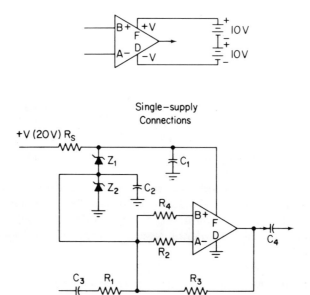

Fig. 2-23. Connections for single-supply operation (without ground reference)

is connected to full Zener supply (-20 V), with the input terminals and IC ground terminal (if any) connected to one-half the Zener supply. Of course, the polarity of the Zener diodes must be reversed.

Series resistance. Figures 2-22 and 2-23 both show a series resistance R_S for the Zener diodes. This is standard practice for Zener operation. The approximate or trial value for R_S is found by:

$$\frac{(\text{ maximum supply voltage} - \text{total Zener voltage}^2\text{ })}{\text{safe power dissipation of Zeners}}$$

For example, assume that the total Zener voltage is 20 V (10 V for each Zener), that the supply voltage may go as high as 24 V, and that 2 W Zeners are used. Under these conditions:

$$\frac{(24-20)^2}{2} = \frac{16}{2} = 8\Omega \text{ for } R_S$$

Effects on circuit operation using a single supply. From a user's standpoint, operation of an IC with a single supply is essentially the same as with the conventional dual power supply. The following notes describe the basic differences in operational characteristics of the IC with both types of power supplies. It is recommended that those readers not already familiar with basic op-amp theory study Chapters 3 through 6.

The normal IC frequency compensation techniques are the same for both types of supplies. The high-frequency limits are essentially the same. However, the low-frequency limit of an IC with a single supply is set by the values of capacitors C_3 and C_4. These capacitors are not required for dual-supply operation. Capacitors C_3 and C_4 are required for single-supply operation since both the input and output of the IC are at a voltage level equal to one-half the total Zener voltage (or 10 V using our example). Thus, the IC op-amp cannot be used as a dc amplifier with the single-supply system. In a dual-supply system, the input and output are at 0 V.

The closed-loop gain is the same for both types of supplies and is determined by the ratio of R_3/R_1. The values of decoupling capacitors C_1 and C_2 are essentially the same for both types of supplies. However, it may be necessary to use slightly larger values with the single-supply system, since the impedance of the Zeners is probably different than that of the power supply (without Zeners).

The value of R_2 should be between 50 and 100 kΩ for a typical IC op-amp. Values of R_2 much higher or lower than these limits can result in decreaseed gain or in an abnormal frequency response. From a practical standpoint, choose trial values using the guidelines and then run gain and frequency response tests. The value of R_4, the input offset resistance, is chosen to minimize offset error caused by impedance unbalance. As an approximate trial value, the resistance of R_4 should be equal to the parallel combination of R_2 and R_3. That is, make R_4 approximately equal to $R_2 R_3 / (R_2 + R_3)$.

3

THE BASIC
IC OP-AMP

An op-amp is essentially a very high-gain, direct-coupled amplifier that uses *feedback* for control of response characteristics. The designation op-amp was originally used for high-performance direct-coupled amplifiers that formed a basic part of analog computers. These op-amps were used to perform mathematical operations applicable to analog computation (summation, scaling, subtraction, integration, etc.). Today's IC op-amp can be used as a replacement for any low-frequency amplifier. For example, the same op-amp used in mathematical operations may be adapted to provide either the broad, flat frequency-gain response required of video amplifiers, or the peaked responses required of various types of shaping amplifiers.

IC op-amps are generally available in DIP and TO-5 style packages, although the flat-pack is used for a few IC op-amp models. All of the problems concerning mounting, power dissipation, power supplies, etc., covered in Chapter 2 are applicable to IC op-amps.

The capabilities and limitations of op-amps are firmly defined by a few simple equations and rules that are based on a certain set of criteria that the op-amp must meet. Effective use of these simple relationships, however, requires knowledge of the conditions under which each is applicable so that errors resulting from various approximations are held to a minimum.

In this chapter we discuss the basic op-amp. (Some typical applications for op-amps are discussed in Chapters 5 and 6.) Here, we shall concentrate on typical op-amp circuits, how to interpret op-amp datasheets, and design considerations for frequency response and gain. Frequency instabilities in the op-amp and the methods used to prevent them are also discussed.

Most of the basic design information for a particular op-amp can be obtained from the datasheet or other catalog information. Likewise, a typical datasheet may describe a few specific applications for the op-amp. However, IC op-amp datasheets generally have two weak points. First, they do not show how the listed parameters relate to design problems. Second, they do not describe the great variety of applications for which a basic op-amp can be used.

In any event, it is always necessary to interpret op-amp data, both from published material and actual test. Each manufacturer has its own system of datasheets and related technical information. It is impractical to discuss all data systems here. Instead, we discuss the typical information found on op-amp datasheets and see how this information affects design and use.

3.1. OP-AMP CIRCUITS

Op-amps generally use at least two (and possibly several) *differential stages* in cascade to provide *common mode rejection* and high gain. Differential amplifiers require both positive and negative power supplies. Since a differential amplifier has two inputs, it provides phase inversion for degenerative feedback, and can be connected to provide either in-phase or out-of-phase amplification.

A conventional op-amp requires that the output be fed back to the input through a resistance or impedance. The output is fed back to the negative or inverting input to produce degenerative feedback (to provide the desired gain and frequency response). As in any amplifier, the signal shifts in phase as it passes from input to output. This phase shift is dependent upon frequency. When the phase shift approaches 180°, the shift adds to (or cancels out) the 180° feedback phase shift. Thus, the feedback is in phase with the input (or nearly so) and will cause the amplifier to oscillate. This condition of phase shift with increased frequency limits the bandwidth of an op-amp. The condition can be compensated by the addition of a phase shift network (usually an *RC* circuit, but sometimes a single capacitor).

Phase shift problems and frequency considerations are discussed in Sec. 3.2. Before going into these subjects, it is necessary to understand the operation of basic op-amp circuits. We start with a review of the basic differential amplifier, and define the characteristics associated with that circuit.

3.1.1. Differential amplifier circuits

The differential amplifier is similar to an emitter-coupled amplifier, except that the two output signals are the result of a *signal difference* between the two inputs. In a theoretical differential amplifier, no output is produced when the signals at the input are identical. That is, an output is produced only when there is a difference in signals at the input. Signals common to both inputs are known as *common-mode signals*. The ability of a differential amplifier to prevent conversion of a common-mode signal into a difference signal (which produces an output) is expressed by the *common-mode rejection ratio* (CMR or CMRR).

One of the main reasons for use of a differential amplifier at the input stage of an op-amp is that the op-amp may be operated in the presence of radiated signals (power line radiation, stray signals from generators, etc.). Leads connected to the input terminals will pick up these radiated signals, even when the leads are shielded. If a single-ended input is used, the undesired signals will be picked up and amplified along with the desired signal input. If the op-amp has a differential input, both leads will pick up the same radiated signals at the same time. Since there is no difference between the radiated signals at the two inputs, there is no amplification of the undesired inputs.

Figure 3-1 is the schematic of a basic differential amplifier. The circuit responds differently to common-mode signals than it does to a single-ended signal. A common-mode signal (such as power-line pickup) drives both bases in-phase with equal-amplitude voltages, and the circuit behaves as though the transistors are in parallel to cancel the output. In effect, one transistor cancels the other.

Normal signals are applied to either of the bases (Q_1 or Q_2). The *inverting input* is applied to the base of Q_2, and the *noninverting input* is applied to the base of Q_1. With signal applied only to the inverting input, and the noninverting input grounded, the output is an amplified and inverted version of the input. For example, if the input is a positive pulse, the output is a negative pulse. If the noninverting input is used with the inverting input grounded, the output is an amplified version of the input (without inversion).

The emitter resistor introduces emitter feedback to both transistors simultaneously. This reduces the common-mode signal gain without reducing the differential signal gain in the same proportion.

Figure 3-2 is the schematic of a more practical differential amplifier. This circuit is typical of those found at the input of an op-amp. The circuit is basically a single-stage differential amplifier (Q_2 and Q_4) with input emitter followers (Q_1 and Q_5) and *constant-current source* Q_3 in the emitter-coupled network. Note that the signal emitter resistor of the circuit in Fig. 3-1 is replaced by Q_3 and its associated circuits in the circuit of Fig. 3-2.

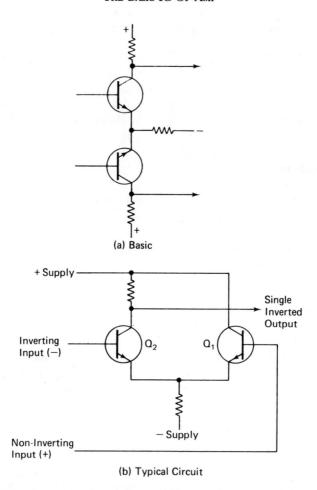

(a) Basic

(b) Typical Circuit

Fig. 3-1. Basic differential amplifier circuits

The use of a transistor such as Q_3 (or some similar circuit) is typical for most differential amplifiers found in op-amps. The circuit of transistor Q_3 is known as a *temperature-compensated* constant-current source. All current for the differential amplifier is fed through Q_3 (an *npn*) connected between the emitters of the differential amplifier and V_{EE} (the negative power supply). If there is an increase in current, a larger voltage is developed across the current-source Q_3 emitter resistor. This larger voltage acts to reverse bias the base-emitter junction, thus reducing current through Q_3. Since all current for the differential amplifier is passed through Q_3, current to the amplifier is also reduced. If there is a decrease in current, the opposite occurs, and the amplifier current increases. Thus, the differential amplifier is maintained at a constant current level.

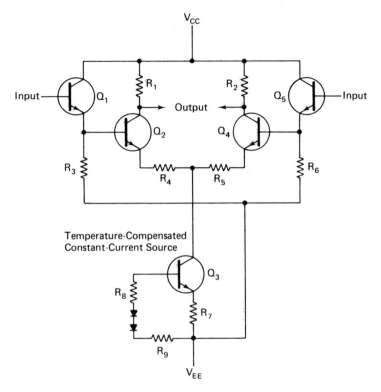

Fig. 3-2. Practical differential amplifier circuit

Transistor Q_3 is also temperature compensated by the diodes connected to the base-emitter bias network. These diodes have the same (approximate) temperature characteristics as the base-emitter junction, and offset any change in Q_3 base-emitter current flow that results from temperature change.

3.1.2. Common-mode definitions

The terms common mode and common-mode rejection are used frequently in op-amp applications. All manufacturers do not agree on the exact definition of common-mode rejection. One manufacturer defines common-mode rejection (CMR, or sometimes listed as CM_{rej}), or the common-mode rejection ratio (CMRR), as the ratio of differential gain (usually large) to common-mode gain (usually a fraction). That is, the amplifier may have a large gain of differential signals (different signals at each input terminal, or one input terminal grounded and the opposite input terminal with a signal), but have little gain (or possibly a loss) of common-mode signals (same signal at both terminals).

Another manufacturer defines CMR as the relationship of change in output voltage to the change in the input common-mode voltage producing it, divided by the open-loop gain (amplifier gain without feedback). For example, using the latter definition, assume that the common-mode input (applied to both terminals simultaneously) is 1 V, the resultant measured output is 1 mV, and the open-loop gain is 100. The CMR is then:

$$\frac{(\text{output}/\text{input})}{\text{open-loop gain}} = CMR$$

$$\frac{(0.001/1)}{100} = 100{,}000 = 100 \text{ dB}$$

Another method used to calculate CMR is to divide the output signal by the open-loop gain to find an *equivalent differential input signal*. Then the common-mode input signal is divided by this equivalent differential input signal. Using the same figures as in the previous CMR calculation:

$$\frac{\text{output signal}}{\text{open-loop gain}} = \text{equivalent differential input signal}$$

$$\frac{0.001 \text{ V}}{100} = 0.00001$$

$$\frac{1 \text{ V}}{0.00001} = 100{,}000 = 100 \text{ dB}$$

No matter what basis is used for calculation CMR is an indication of the *degree of circuit balance* of the differential stages, since common-mode input signals should be amplified identically in both halves of the circuit. A large output for a given common-mode input is an indication of large unbalanced or poor CMR. If there is an unbalance, a common-mode signal becomes a differential signal after it passes the first stage.

As with amplifier gain, CMR usually decreases as frequency increases. However, as a guideline, the CMR of any differential amplifier should be at least 20 dB greater than the open-loop gain at any frequency (within the limits of the op-amp).

3.1.3. Floating inputs and ground currents

Since a differential amplifier is sensitive only to the difference between two input signals (in theory), the signal source need not be grounded and can be *floating*. Thus, op-amps with differential inputs are often used in applications where the signal source is from a bridge (such as a bridge-type transducer) and the signal source is grounded.

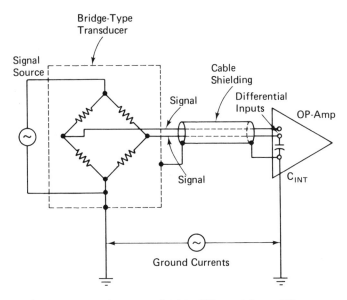

Fig. 3-3. Bridge-type transducer used with differential amplifier

A floating-input circuit can create problems. When the input is floating, cable shielding between the op-amp and signal source may be connected to chassis ground rather than to signal ground. However, both ac and dc voltages can exist between two widely separated earth grounds, causing current to flow. (Such currents are known as *ground currents*, and the circuits producing the current flow are known as *ground loops*.) The condition is shown in Fig. 3-3 where a bridge-type transducer is used with a differential op-amp.

Note that the signal source is connected to the transducer earth ground (local ground or physical ground, as it may sometimes be called). This ground point is connected to the op-amp ground through the cable shielding. The op-amp ground is connected to one of the differential inputs through the internal capacitance (represented as C_{INT}) of the op-amp, even though there may be no dc connection between ground and the input terminal of the floating-input op-amp.

The same differential input terminal is connected to the signal source through the signal leads and the transducer elements (bridge resistors in this case). Thus, the ac ground currents are mixed with the signal currents. This can result in an unbalance of the differential amplifier. Also, radiated signals picked up by the shield appear as undesired differential signals, rather than common-mode signals, and produce an undesired output.

One method used to minimize this condition is shown in Fig. 3-4. Here, a guard shield is placed around the input circuits of the differential

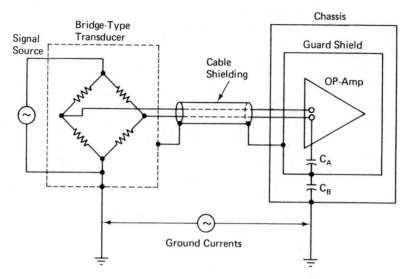

Fig. 3-4. Floated-input op-amp with guard shield to reduce capacitance between signal leads and ground

amplifier. This not only shields the differential amplifier from radiated signals but also provides an electrostatic shield to break the internal capacitance C_{INT} into two series capacitances C_A and C_B. A much higher impedance is then presented to the flow of ac ground signals. This type of op-amp is termed a *floated-input* and *guarded* op-amp.

3.1.4.   Universal op-amp circuit

Figure 3-5 shows the circuit of a universal op-amp, described by the manufacturer (RCA) as a BiMOS IC. The term BiMOS is used since the IC includes a combination of bipolar and MOS techniques. The basic purpose of such a circuit is to produce an output signal that is linearly proportional to the difference between two signals applied to the input. The circuit shown provides an open-loop voltage gain of approximately 110 dB.

The following is a brief description of the circuit shown in Fig. 3-5. As illustrated, the op-amp consists of four stages: an input stage, a second stage, an output stage, and a bias circuit.

Input stage.   The inputs to the op-amp are applied to the gates of MOS differential amplifiers Q_6 and Q_7. Diodes D_5 through D_8 provide gate-oxide protection for the MOS input stage, as described in Sec. 2.3.3. The drain-source currents for Q_6 and Q_7 are drawn through the emitter-collector paths of bipolar amplifiers Q_9 and Q_{10}. The single-ended output of the input stage is taken from the collector of Q_{10} and applied to the base of second-stage transistor Q_{11}.

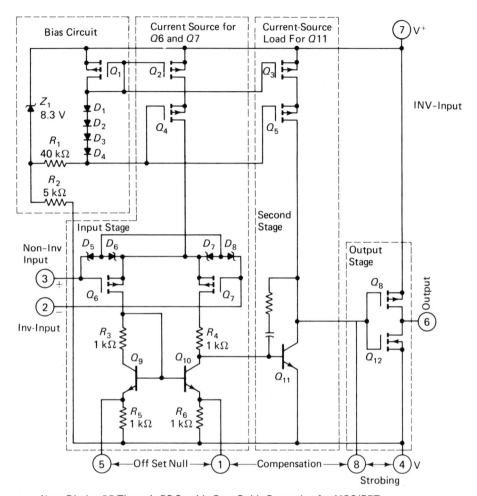

Note Diodes $D5$ Through $D8$ Provide Gate-Oxide Protection for MOS/FET Input Stage

Fig. 3-5. RCA CA3130/CA3160 BiMOS universal op-amp

Note that the emitters of Q_9 and Q_{10} are connected to terminals 5 and 1 of the IC. This permits connection to an external *offset null voltage*. Any differential amplifier will have some built-in unbalance, resulting in offset of the input and output. That is, the input and output of the IC will not be exactly at zero under no-signal conditions. When it is critical that offset be at a minimum for some particular design, a variable external voltage is applied to the input circuit through terminals 1 and 5. The external voltage is adjusted until the unbalance is eliminated (or at an absolute minimum) as indicated by a minimum offset at the input and output. (Refer to Sec. 3.3.9 for a further discussion of offset.)

MOS transistors Q_2 and Q_4 provide constant-current sources for Q_6 and Q_7. The gates of Q_2 and Q_4 are held constant by voltages from a bias circuit.

Bias circuit.   The circuit composed of Zener diode Z_1, series diodes D_1 through D_4, and MOS transistor Q_1 provides bias voltages to compensate for variations in supply voltage, as well as variations in temperature. Zener Z_1 provides the voltage regulation in the normal manner. Diodes D_1 through D_4, and Q_1, provide temperature compensation as described in Sec. 3.1.1.

Second stage.   Bipolar transistor Q_{11} amplifies the signal from the differential first stage, and applies the amplified signal to the output stage Q_8 and Q_{12}. Transistors Q_3 and Q_5 provide a constant current source for Q_{11} (in essentially the same manner as Q_2 and Q_4 provide for the differential input stages).

The series resistor-capacitor network connected between the collector and base of Q_{11} provides feedback to control the phase-gain frequency characteristics of the amplifier circuit, as described in Sec. 3.2. This internal phase compensation is not used in one version of the IC. Instead, an external phase compensation network is connected at terminals 1 and 8.

Output stage.   Transistors Q_8 and Q_{12} are connected in a complementary MOS (CMOS) configuration similar to that discussed in Sec. 1.2.2. By using CMOS the output voltage can swing to within 10 mV of the full-power source voltage, in either polarity. Note that when a positive signal is applied to noninverting input terminal 3, the output at terminal 6 is also positive. A positive at the inverting input (terminal 2) produces a negative signal at output terminal 6.

The input to the gates of Q_8 and Q_{12} from terminal 8 provides a means of controlling the op-amp from an external source. The op-amp is designed for systems in which the circuit must be disabled temporarily, by a "strobe," or "squelch," or other disable pulse. For example, pulses applied to terminal 8 (or between terminals 8 and 4) can completely disable (or enable, depending on polarity) the output stage, and thus the entire op-amp signal path.

3.1.5.   Variable bias (micropower) op-amps

Variable-bias op-amps are similar to conventional op-amps, except that standby power can be controlled by an external bias voltage. The bias input is generally introduced at the constant-current transistor for the input differential amplifier. This is shown in Fig. 3-6, which is the schematic for a variable-bias op-amp in IC form. The bias sets the quiescent current

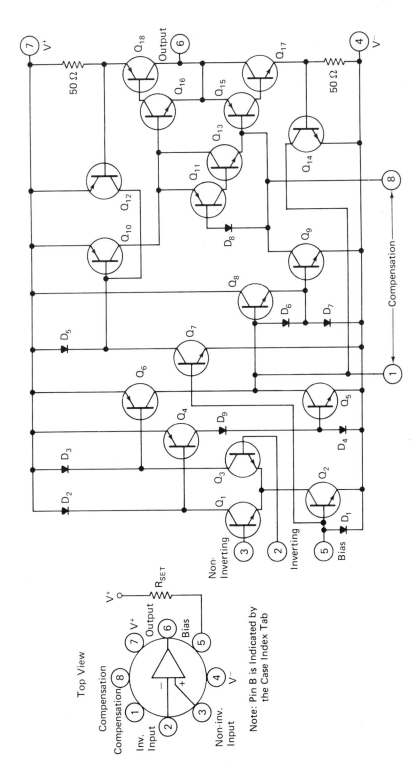

Fig. 3-6. RCA CA3078T micropower op-amp

71

(no-signal current, standby current, operating current, etc.) of the entire op-amp. Thus, the operating point of the op-amp can be controlled by external means.

An increase in the bias current produces an increase in quiescent current. For example, a variation from 1 to 400 nA of input bias produces a corresponding variation in quiescent current (I_Q) of 1 to 800 μA in the circuit of Fig. 3-6. Thus, an IC op-amp with variable bias can deliver milliamperes of current, yet only consume microwatts of standby power. For this reason, some variable-bias op-amps are known as *micropower* op-amps.

The variation in I_Q produces changes in other op-amp characteristics (which are defined in Sec. 3-3). For example, in the op-amp of Fig. 3-6, input offset current increases directly with I_Q. However, input offset voltage is not greatly affected by I_Q. The maximum output capability of the op-amp is affected by I_Q, but not to the extent of the input offset current. In the circuit of Fig. 3-6, maximum output current capability increases up to a point for an increase in I_Q. Once that point is reached (at about 6 μA) further increases in I_Q produce no further increases in maximum output current capability.

The I_Q has little effect on voltage gain of the op-amp. This is especially true when the load resistance is large. Load resistance is also the dominating factor in output voltage swing. The bias used to control the op-amp can be

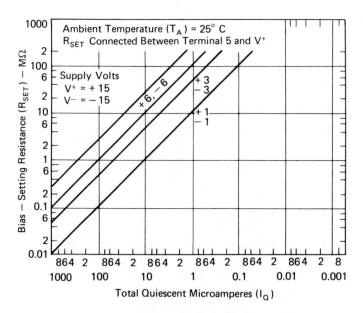

Fig. 3-7. Bias-setting resistance versus total quiescent current for RCA CA3078T

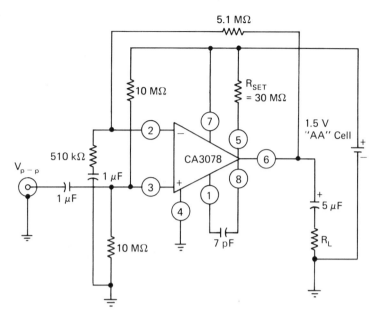

Fig. 3-8. Noninverting 20 dB amplifier using RCA CA3078 micropower op-amp

obtained from any source. Generally, the bias is taken from the V+ supply through a fixed resistance. Figure 3-7 gives the required values (for the op-amp of Fig. 3-6) using an external resistor R_{SET}. Figure 3-7 also shows a graph used to determine the value of the bias resistance R_{SET}.

Because of the low power consumption possible with variable-bias op-amps, battery operation is practical. Figure 3-8 shows an IC variable-bias op-amp used as a 20 dB amplifier powered from a 1.5 V AA battery cell.

3.2. FREQUENCY RESPONSE (BANDWIDTH) AND GAIN

Most of the design problems for op-amps are the result of trade-offs between gain and frequency response (or bandwidth). The open-loop (without feedback) gain and frequency response are characteristics of the basic op-amp circuit, but they can be modified with *phase compensation* networks. The *closed-loop* (with feedback) gain and frequency response are primarily dependent upon *external feedback* components. Although op-amps are generally used in the closed-loop operating mode, the open-loop characteristics have considerable effect on operation, and must be considered when designing a feedback system for an op-amp.

3.2.1. Inverting and noninverting feedback

The two basic op-amp feedback systems, inverting feedback and non-invertingfeedback, are shown in Figs. 3-9 and 3-10, respectively. In both cases, the op-amp output is fed back to the inverting (or minus) input through an impedance Z_F to control frequency response and gain. In the inverting feedback system of Fig. 3-9, the input signal is also applied to the inverting input, resulting in an inverted output. In the noninverting system of Fig. 3-10, the input signal is applied to the noninverting input, producing a noninverted output (a positive input produces a positive output). The inverting feedback system of Fig. 3-9, in which a positive input produces a negative output, is the more commonly used.

The equations shown in Figs. 3-9 and 3-10 are classic guidelines. The equations do not take into account the fact that open-loop gain is not infinitely high and output impedance is not infinitely low. Thus, the equations contain built-in inaccuracies and must be used as guides only.

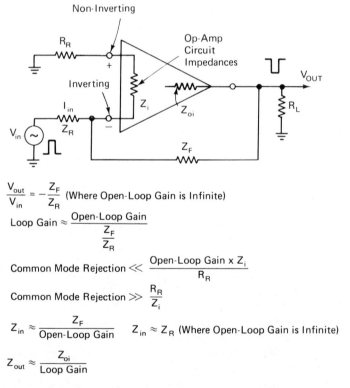

$$\frac{V_{out}}{V_{in}} = -\frac{Z_F}{Z_R} \quad \text{(Where Open-Loop Gain is Infinite)}$$

$$\text{Loop Gain} \approx \frac{\text{Open-Loop Gain}}{\dfrac{Z_F}{Z_R}}$$

$$\text{Common Mode Rejection} \ll \frac{\text{Open-Loop Gain} \times Z_i}{R_R}$$

$$\text{Common Mode Rejection} \gg \frac{R_R}{Z_i}$$

$$Z_{in} \approx \frac{Z_F}{\text{Open-Loop Gain}} \quad Z_{in} \approx Z_R \quad \text{(Where Open-Loop Gain is Infinite)}$$

$$Z_{out} \approx \frac{Z_{oi}}{\text{Loop Gain}}$$

Fig. 3-9. Inverting feedback op-amp (theoretical relationships)

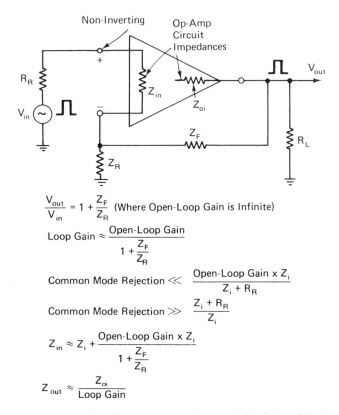

$$\frac{V_{out}}{V_{in}} = 1 + \frac{Z_F}{Z_R} \quad \text{(Where Open-Loop Gain is Infinite)}$$

$$\text{Loop Gain} \approx \frac{\text{Open-Loop Gain}}{1 + \dfrac{Z_F}{Z_R}}$$

$$\text{Common Mode Rejection} \ll \frac{\text{Open-Loop Gain} \times Z_i}{Z_i + R_R}$$

$$\text{Common Mode Rejection} \gg \frac{Z_i + R_R}{Z_i}$$

$$Z_{in} \approx Z_i + \frac{\text{Open-Loop Gain} \times Z_i}{1 + \dfrac{Z_F}{Z_R}}$$

$$Z_{out} \approx \frac{Z_{oi}}{\text{Loop Gain}}$$

Fig. 3-10. Noninverting feedback op-amp (theoretical relationships)

3.2.2. Gain/frequency characteristics of a theoretical op-amp

The gain/frequency relationships shown in Fig. 3-11 are based on a theoretical op-amp. Thus, the open-loop gain remains flat as frequency increases, and then begins to roll off at some particular frequency. The point at which the rolloff starts is sometimes referred to as a *pole*. The curve of Fig. 3-11 is known as a *one-pole* plot (and may also be known as a Bode plot, gain/bandwidth plot, or a frequency response curve). At the first (and only) pole, the open-loop gain rolls off at 6 dB per octave, or 20 dB per decade. The term 6 dB/octave means that the gain drops by 6 dB each time frequency is doubled. This is the same as a 20 dB drop each time the frequency is increased by a factor of 10.

If the open-loop gain of an amplifier is as shown in Fig. 3-11, any stable closed-loop gain could be produced by proper selection of feedback components, provided the closed-loop gain is less than the open-loop gain.

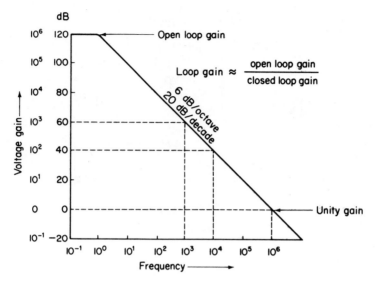

Fig. 3-11. Frequency response curve of theoretical op-amp

The only concern would be a trade-off between gain and frequency response.

For example, if a gain of 40 dB (10^2) is desired, a feedback resistance Z_F that is 10^2 times larger than the input resistance Z_R is used (such as a Z_F of 1000 and a Z_R of 10). The closed-loop gain is then flat to 10^4 Hz, and rolls off at 6 dB/octave to unity gain (a gain of 1) at 10^6 Hz. If a 60 dB (10^3) gain is required instead, the feedback resistance Z_F is raised to 10^3 times the input resistance Z_R (Z_F of 10,000 and a Z_R of 10). This reduces the frequency response. Gain is flat to 10^3 Hz (instead of 10^4), followed by rolloff of 6 dB/octave down to unity gain. It should be noted that the gain/frequency response curves of practical op-amps rarely look like that in Fig. 3-11. A possible exception is when internally compensated IC op-amps are used.

3.2.3. Gain/frequency characteristics of a practical op-amp

The open-loop frequency response curve of a practical op-amp more closely resembles that shown in Fig. 3-12. This curve is a three-pole plot. The first pole occurs at about 0.2 MHz, the second pole at 2 MHz, and the third pole at 20 MHz. (In a truly practical frequency response curve there will be no sharp breaks at the poles. Instead, the pole "corners" will be rounded and often difficult to distinguish. However, the curve of Fig. 3-12 is given here to illustrate certain frequency response characteristics.)

In the curve of Fig. 3-12, gain is flat at 60 dB to the first pole, then rolls off to 40 dB at the second pole. Since there is a decade between the first

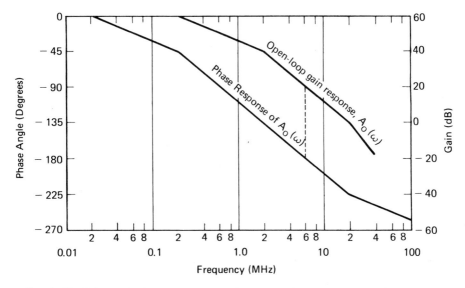

Fig. 3-12. Gain and phase response of an open-loop op-amp without phase compensation

and second poles (0.2 MHz and 2 MHz), the rolloff is 20 dB per decade, or 6 dB/octave. As frequency increases, rolloff continues from the second pole to the third pole, where gain drops from 40 dB to 0 dB. Thus, the rolloff is 40 dB/decade or 12 dB/octave. At the third pole, gain drops below unity as frequency increases at a rate of 60 dB/decade or 18 dB/octave.

Some op-amp datasheets provide a curve similar to that shown in Fig. 3-12. If the datasheets are not available, it is possible to test the op-amp under laboratory conditions, and draw an actual response curve (frequency response and phase shift).

3.2.4. Phase shift problems

Note that the curve of Fig. 3-12 also shows phase shift of an op-amp. As frequency increases, the phase shift between input and output signals of the op-amp increases. At frequencies up to about 0.02 MHz, the phase shift is zero. That is, the output signals are in phase with the noninverting input, and exactly 180° out of phase with the inverting input. As frequency increases up to about 0.2 MHz, phase shift increases by about 45°. That is, the output signals are 45° out of phase from the noninverting input and 45° from 180° (or 135°) out of phase with the inverting input. At about 6 MHz, the output signals are 180° out of phase with the noninverting input, and in phase with the inverting input. Since the output is fed back to the inverting input through Z_F, the input and output are in phase. If the output amplitude is large enough, the op-amp will then oscillate.

Oscillation can occur if the output is shifted near to 180° (and fed back to the inverting input). Even if oscillation does not occur, op-amp operation can become unstable. For example, the gain will not be flat. Usually a *peaking condition* will occur, in which output remains flat up to a frequency near the 180° phase-shift point, then gain will increase sharply to a peak. This is caused by the output signals being nearly in phase with input signals and reinforcing the input signals. At higher frequencies, gain will drop off sharply and/or oscillation will occur.

As a guideline, the op-amp should never be operated at a frequency at which phase shift is near 180° (above 170°) without compensation. In the op-amp of Fig. 3-12, the maximum uncompensated frequency is about 5 MHz. This is the frequency at which open-loop gain is about +25 dB.

One guideline often mentioned in op-amp literature is based on the fact that the 180° phase-shift point almost always occurs at a frequency at which the open-loop gain is in the 12 dB/octave slope. Thus, the guideline states that a closed-loop gain should be selected so that the unity gain is obtained at some frequency near the beginning of the 12 dB/octave slope (such as at 2 MHz in the op-amp of Fig. 3-12). However, this usually results in a very narrow bandwidth.

A more practical guide can be stated as follows: when a selected closed-loop gain is equal to or less than the open-loop gain at 180° phase-shift point, the op-amp will be unstable. For example, if a closed-loop gain of 20 dB or less is selected, an op-amp with open-loop, uncompensated curves similar to Fig. 3-12 will be unstable.

To find the minimum closed-loop gain, simply note where the −180° phase angle intersects the phase shift line. Then draw a vertical line up to cross the open-loop gain line. The closed-loop gain must be more than the open-loop gain at the frequency where the 180° phase shift occurs, but less than the maximum open-loop gain. Using Fig. 3-12 as an example, the closed-loop gain would have to be greater than 20 dB, but less than 60 dB.

Keep in mind that the guidelines discussed thus far apply to an uncompensated op-amp. With proper phase compensation, bandwidth (frequency response) and/or gain can be extended.

3.2.5. Op-amp phase compensation methods

Op-amp design problems created by excessive phase shift can be solved by compensating techniques that alter response so that excessive phase shifts no longer occur in the desired frequency range. The following are the basic methods of phase compensation.

Closed-loop modification.   The closed-loop gain of an op-amp can be altered by means of capacitors and/or inductances in the external feedback circuit (in place of fixed resistors). Capacitors and inductances change impedance with changes in frequency. This provides a different amount of feedback at different frequencies, and changes the amount of phase shift in the feedback signal. The capacitors and inductances can be arranged to offset the undesired open-loop phase shift.

Phase-shift compensation by closed-loop modification is generally not recommended since the method can create impedance problems at both the high- and low-frequency limits of operation. However, closed-loop modification is used for applications in which the op-amp is part of a bandpass, bandrejection, or peaking filter, as are described in Chapter 5.

Input impedance modification. The open-loop input impedance of an op-amp can be altered by means of resistors and capacitors connected at the op-amp input terminals. The impedance presented by the *RC* combination changes with frequency, thus altering the input impedance of the op-amp. In turn, the change in input impedance (with frequency) changes the bandwidth and phase-shift characteristics of the op-amp. Such an arrangement causes the rolloff to start at a lower frequency than the normal open-loop response of the op-amp, but produces a stable rolloff similar to that of the "ideal" curve in Fig. 3-11. With an op-amp properly compensated by an *RC* circuit at the input, the desired closed-loop gain can be produced by selection of external feedback resistors in the normal manner.

Phase compensation techniques that alter the open-loop input impedance permit the introduction of a *zero* into the response. This zero can be designed to cancel one of the poles in the open-loop response. Typically, the first pole is cancelled, and the open-loop gain drops to zero at the second pole. That is, after modification, the response drops to zero at a frequency where the uncompensated response changes from 6 dB/octave to 12 dB/octave. This is shown in Fig. 3-13. In another, less-frequently used input modification design, the response drops to zero at the frequency where the uncompensated response changes from 12 dB/octave to 18 dB/octave. Both input impedance modification designs are discussed in Sec. 3.2.6.

Phase-lead compensation. The open-loop gain and phase shift characteristics of an op-amp can be modified by means of a capacitor (or capacitors) connected to stages in the op-amp. Usually, the capacitors are connected between collectors in one of the high-gain differential stages. In other cases, the capacitors are connected from the collectors to ground. Generally, the capacitors are external to the op-amp, and are connected to the internal stages by means of terminals provided on the package (such as the compensation terminals shown in Fig. 3-5).

Phase-lead compensation requires a knowledge of the op-amp circuit characteristics. Usually, information for phase-lead compensation is provided on the op-amp datasheet. Typical phase-lead systems are described in Sec. 3.2.7.

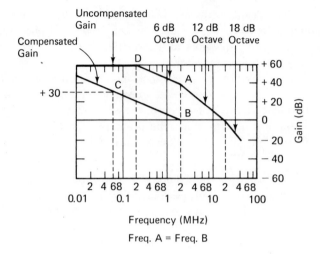

Frequency (MHz)

Freq. A = Freq. B

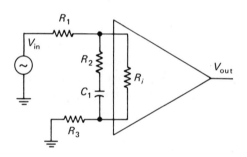

R_i = Input Impedance
    of Op-Amp

$R_1 = R_3$

$$R_1 + R_3 = \left(\frac{\text{Uncompensated Gain (dB)}}{\text{Compensated Gain dB}} - 1 \right) R_i$$

$$R_2 = \frac{R_1 + R_3}{\left(\dfrac{\text{Freq. D}}{\text{Freq. C}} - 1 \right) \left(1 + \dfrac{R_1 + R_3}{R_i} \right)} \qquad C_1 = \frac{1}{6.28 \times \text{Freq. D} \times R_2}$$

$$\text{Compensated Gain} = \frac{\text{Uncompensated Gain} \times R_i}{R_i + R_1 + R_3}$$

$$\text{Freq. D} = \frac{1}{6.28 \times R_2 \times C_1}$$

Fig. 3-13. Phase compensation by modification of input impedance (early rolloff method)

Phase-lag compensation. The open-loop gain and phase shift characteristics of an op-amp can be modified by means of a series capacitor and resistor connected to stages in the op-amp. There are two basic phase-lag compensation systems.

In one system, generally known as *RC rolloff, straight rolloff,* or *phase-lag rolloff compensation,* the open-loop response is altered by means of an *RC* network connected across a circuit component, such as across the input or output of an op-amp gain stage. In the other system, generally known as *Miller-effect rolloff* or *Miller-effect phase-lag compensation,* the open-loop response is altered by means of an *RC* network connected between input and output of an inverting gain stage in the op-amp. (The internal compensation *RC* network shown in Fig. 3-5 is an example of Miller-effect phase compensation.) The impedance of the compensating *RC* network then appears to be divided by the gain of that stage.

With either method, the rolloff starts at the pole frequency produced by the *RC* network. The Miller-effect rolloff technique requires a much smaller phase-compensating capacitor than that which must be used with the straight-rolloff method. Also, the reduction in output swing capability that is inherently in the straight rolloff is delayed significantly when the Miller-effect rolloff is used.

As with phase-lead, either method of phase-lag compensation requires a knowledge of the op-amp circuit characteristics. Usually, information for phase-lag compensation is provided on the op-amp datasheet, when such methods are recommended by the manufacturer. A typical straight-rolloff system is described in Sec. 3.2.7. Miller-effect rolloff is discussed in Sec. 3.2.8.

How to select a phase compensation method. A comprehensive op-amp datasheet will recommend one or more methods of phase compensation and will show the relative merits of each method. Usually this is done by means of response curves for various values of the compensating network. Several examples are described in Secs. 3.2.6 through 3.2.8.

The recommended phase compensation methods and values should be used in all cases. Proper phase compensation of an op-amp is at best a difficult, trial-and-error job. By using the datasheet values it is possible to take advantage of the manufacturer's test results on production quantities of a given op-amp. If the datasheet is not available or if the datasheet does not show the desired information, it is still possible to design a phase-compensating network using guideline equations.

The first step in phase compensation (when not following the datasheet) is to test the op-amp for open-loop frequency response and phase shift. Then draw a response curve similar to that in Fig. 3-12. On the basis of actual open-loop response, and the information in Secs. 3.2.5 through 3.2.8, select trial values for the phase-compensating network, then repeat the frequency-response and phase-shift tests. If the response is not as desired, change the values as necessary.

Each method of phase compensation has its advantages and disadvantages. The main advantage of open-loop input impedance modification (Sec. 3.2.6) is that it can be accomplished without datasheet information (or with limited information). The only op-amp characteristic required is input impedance. This is almost always available in datasheet or catalog form. If not, input impedance can be found by a simple test.

Phase lead and phase lag are the most widely accepted techniques for op-amps (particularly IC op-amps). These methods have an advantage (over input modification) in that the phase-compensation network is completely isolated from the feedback network. In the case of input modification, resistance in the phase compensation network forms part of the feedback network.

Phase lead and phase lag compensation have certain disadvantages. A careful inspection of the information in Secs. 3.2.7 through 3.2.9 will show that it is necessary to know certain internal characteristics of the op-amp before an accurate prediction of the compensated frequency response can be found. In the case of phase lead (Sec. 3.2.7), the compensated response is entirely dependent upon the value of the capacitors, and must be found by actual test. In the case of phase lag compensation (Secs. 3.2.8 and 3.2.9), the values for R and C of the compensation network are based on the uncompensated open-loop frequency at which gain changes from a 6 dB/octave drop to a 12 dB/octave drop. This can be found by test of the uncompensated op-amp. However, to predict the frequency at which the compensated response will start to roll off (or the gain after compensation) requires a knowledge of internal-stage transconductance (or gain) and stage load. This information is usually not available and cannot be found by simple test.

To sum up, if the datasheet is available, use the recommended phase compensation method. It will probably be phase lead or phase lag. If no phase compensation information is available, use input impedance modification.

3.2.6. Phase compensation by modification of input impedance

There are two accepted methods for phase compensation using input impedance modification. The first method, shown in Fig. 3-13, is the most widely used since it provides a straight rolloff similar to the ideal curve of Fig. 3-11. Once the input circuit is modified, conventional feedback can be used to select any point along the rolloff. That is, any combination of gain and frequency can be produced as described in Sec. 3.2.2. The main disadvantage to the method of Fig. 3-13 is early rolloff. Thus, if high gain is required, the bandwidth will be very narrow.

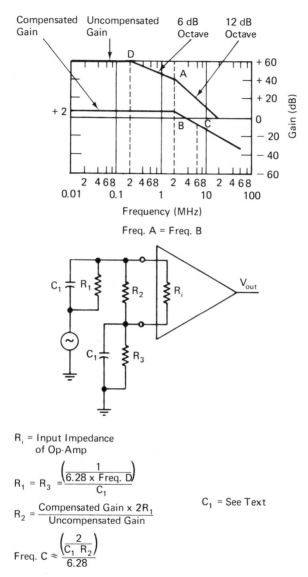

R$_i$ = Input Impedance of Op-Amp

$$R_1 = R_3 = \frac{\left(\dfrac{1}{6.28 \times Freq. D}\right)}{C_1}$$

$$R_2 = \frac{Compensated\ Gain \times 2R_1}{Uncompensated\ Gain}$$

C$_1$ = See Text

$$Freq.\ C \approx \frac{\left(\dfrac{2}{C_1\ R_2}\right)}{6.28}$$

Fig. 3-14. Phase compensation by modification of input impedance (extended bandwidth method)

The method shown in Fig. 3-14 is used only where bandwidth is of greatest importance, and gain can be sacrificed. As shown, rolloff does not start until the breakpoint between 6 dB/octave and 12 dB/octave is reached, and gain is flat up to that point (no peaking condition). However, the method of Fig. 3-14 usually results in little gain across the operating frequency range.

Early rolloff method. Assume that the method of Fig. 3-13 is to be used with an op-amp having the characteristics shown in Fig. 3-12. That is, the uncompensated, open-loop gain is 60 dB, the 6 dB/octave rolloff starts at about 0.2 MHz, the 12 dB/octave starts at 2 MHz and the 18 dB/octave starts at 20 MHz (which is also the point at which the open-loop gain drops to zero).

The first step in using the method shown in Fig. 3-13 is to note the frequency at which the uncompensated rolloff changes from 6 dB to 12 dB (point A of Fig. 3-13). The compensated rolloff should be zero (unity gain, point B) at the same frequency. Draw a line up to the left from point B that *increases* at 6 dB/octave. For example, with point B at 2 MHz, the line should intersect 0.2 MHz as it crosses the 20 dB gain point, should intersect 0.02 MHz as it crosses the 40 dB point. Any combination of compensated gain and rolloff starting frequency (point C) can be selected along the line. For example, if the rolloff starts at 0.2 MHz, the gain is about 20 dB, and vice versa.

Assume that the circuit of Fig. 3-13 is used to produce a compensated gain of 30 dB, with rolloff starting at about 0.06 MHz and dropping to zero (unity gain) at 2 MHz. The typical input impedance R_i is 10 kΩ. (Uncompensated gain, similar to that of Fig. 3-12, and typical input impedance can be found by referring to the datasheet.)

Using the equations of Fig. 3-13;

$$R_1 + R_3 = \left(\frac{60}{30} - 1\right)10,000$$

$$R_1 + R_3 = 1 \times 10,000 = 10,000$$

$$\text{If } R_1 = R_3, R_1 = R_3 = 5000$$

Using the equation in Fig. 3-13, the value of R_2 is:

$$R_2 = \frac{10,000}{\left(\dfrac{0.2}{0.06} - 1\right)\left(1 + \dfrac{10,000}{10,000}\right)}$$

$$R_2 = \frac{10,000}{2.3 \times 2}$$

$$R_2 \approx 2100 \ \Omega \text{ (nearest standard value)}$$

The value of C_1 is:

$$C_1 = \frac{1}{(2100)(6.28)(0.2 \text{ MHz})} \approx 0.0004 \ \mu\text{F(nearest standard value)}$$

If the circuit of Fig. 3-13 shows any instability in the open-loop or closed-loop condition, try increasing the values of R_1 and R_3 (to reduce gain); then select new values for R_2 and C_1.

Extended bandwidth method. Assume that the method of Fig. 3-14 is to be used with an op-amp having the characteristics shown in Fig. 3-12. The first step in using the method shown in Fig. 3-14 is to note the frequency at which the uncompensated rolloff changes from 6 dB to 12 dB (point A of Fig. 3-14, 2 MHz). The compensated rolloff (point B) should start at the same frequency.

Assume that the circuit of Fig. 3-14 is used to produce a flat, compensated gain of 2 dB, with rolloff starting at 2 MHz. Find the approximate values for R_1 through R_3, and C_1. Also find the approximate operating frequency (point C).

Assume a convenient value for each capacitor C_1, say 0.001 μF. Using this value, and the frequency (Freq. D) at which the uncompensated 6 dB rolloff starts (0.2 MHz), find the value of R_1 and R_3 as follows:

$$R_1 = R_3 = \frac{\left(\dfrac{1}{6.28 \times 0.2^{-6}}\right)}{0.001^{-6}}$$

$$\approx \frac{0.8^{-6}}{0.001^{-6}} \approx 800 \ \Omega$$

Using a desired compensated gain of 2 dB, an uncompensated gain of 60 dB, and a value of 1600 for $2R_1$, the value of R_2 is:

$$R_2 = \frac{2 \times 1600}{60} \approx 53 \ \Omega$$

Using 0.001 μF for C_1 and 53 ohms for R_2, the approximate maximum operating frequency (point C) is:

$$\text{Freq. C} = \frac{\left(\dfrac{2}{0.001^{-6} \times 53}\right)}{6.28} \approx 6 \ \text{MHz}$$

3.2.7. Phase-lead compensation examples

Phase-lead compensation is accomplished by the addition of a capacitor (or capacitors) to the basic op-amp circuit, and requires a knowledge of internal op-amp circuit characteristics. As a result, the manufacturer's datasheet or some similar information must be used. The alternate method is to use typical values for the compensation capacitor and test the results. Of course, this is time consuming, and may not prove satisfactory, even after tedious testing. The following are some examples of manufacturer's data on phaselead compensation.

Figure 3-15 shows typical phase lead compensation characteristics for an op-amp. The two compensating capacitors C_X and C_Y are connected from the collectors of the first differential amplifier (of the op-amp) to ground. The dashed lines in Fig. 3-15 illustrate the use of the curves for design of a 60 dB amplifier. First, the intersection of the various gain-frequency curves is followed out along the 60 dB line to the curve for a capacitor value of 0.001 μF. The intersection occurs at approximately 230 kHz. This means that if a 0.001 μF phase lead capacitor is used, the op-amp response should be flat at 60 dB (within a 3 dB range) up to a frequency of 230 kHz. At higher frequencies, the op-amp output will drop off. Thus, any gain up to 60 dB can be selected by proper choice of feedback and input resistances.

Next, follow the 230 kHz line vertically until it intersects the phase curve. The intersection occurs at approximately 118°. This means that if a 0.001 μF capacitor is used, and the op-amp is operated at 230 kHz, the phase shift will be 118°. That is, the output will be 118° from the noninverting input, and 62° from the inverting. Thus, there is a 62° *phase margin* (180°-118°) between input and output (assuming that the input

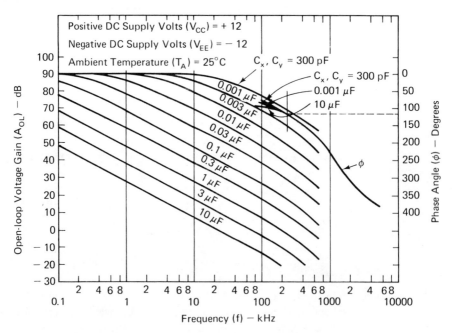

Fig. 3-15. Phase-compensation characteristics for RCA op-amp

signal is applied to the inverting input in the usual manner). A 62° phase margin should provide a very stable operation.

Now assume that it is desired to operate at a higher frequency, but still provide the 60 dB gain. Follow the 60 dB line to intersect with the 300 pF curve (300 pF is the smallest recommended capacitor). The intersection occurs at approximately 600 kHz. However, the 600 kHz line intersects the phase curve at about 175°, resulting in a phase margin of about 5°. This will probably produce unstable operation. Thus, if the 60 dB gain must be obtained, the capacitor value must be larger than 300 pF.

The curves of Fig. 3-15 show that, for a given gain, a larger value of phase lead capacitance reduces frequency capability, and vice versa. However, a reduction in frequency increases stability. For a given frequency of operation, capacitor size has no effect on stability, only on gain.

3.2.8 Conventional phase-lag compensation examples

As shown in Fig. 3-16, conventional phase lag compensation requires the addition of a capacitor and resistor to the basic op-amp circuit. This *RC* network is connected across a circuit component (such as across a stage output or input, or across the op-amp output). Conventional phase-lag compensation is generally external to the op-amp circuit. Many IC op-amps have terminals provided for connection of external phase compensation circuits to the internal circuit.

As in the case of phase lead compensation, the manufacturer's information must be used for phase lag compensation. The only alternative is to use the equations shown in Fig. 3-16, and test the results. This is not recommended unless the manufacturer's information is not available. Always use the published information, at least as a first trial value.

As shown in Fig. 3-16, the values of the external phase compensating resistor and capacitor are dependent upon the frequency at which the uncompensated gain changes from 6 dB/octave to 12 dB/octave (generally known as the *second pole* of the uncompensated gain curve). Thus, it is relatively simple to find values for *R* and *C* if the uncompensated gain characteristics are known (or can be found by test). However, note that the frequency at which the compensated rolloff will start (the start of the compensated 6 dB/octave rolloff) is dependent upon the value of the external capacitor *C* and internal characteristics of the op-amp circuits. Thus, even though satisfactory values of *R* and *C* can be found to produce a straight rolloff, there is no way of determining the frequency at which rolloff will start (using the equations).

If conventional phase lag rolloff is to be used, and the values must be found by test (no manufacturer's data), assume a convenient value for *R*, and find *C* using the equations of Fig. 3-16. Then test the compensated

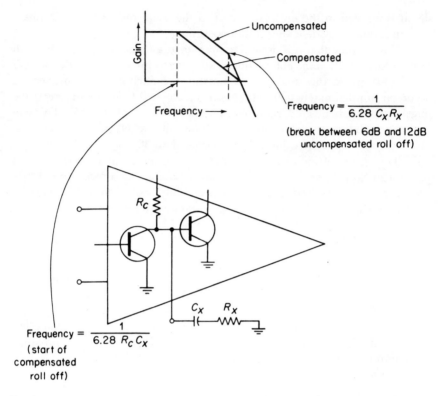

Fig. 3-16. Frequency-response compensation with external capacitor and resistor (conventional phase lag)

op-amp. If the compensated rolloff starts at too low a frequency, decrease the value of C, and find a new corresponding value for R.

As an example, assume that the uncompensated rolloff changes from 6 dB/octave to 12 dB/octave at 10 MHz, and that it is desired to have the compensated rolloff at 300 kHz. Assume that R_X is 1000 ohms (a convenient value). Using the equation of Fig. 3-16, a frequency of 10 MHz, and an R_X of 1000 ohms, the value for C_X is:

$$C_X = \frac{1}{6.28 \times 10 \text{ MHz} \times 1000} \approx 16 \text{ pF}$$

Test the op-amp with C_X at 10 pF and R_X at 1000 ohms. If the compensated rolloff starts at some frequency lower than 300 kHz, increase the value of C_X and find a new value for R_X. If the rolloff starts above 300 kHz, decrease the value of C, and use another value for R.

For example, if compensated rolloff starts at 100 kHz instead of the desired 300 kHz, increase the value of C_X to 30 pF. Using the equation of

Fig. 3-16, the corresponding value of R_X is:

$$R_X = \frac{1}{6.28 \times 10 \text{ MHz} \times 30 \text{ pF}} \approx 530 \text{ ohms}$$

3.2.9. Miller-effect phase lag compensation examples

As shown in Fig. 3-17, the problems in determining Miller-effect phase lag values are essentially the same as for conventional phase lag. That is, the values of the compensating resistor and capacitor are dependent upon the frequency at which uncompensated rolloff changes from 6 to 12 dB. The frequency at which compensated rolloff starts is dependent upon compensating RC values and internal op-amp values. (This is also true for the frequency at which compensated rolloff reaches unity gain.)

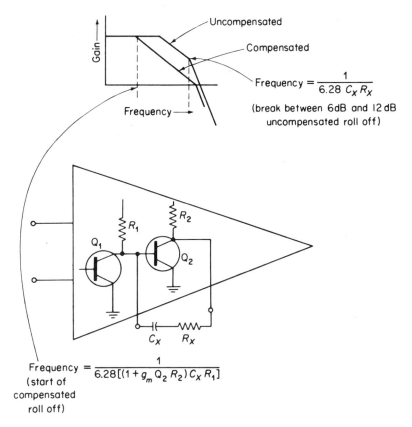

$$\text{Frequency} = \frac{1}{6.28\, C_X\, R_X}$$

(break between 6dB and 12 dB uncompensated roll off)

$$\text{Frequency} = \frac{1}{6.28[(1 + g_m\, Q_2\, R_2)\, C_X\, R_1]}$$

(start of compensated roll off)

Fig. 3-17. Frequency-response compensation with external capacitor and resistor (Miller-effect rolloff)

Again, always use manufacturer's information. If Miller-effect phase compensation values must be found without manufacturer's data, assume a convenient value for R, and find C using the equations of Fig. 3-17. Decrease the value of C if compensated rolloff starts at too low a frequency, and vice versa.

3.2.10. Internal Miller-effect compensation

As discussed in Sec. 3.1.4, one version of the op-amp shown in Fig. 3-5 is provided with internal compensation. The Miller-effect technique is used for compensation in this case. The resistor-capacitor network connected between the collector and base of Q_{11} provides the desired phase shift. This results in a fixed rolloff similar to that shown in Fig. 3-18. Note that this rolloff approaches that of the "ideal" op-amp shown in Fig. 3-11.

Any closed-loop gain (less than open-loop gain) can be selected by feedback resistance in the normal manner. Likewise, any combination of gain/frequency response can be selected. For example, if it is desired to start the rolloff at 10^4 Hz, choose a feedback resistance Z_F that is 10^2 times larger than the input resistance Z_R (such as $Z_F = 10,000$ and $Z_R = 100$). The closed-loop gain is then flat to 10^4 Hz, and rolls off at 6 dB/octave to unity gain at 10^6 Hz.

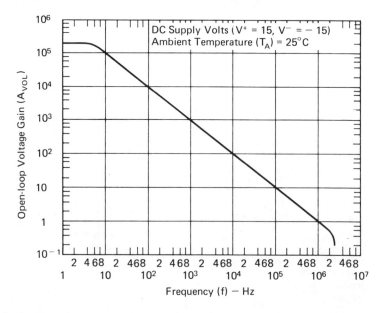

Fig. 3-18. Open-loop gain versus frequency characteristics of RCA op-amp with internal phase compensation

Keep in mind that the gain/frequency characteristics of an internally compensated op-amp cannot normally be altered. Thus, if a different gain/frequency response is required (say 10^2 gain at 10^5 Hz), an internally compensated op-amp is unsuitable.

3.3. OP-AMP CHARACTERISTICS

The designer must have a thorough knowledge of op-amp characteristics to get the best possible results from the op-amp in any system. From a user's standpoint, op-amp characteristics provide a good basis for op-amp system design. However, commercial IC op-amps are generally designed for specific applications. Although some commercial op-amps are described as general purpose, it is impossible to design an op-amp with truly universal characteristics.

For example, certain op-amps are designed to provide high-frequency gain at the expense of other performance characteristics. Other op-amps provide very high gain or high input impedance in low-frequency applications. IC op-amps that are fabricated by the diffusion process can be made suitable for comparator applications (where both halves of the differential amplifier circuits must be identical). Likewise, IC op-amps can be processed to provide high gain at low dissipation levels. For this reason, any description of op-amp characteristics must be of a general nature, unless a specific application is being considered.

Most of the op-amp characteristics required for proper use of the op-amp in any application can be obtained from the manufacturer's datasheet or similar catalog information. There are some exceptions to this rule. For certain applications it may be necessary to test the op-amp under simulated operating conditions.

In using datasheet information or test results, or both, it is always necessary to interpret the information. Each manufacturer has its own system of datasheets. It is impractical to discuss all datasheet formats here. Instead, we discuss typical information found on op-amp datasheets, as well as test results, and see how this information affects the op-amp user.

3.3.1. Open-loop voltage gain

The open-loop voltage gain (A_{VOL} or A_{OL}) is defined as the ratio of a change in output voltage to a change in input voltage at the input of the op-amp. Open-loop gain is always measured without feedback, and usually without phase-shift compensation.

Open-loop gain is *frequency dependent* (gain decreases with increased frequency). This is shown in Fig. 3-12. As shown, the gain is flat (within about ± 3 dB) up to frequencies of about 0.2 MHz. Then the gain rolls off to unity at frequencies above 20 MHz. The open-loop gain is also *temperature dependent*, and dependent upon supply voltage. Generally, gain increases with supply voltage and temperature, but not necessarily in the same ratio.

Ideally, open-loop gain should be infinitely high since the primary function of an op-amp is to amplify. In general, the higher the gain the higher the accuracy of op-amp transfer function (relationship of output to input). However, there are practical limits to gain magnitude, and also levels at which an increase in magnitude buys little in the way of increased performance. The true significance of open-loop gain is many times misapplied in op-amp operation where in reality open-loop gain determines closed-loop accuracy limits, rather than ultimate accuracy.

The numerical values of the open-loop gain (and the bandwidth) of an op-amp are of relatively little importance in themselves. The important requirement is that the open-loop gain must be greater than the closed-loop gain over the frequency of interest if an accurate transfer function is to be maintained. For example, if a 40 dB op-amp and a 60 dB op-amp are used in a 20 dB closed-loop gain configuration, and the open-loop gain is decreased 50 percent in each case (say because of component aging), the closed-loop gain of the 40 dB op-amp varies 9 percent, and that of the 60 dB op-amp varies only 1 percent.

The *frequency rolloff characteristics* are the prime determinants of the op-amp frequency response. The greater the rate of rolloff prior to the intersection of feedback ratio (closed-loop) frequency characteristics with the open-loop response (in the active region), the more difficult phase compensation of the op-amp becomes.

An 18 dB/octave rolloff is generally considered the maximum slope that can occur in the active region before proper phase compensation becomes extremely difficult or impossible to achieve. In addition, because op-amps have useful application down to and including unity gain, the active region of the op-amp may be considered as the entire portion of the frequency characteristic above the 0 dB bandwidth. Thus, a well-designed op-amp should roll off at no greater than 18 dB/octave until well below unity gain.

As discussed, open-loop gain can be modified by several compensation methods. A typical op-amp datasheet will show the results of such compensation, usually by means of graphs such as the one shown in Fig. 3-15.

After compensation is applied, the op-amp can be connected in the closed-loop configuration. The voltage gain under closed-loop conditions is dependent upon external components (the ratio of feedback resistance to input resistance). Thus, closed-loop gain is usually not listed as such on op-amp datasheets. However, the datasheet may show some typical gain curves with various ratios of feedback. If available, such curves can be used directly to select values of feedback components (as well as phase compensation components).

3.3.2. Phase shift

Figure 3-12 shows the open-loop phase shift curve of a typical op-amp. Because a closed-loop gain of unity allows the highest frequency response for the loop gain, the closed-loop unity gain frequency is considered the worst case for phase shift.

One figure of merit commonly used in evaluating the stability of an op-amp is *phase margin*. As discussed in Sec. 3.2, oscillations can be sustained if the total phase shift around the loop (from input to output, and back to input) can reach 360° before the total gain around the loop drops below unity (as the frequency is increased). Because an op-amp is normally used in the inverting mode, 180° of phase shift is available to begin with. Additional phase shift is developed by the op-amp because of internal circuit conditions.

Phase margin represents the difference between 180° and the phase shift introduced by the op-amp at the frequency at which loop-gain is unity. A value of 45° phase margin is considered quite conservative to provide a guard against production variations, temperature effects and other stray factors. This means that the op-amp should not be operated at a frequency where the phase shift exceeds 135° (180° − 45°). However, it is possible to operate op-amps at frequencies where the phase shift is kept within the 160° to 170° region. Of course, the ultimate stability of an op-amp must be established by tests.

3.3.3. Bandwidth, slew rate and output characteristics

The bandwidth, slew rate, output voltage swing, output current, and output power of an op-amp are all interrelated. These characteristics are frequency dependent, and depend upon phase compensation. The characteristics are also temperature and power supply dependent, but to a lesser extent. Before discussing the interrelationship, let us define each of the characteristics.

Bandwidth for an op-amp is usually expressed in terms of open-loop operation. The common term is BW_{OL} at − 3 dB. A BW_{OL} of 800 kHz indicates that the open-loop gain of the op-amp drops to a value of 3 dB below the flat or low-frequency level at a frequency of 800 kHz.

Frequency range is sometimes used in place of open-loop bandwidth. The frequency range of an op-amp is often listed as "useful frequency range" (such as dc up to 18 MHz). Useful frequency range for an op-amp is similar to the F_T (total frequency) term used with discrete transistors (Chapter 1). Generally, this high-frequency limit specified for an op-amp is the frequency at which gain drops to unity.

Power bandwidth is a more useful characteristic since it represents the bandwidth of the op-amp in closed-loop operation connected to a normal load. As shown in Fig. 3-19, power bandwidth is given as the peak-to-peak output voltage capability of the op-amp (working into a given load) across a

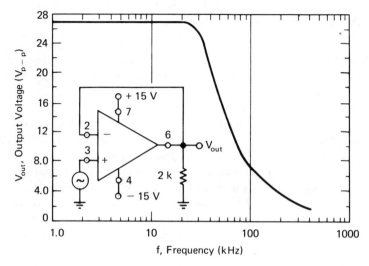

Fig. 3-19. Power bandwidth for Motorola MC1556

band of frequencies. Power bandwidth figures usually imply that the output indicated is free of distortion, or that distortion is within limits (such as total harmonic distortion less than 5 percent). In the op-amp of Fig. 3-19, the output voltage is about 27 V (peak-to-peak) up to about 20 kHz, and drops to near zero at 300 kHz.

Power output of an op-amp is generally listed in terms of power across a given load (such as 250 mW across 500 ohms). However, power output is usually listed at only one frequency. The same is true of *output current* or *maximum output current* characteristics found on some datasheets. Thus, power bandwidth is the more useful characteristic.

Output voltage swing is defined as the peak or peak-to-peak output voltage swing (referred to zero) that can be obtained without clipping. A symmetrical voltage swing is dependent upon frequency, load current, output impedance, and slew rate. Generally, an increase in frequency will decrease the possible output voltage swing. For a given frequency, an increase in phase compensation capacitance decreases output voltage swing. However, an increase in load resistance generally increases output voltage swing.

Slew rate of an op-amp is the maximum rate of change of the output voltage, with respect to time, that the op-amp is capable of producing while maintaining linear characteristics (symmetrical output without clipping).

Slew rate is expressed in terms of:

$$\frac{\text{difference in output voltage}}{\text{difference in time}} \quad \text{or} \quad \frac{dV_O}{dt}$$

Usually, slew rate is listed in terms of volts per microsecond. For example, if the output voltage from an op-amp is capable of changing 7 V in 1 μs, then the slew rate is 7. If, after compensation or other change, the op-amp changes a maximum of 3 V in 1 μs, the new slew rate is 3.

Slew rate of an op-amp is the direct function of the phase-shift compensation capacity. At higher frequencies, the current required to charge and discharge a compensating capacitor can limit available current to succeeding stages or loads, and thus result in lower slew rates. This is one reason why op-amp manufacturers usually provide for compensation of early stages in the op-amp (near the input) where signal levels are small and little current is required.

Slew rate decreases as compensation capacitance increases. Thus, where high frequencies are involved, the lowest values of compensation capacitor should be used. The major effect of slew rate in op-amp applications is on output power. All other factors being equal, a lower slew rate results in lower power output. Slew rate and the term *full power response* of an op-amp are directly related. Full power response is the maximum frequency measured in a closed-loop unity-gain configuration for which rated output voltage can be obtained for a sinewave signal with a specified load and without distortion because of a limiting slew rate.

The slew rate versus full power response relationship can be shown as:

$$\text{slew rate (in volts/second)} = 6.28 \times F_M \times E_O$$

where F_M is the full power response frequency (in Hz), and E_O is the peak output voltage (one-half the peak-to-peak voltage).

For example, using the characteristics shown in Fig. 3-19, the output voltage E_O is about 13 V (one-half the peak-to-peak of 26 V) at a frequency of 30 kHz. Thus, the slew rate is:

$$\text{slew rate} = 6.28 \times 30,000 \times 13 \approx 2,449,200 \text{ V/s} \approx 2.45 \text{ V/}\mu\text{s}$$

The equation can be turned around to find the full power response frequency. For example, assume that an op-amp is rated as having a slew rate of 2.5 V/μs and a peak-to-peak output of 20 V ($E_O = 10$ V). Find the full power response frequency F_M as follows:

$$F_M = \frac{2.5 \text{ V/}\mu\text{s}}{6.28 \times 10} = \frac{2,500,000 \text{ V/s}}{62.8} \approx 40,000 \text{ Hz} \approx 40\text{k Hz}$$

Of course, if curves such as shown in Fig. 3-19 are available, it is not necessary to calculate the maximum frequency for a given output. Simply follow the 20 V line until it crosses the curve at 40 kHz.

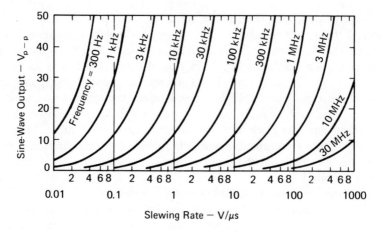

Fig. 3-20. Slewing rate curve for RCA op-amp

Slew rate versus output voltage. The graph of Fig. 3-20 shows the relationship among slew rate, full power response frequency, and output voltage. For example, if slew rate is 6 V/μs, the maximum output voltage (peak-to-peak) is about 20 V at a frequency of 100 kHz, and vice versa.

Power output versus slew rate. With a constant output load, the power output of an op-amp is dependent upon output voltage. In turn, all other factors being equal, output voltage is dependent upon the slew rate. Since slew rate depends upon phase compensation capacitance, op-amp power output is also dependent upon compensation. Some datasheets omit slew rate, but provide a graph that shows the direct relationship between full power output frequency and phase compensation.

3.3.4. Settling time

Some op-amp datasheets show a *settling time* factor instead of, or in addition to, the slew rate. Settling time indicates the amount of time required for the output voltage to swing from the maximum (usually at or near the supply voltage) to zero or to some specific voltage near zero. For example, the op-amp of Fig. 3-5 is described as having a settling time of 1.2 μs to 10 mV. This means that the output voltage will swing from the maximum to within 10 mV of zero in 1.2 μs. Since the op-amp is normally operated at 12 V, the output voltage will swing about 12 V in 1.2 μs, or about 10 V per microsecond. Thus, the slew rate is 10 V/μs. From a design standpoint, the shortest possible settling time is the most desirable.

3.3.5. Input and output impedance

Input impedance is defined as the impedance seen by a source looking into one input of the op-amp with the other input grounded, as shown in Fig. 3-21. The primary effect of input impedance on design is to reduce amplifier loop gain. If the input impedance is quite different from the impedance of the device driving the op-amp (source impedance), there will be a loss of input signal caused by the mismatch. However, in practical terms it is not possible to alter the op-amp impedance. Thus, if impedance match is critical, either the op-amp or driving source must be changed to effect a match.

Output impedance is defined as the impedance seen by a load at the output of the op-amp, as shown in Fig. 3-22. Excessive output impedance can reduce the gain since, in conjunction with the load and feedback resistors, output impedance forms an attenuator network. In general, output impedance of op-amps is less than 200 ohms. Generally, input resistances are at least 1000 ohms, with feedback resistance several times higher than 1000 ohms. Thus, the output impedance of a typical op-amp will have little effect on gain.

If the op-amp is serving primarily as a voltage amplifier (as is usually the case) the effect of output impedance will be at a minimum. Output

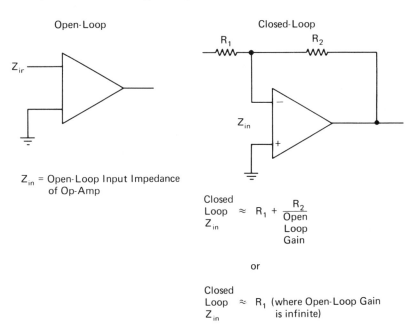

Fig. 3-21. Input impedance relationships

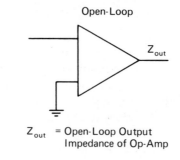

Open-Loop

Z_{out} = Open-Loop Output
Impedance of Op-Amp

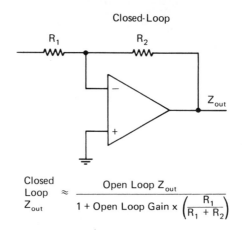

Closed-Loop

$$\text{Closed Loop } Z_{out} \approx \frac{\text{Open Loop } Z_{out}}{1 + \text{Open Loop Gain} \times \left(\frac{R_1}{R_1 + R_2}\right)}$$

Fig. 3-22. Output impedance relationships

impedance has a more significant effect in design of power applications where the op-amp must supply large amounts of load current.

Closed-loop output impedance is found by using the equation of Fig. 3-22. Thus, it will be seen that output impedance increases as frequency increases, since open-loop gain decreases. Both input and output impedance change with temperature, as well as frequency. Generally both characteristics are listed on datasheets at 25° C and 1 kHz, unless a graph is provided. Typically, input impedance increases, and output impedance decreases (slightly) with increases in temperature.

3.3.6.   Input common-mode voltage swing

Input common-mode voltage (V_{ICM}) is defined as the maximum peak input voltage that can be applied to either input terminal of the op-amp without causing abnormal operation or damage, as shown in Fig. 3-23. Some op-amp datasheets list a similar term: *common mode input voltage range* (V_{CMR}). Usually, V_{ICM} is listed in terms of peak voltage, with positive or negative peaks being

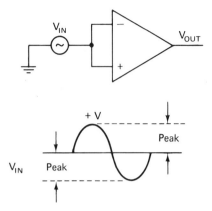

Fig. 3-23. Input common-mode voltage swing relationships

equal. V_{CMR} is often listed for positive and negative voltages of different values (such as $+1$ V and -3 V).

In practical use, either of these parameters limits the differential signal amplitude that can be applied to the op-amp input. So long as the input signal does not exceed the V_{ICM} or V_{CMR} values (in either the positive or negative directions), there should be no problem.

Note that some op-amp datasheets list "single-ended" input voltage signal limits where the differential input is not to be used.

3.3.7. Common-mode rejection ratio

Common-mode rejection terms are defined in Sec. 3.1.2. Under differential drive conditions, the common-mode rejection has no drastic effects on the performance of an op-amp, unless the rejection ratio is extremely low. However, in a common-mode drive application, such as in a comparator, high common-mode rejection is essential. For example, if an op-amp with a 60 dB differential gain, and a 50 dB common-mode rejection, is used to compare a 1 V signal against a 1 V reference, the output will be 3.2 V, when it should be zero. Such results would be totally unacceptable for a comparator type application. This is why the common-mode rejection should be at least 20 dB greater than the differential gain.

3.3.8. Input bias current

Input bias current is defined as the average value of the two input bias currents of the op-amp differential input stage. This is shown by the equation

$$I_b = \frac{I_1 + I_2}{2}$$

where I_b is input bias, I_1 and I_2 are currents of the inverting and noninverting inputs, respectively.

Input bias current is essentially a function of the large signal current gain of the input stage, and typically decreases as temperature increases.

In use, the significance of input bias current is the resultant voltage drop across input resistors or other source resistances. This voltage drop can restrict the input common-mode voltage range at higher impedance levels. The voltage drop must be overcome by the input signal. Also, a large input bias current is undesirable in applications where the source cannot accommodate a significant direct current. Examples of such applications are those in which the source resistance is very large (resulting in a large voltage drop), or sources of a magnetic nature that can be severely unbalanced by a flow of direct current (such as transducers that operate on magnetic principles).

Some op-amps have very low input bias current, and are thus well suited to these applications. Where very low input bias current is required, the input differential stages of the op-amp often use FETs or other MOS devices that draw very little input current. This is the case with the op-amp shown in Fig. 3-5 where the input current is 5 pA. Also note that when input current is low, input impedance (Sec. 3.3.5) is high. For example, the input impedance of the op-amp in Fig. 3-5 is 1.5 Terra Ω (1,500,000 MΩ).

3.3.9.   Input offset voltage and current

Input offset voltage is defined as the voltage that must be applied at the input terminals to obtain zero output voltage, as shown in Fig. 3-24. Input offset voltage indicates the *matching tolerance* in the differential amplifier stages. A perfectly matched amplifier requires zero input voltage to provide zero output voltage. Typically, input offset voltage is on the order of a few millivolts for an IC op-amp, as shown in Fig. 3-24. (Note that the input offset voltage for the op-amp shown in Fig. 3-5 is 8 mV.) The offset voltage for an op-amp can also be defined as the deviation of the output dc level from the arbitrary input-output level, usually taken as ground reference when both inputs are shorted together.

Input offset current is defined as the difference in input bias current into the input terminals of an op-amp, as shown in Fig. 3-24c. Input offset current is an indication of the *degree of matching* of the input differential stage. Typically, input offset current is on the order of 1 or 2 μA for an IC op-amp (particularly where the input stages use bipolar transistors), as shown in Fig. 3-24a. (Note that the input offset current for the op-amp shown in Fig. 3-5 is 0.5 pA, since the input stages use FETs.) The offset current for an op-amp can also be defined as the deviation when the inputs are driven by two identical dc input bias current sources.

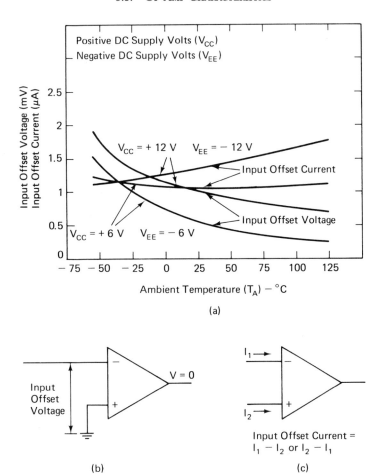

Fig. 3-24. Input offset voltage and current

Offset voltage and current are usually referred back to the input because their output values are dependent on feedback. (That is, datasheets rarely list output offset characteristics.) In normal use, the offset in an op-amp results from a combination of offset voltage and current. For example, if an op-amp has a 1 mV input offset voltage, and a 1 μA input offset current, with the inputs returned to ground through 1000 Ω resistors, the total input offset is either zero or 2 mV, depending upon the phase relationship between the two offset characteristics.

The offset of an op-amp is a direct-current error that should be minimized for numerous reasons, including the following:

1. The use of an op-amp as a true dc amplifier is limited to signal levels much greater than the offset;

UNIVERSITY OF PITTSBURGH

BRADFORD CAMPUS LIBRARY

2. Comparator applications require that the output voltage be zero (within limits) when the two input signals are equal and in phase;

3. In a direct-current cascade, the offset of the first stage determines the offset characteristics of the entire system.

Thus, any offset at the input of an op-amp is multiplied by the gain at the output. If the op-amp serves to drive additional amplifiers, the increased offset at the op-amp output will be multiplied even further. The gain of the entire system must then be limited to a value that is insufficient to cause limiting in the final output stage.

The effect of input offset voltage on op-amp use is that the input signal must overcome the offset voltage before an output will be produced. For example, if an op-amp has a 1 mV input offset voltage, and a 1 mV signal is applied, there is no output. If the signal is increased to 2 mV, the op-amp will produce only the peaks (where the input signal exceeds the 1 mV input offset). Since input offset voltage is increased by gain, the effect of input offset voltage is increased by the ratio of feedback resistance to input resistance, plus one, in the closed-loop condition. For example, if the ratio is 100 to 1 (for a gain of 100), the effect of input offset voltage is increased by 101.

Input offset current can be of greater importance than input offset voltage when high impedances are used in design. If the input bias current is different for each input, the voltage drops across the input resistors (or input impedance) will not be equal. If the resistance is large, there will be a large unbalance in input voltages. This condition can be minimized by means of a resistance connected between the noninverting input and ground, as shown in Fig. 3-25. The value of this resistor R_3 should equal the parallel equivalent of the input and feedback resistors R_1 and R_2, as shown by the equations of Fig. 3-25. In practical design, the trial value for R_3 is based on the equation of Fig. 3-25. The value of R_3 is then adjusted for minimum voltage difference at both terminals (under normal operating conditions, but with no signal).

Some IC op-amps include provisions to neutralize any offset. Typically, an external voltage is applied through a potentiometer to terminals on the op-amp. The voltage is adjusted until the offset, at the *input and output*, is zero. For example, note the terminals marked "offset null" (terminals 1 and 5) on the schematic of Fig. 3-5. The terminals are connected to the emitters of the differential amplifier stage that follows the input stage. Figure 3-26 shows a typical external offset null or neutralization used with the op-amp of Fig. 3-5.

For op-amps without offset compensation, the effects of input offset can be minimized by an external circuit. Figure 3-27 shows two such circuits, one for inverting and the other for noninverting op-amps. The equations shown in Fig. 3-27 assume that resistor R_B must be of a value to produce a

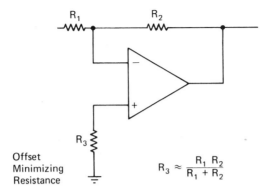

Fig. 3-25. Minimizing input offset current (and input offset voltage) by means of resistor at noninverting input

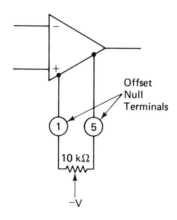

Fig. 3-26. Typical offset null or neutralization circuit

null range of ± 7.5 mV. This is generally sufficient for most IC op-amps. However, if a different input offset voltage range is required, simply substitute the desired range for ± 7.5 mV.

In addition to the basic circuits of Fig. 3-27, some of the applications described in Chapters 5 and 6 require special offset null circuits. One reason for an offset null is that the input and output dc levels of an op-amp should be equal, or nearly equal. This condition is desirable to assure that the resistive feedback network can be connected between the input and output (such as R_2 in Fig. 3-25) without upsetting either the differential or the common-mode direct current bias.

The average temperature coefficient of input offset voltage, listed on some datasheets as TCV_{IO}, is dependent upon the temperature coefficients of various components within the op-amp. Temperature changes affect stage

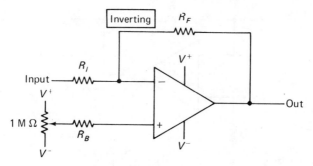

Value of R_B required to have a null adjustment
range of ± 7.5 mV:

$$R_B \approx \frac{R_I\, V^+}{7.5 \times 10^{-3}} \text{ assuming } R_B \gg R_I$$

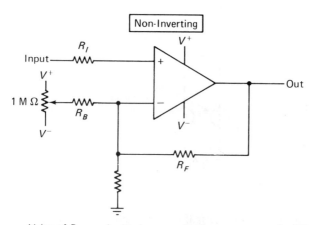

Value of R_B required to have a null adjustment range of ± 7.5 mV:

$$R_B \approx \frac{R_i R_F\, V^+}{(R_i + R_F)\, 7.5 \times 10^{-3}}$$

$$\text{Assuming } R_B \gg \frac{R_I R_F}{R_I + R_F}$$

Fig. 3-27. Input offset minimizing circuits

gain, match of differential amplifiers, and so forth, and thus change input offset voltage. From a user's standpoint TCV_{IO} need be considered only if the parameter is large, and the op-amp must be operated under extreme temperatures. For example, if input offset voltage doubles with an increase to a temperature that is likely to be found during normal operation, the higher input offset voltage should be considered the "normal" value for design.

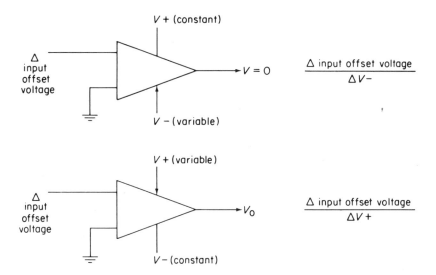

Fig. 3-28. Power supply sensitivity

3.3.10. Power supply sensitivity

Power supply sensitivity is defined as the ratio of change in input offset voltage to the change in supply voltage producing it, with the remaining supply held constant, as shown in Fig. 3-28. Some op-amp datasheets list a similar characteristic: *input offset voltage sensitivity*. In either case, the characteristic is expressed in terms of mV/V or μV/V, representing the change (in mV or μV) of input offset voltage to a change (in volts) of *one power supply*. Usually, there is a separate sensitivity characteristic for each power supply, with the opposite power supply assumed to be held constant. For example, a typical listing is 0.1 mV/V for a positive supply. This implies that with the negative supply held constant, the input offset voltage will change by 0.1 mV for each 1 V change in positive supply voltage.

The effects of power supply sensitivity (or input offset voltage sensitivity) are obvious. If an op-amp has considerable sensitivity to power supply variations, overall performance is affected by each supply voltage change. The power supply regulation must be increased to provide correct operation with minimum input signal levels.

3.3.11. Noise voltage

There are many systems for measuring noise voltage in an op-amp, and equally as many methods used to list the value on datasheets. Some datasheets omit the value entirely. In general, noise is measured with the op-amp in the open-loop condition, with or without compensation, and with the input shorted or with a fixed resistance load at the input terminals.

The input and/or output voltage is measured with a sensitive voltmeter or oscilloscope. Input noise is typically on the order of a few microvolts; output noise is usually less than 100 mV. Output noise is almost always greater than input noise (because of the amplifier gain).

Except in cases where the noise value is very high or the input signal is very low, op-amp noise can be ignored (unless a particular application requires extremely low noise). Obviously, a 10 μV noise at the input will mask a 10 μV signal. If the signal is raised to 1 mV with the same op-amp, the noise will be unnoticed. Noise is dependent upon temperature as well as upon the method of compensation used.

3.3.12. Power dissipation

An IC op-amp datasheet usually lists two power dissipation ratings. The same is true of some discrete component op-amps. One value is the *total device dissipation*, which includes any load current. The other value is *device dissipation*, which is defined as the dc power dissipated by the op-amp itself (with output at zero and no load). The device dissipation must be subtracted from the total dissipation to calculate the load dissipation.

For example, if an IC op-amp can dissipate a total of 300 mW (at a given temperature and supply voltage, and with or without a heat sink) and the IC op-amp itself dissipates 100 mW, the load cannot exceed 200 mW ($300 - 100 = 200$).

4

OPERATIONAL TRANSCONDUCTANCE AMPLIFIERS (VARIABLE OP-AMPS)

An operational transconductance amplifier (OTA) is similar in form to a conventional op-amp such as described in Chapter 3. However, OTAs and op-amps are not always interchangeable. For that reason an explanation of the unique characteristics found in OTAs is in order. The OTA not only includes the usual differential inputs of an op-amp, but also contains an additional *control input* in the form of an *amplifier bias current* (or I_{ABC}). This control input increases the OTA's flexibility for use in a wide range of applications. In effect, the OTA's characteristics can be varied or programmed to meet specific design needs.

The characteristics of an ideal OTA are similar to those of an ideal op-amp except that the OTA has an extremely high output impedance. Because of this basic difference, the output signal of an OTA is best described in terms of current that is proportional to the difference between the voltages of the two inputs. Thus, the transfer characteristic is best defined in terms of *transconductance* (similar to those of a vacuum tube) rather than voltage gain. Transconductance, or g_m, is the difference in output current, divided by the difference in input voltage, and is expressed in mhos or millimhos (mmhos). Except for the high output impedance, and the definition of input/output relationships, the characteristics of a typical OTA are similar to those of a typical op-amp.

This chapter describes operation of the OTA and features various circuits using the OTA. As is discussed, the OTA provides the equipment designer with a wider variety of circuit arrangements than does the conventional op-amp. This is because the user can select the optimum circuit conditions for a specific application simply by varying the bias (I_{ABC}) conditions of the OTA. For example, if low power consumption, low bias and low offset current, or high input impedance are desired, then low I_{ABC} current is applied to the OTA. On the other hand, if operation into a moderate load impedance is the main consideration, then higher levels of I_{ABC} bias may be used.

4.1. CIRCUIT DESCRIPTION OF TYPICAL OTA

Figure 4-1 shows the equivalent circuit for the OTA. The output signal is a "current" that is proportional to the transconductance (g_m) of the OTA, established by the I_{ABC} and the differential input voltage. The OTA can either source or sink current at the output, depending upon the polarity of the input signal.

Figure 4-2 is a simplified block diagram of an OTA. Transistors Q_1 and Q_2 form the usual differential input amplifier found in most op-amps. The lettered circles (with arrows leading either into or out of the circles)

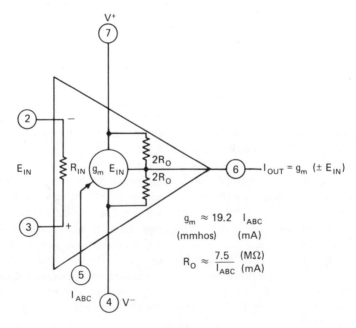

Fig. 4-1. Basic equivalent circuit of the OTA

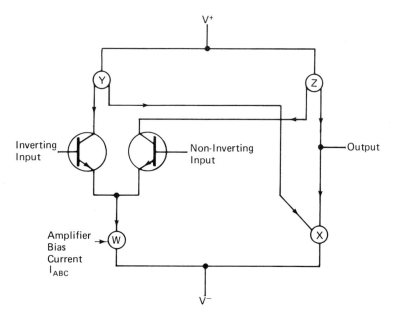

Fig. 4-2. Simplified diagram of OTA

indicate *current-mirrors*. The use of current-mirrors is essential to OTA operation. There are two basic types of current-mirror, as indicated in Figs. 4-3 and 4-4.

Figure 4-3 shows the basic type of current-mirror composed of two transistors, one of which (Q_2) is diode-connected. Because diode-connected transistor Q_2 is not in saturation, and is "active," the "diode" formed by the connection may be considered as a transistor with 100 percent feedback. Therefore, the base current still controls the collector current, as is the case in normal transistor action. That is, collector current I_C equals beta times base current I_b. If a current I_1 is forced into the diode-connected transistor Q_2, the base-to-emitter voltage will rise until the total current being supplied is divided between the collector and base regions. Thus, a base-to-emitter voltage is established in Q_2 such that Q_2 "sinks" the applied current I_1.

If the base of a second transistor Q_1 is connected to the base-to-collector junction of Q_2, then Q_1 is able to sink a current approximately equal to that flowing in the collector lead of the diode-connected transistor Q_2. This assumes that both transistors have identical characteristics (which is usually the case with IC fabrication where all components are formed on the same chip).

The difference in current between the input I_1 and the collector current I_2 of transistor Q_1 is owed to the fact that the base-current for both transistors is supplied from I_1. The ratio of the sinking current I_2 to the input

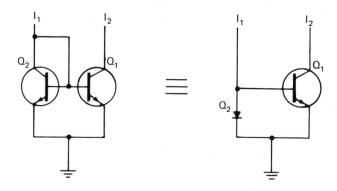

Fig. 4-3. Current-mirror using diode-connected transistor

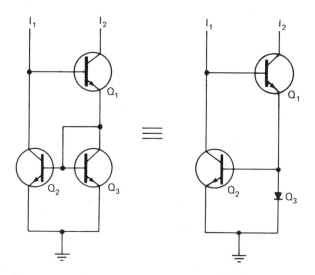

Fig. 4-4. Improved current-mirror with additional transistor

current I_1 is equal to $I_2/I_1 = \text{beta}/(\text{beta}+2)$. As beta increases, the output sinking current I_2 level approaches that of the input current I_1. Thus, the basic current-mirror of Fig. 4-3 is sensitive to transistor beta.

Circuit sensitivity to beta can be decreased, and an improvement in circuit output resistance characteristics can be made by the insertion of a diode-connected transistor in series with the emitter of Q_1. Such an arrangement is shown in Fig. 4-4. The diode-connected transistor Q_3 can be considered as a current-sampling diode that senses the emitter current of Q_1 and adjusts the base current of Q_1 (via Q_2) to maintain a constant current in I_2.

The current-mirror in Fig. 4-2 uses the basic configuration of Fig. 4-3. Current-mirrors X, Y, and Z are basically the constant-current version of Fig. 4-4. Mirrors Y and Z use *pnp* transistors, as shown by the arrows pointing outward from the mirrors.

4.1.1. Typical OTA circuits

Figure 4-5 is a complete schematic diagram of an OTA. The example shown is an RCA type CA3080 OTA available in IC form. The OTA shown uses only active devices (transistors and diodes, no resistors). Current applied to the I_{ABC} input established the emitter current of the input differential amplifier Q_1 and Q_2. This provides effective control of the differential transconductance (g_m).

In the case of the RCA CA3080, the transconductance $g_m = 19.2 \times I_{ABC}$, where g_m is in millimhos and I_{ABC}, is in mA. The temperature coefficient of g_m is approximately $-0.33\%/\,°C$ (at room temperature).

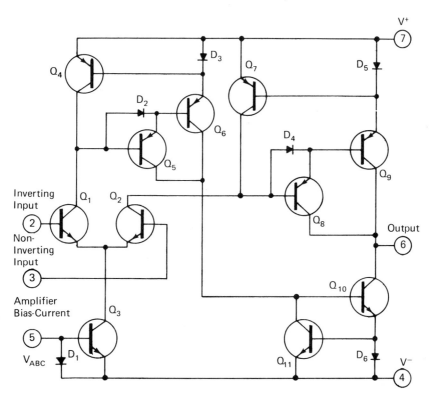

Fig. 4-5. RCA CA3080 and CA3080A operational transconductance amplifier (OTA)

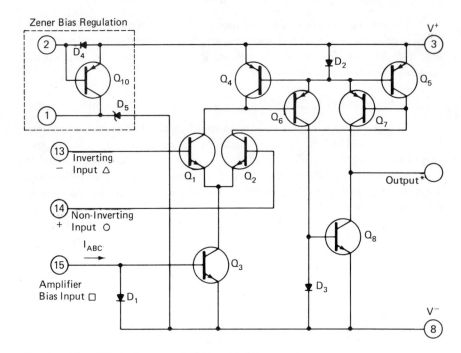

Fig. 4-6. Simplified diagram of OTA with bias regulator

Transistor Q_3 and diode D_1 of Fig. 4-5 form the current-mirror W of Fig. 4-2. Similarly, transistors Q_7, Q_8 and Q_9, and diode D_5, make up the current-mirror Z of Fig. 4-2. Darlington-connected transistors are used in mirrors Y and Z to reduce the voltage sensitivity of the mirror, but an increase in mirror output impedance.

Transistors Q_{10}, Q_{11}, and diode D_6 of Fig. 4-5 make up the current-mirror X of Fig. 4-2. Diodes D_2 and D_4 are connected across the base-emitter junctions of Q_5 and Q_8, respectively, to improve the circuit speed. The amplifier output signal is derived from the collectors of the Z and X current-mirror of Fig. 4-2, providing a push-pull class A output stage that produces full differential g_m.

Figure 4-6 is the schematic diagram of another RCA OTA. This device, available in IC form, is designated as the CA3060. The OTAs in the CA3060 family incorporate a unique Zener diode regulator system (D_4, D_5, Q_{10}) that permits current regulation below supply voltages normally associated with such a system.

4.2. DEFINITION OF OTA TERMS

The following terms apply to all types of OTA circuits. However, the terms were first applied to IC OTA devices developed by RCA.

Amplifier Bias Current (I_{ABC}): The current supplied to the amplifier bias terminal to establish the operating point (such as the I_{ABC} current at the base of Q_3 in Fig. 4-6).

Amplifier Supply Current (I_A): The current drawn by the amplifier from the positive supply source. The total supply current—which includes the sum of the amplifier supply current, the amplifier bias current, and the regulator bias current—is not to be mistaken for the amplifier supply current.

Bias Regulator Current: The current flowing from the Zener bias regulator (such as at terminal 2 of Fig. 4-6), set by an external source, that establishes the operating conditions of the bias regulator.

Bias Terminal Voltage (V_{ABC}): The voltage existing between the amplifier bias terminal and the negative supply voltage terminal (such as between the I_{ABC} terminal and terminal 8 of Fig. 4-6).

Peak Output Current (I_{OM}): The maximum current that is drawn from a short circuit on the output of the amplifier (positive I_O) or the maximum current delivered into a short circuit load (negative I_O). Peak-to-peak current swing is twice the peak output current (I_{OM}).

Peak Output Voltage (V_{OM}): The maximum positive voltage swing (V_{OM} +) or the maximum negative voltage swing (V_{OM}−) for a specific supply voltage and amplifier bias.

Power Consumption (P): The product of the sum of the supply voltages and the supply current, or $((V+) + (V-)) \times (I_A)$. This is not the total power consumed by an operating circuit. The power in the regulator must also be included for total power consumed.

Zener Regulator Voltage (V_Z): The regulator voltage (such as across terminals 1 and 8 of Fig. 4-6), measured with current flowing in the bias regulator.

4.3. EFFECTS OF CONTROL BIAS ON OTA CHARACTERISTICS

Unlike conventional op-amps, the characteristics of OTAs can be altered by adjustment of I_{ABC}. In effect, many of the OTA characteristics can be programmed to meet specific design problems. The following is a summary of the effects of bias on typical OTA units. Although the absolute values given in the following examples apply to RCA OTAs, the range of characteristic changes shows the possible effects on operation of any OTA with variations in I_{ABC}. (Note that the characteristics listed here for OTAs are the same as for conventional op-amps described in Chapter 3).

Input Offset Voltage (V_{io}) is not drastically affected by variations in I_{ABC}. A possible exception is when the OTA is operated at high temperatures.

Input Offset Current (I_{io}) is directly affected by I_{ABC}. At +25 °C, I_{io} increases almost in direct proportion with increases in I_{ABC}.

Input Bias Current (I_{ib}) also increases directly with increases in I_{ABC}.

Peak Output Current (I_{OM}^{+} or I_{OM}^{-}) is another characteristic that increases with I_{ABC}.

Peak Output Voltage (V_{OM}) is not drastically affected by variations in I_{ABC}. Instead, V_{OM} is set (primarily) by supply voltage, as is the case with a conventional op-amp.

Amplifier Supply Current (I_A), *Device Dissipation* (P_D), and *transconductance* (g_m) all increase directly with I_{ABC}.

Input Resistance (R_i) *and Output Resistance* (R_o) both decrease with increases in I_{ABC}.

Input and Output Capacitance (C_I and C_O), as well as *Amplifier Bias Voltage* (V_{ABC}) all increase with I_{ABC}. However, these characteristics do not increase in direct proportion to I_{ABC}. That is, a large increase in I_{ABC} occurs for a small increase in C_I, C_O and V_{ABC}.

4.4. BASIC DESIGN CONSIDERATIONS FOR OTA UNITS

The basic function of an OTA is as a substitute for an op-amp. However, the OTA allows the designer to select and control operating conditions of the op-amp circuit by adjusting the I_{ABC}. This permits the designer to have complete control over circuit transconductance, peak output current, and total power consumption, relatively independent of supply voltage. In addition, the high output impedance makes the OTA ideal for applications in which current summing is involved.

The following steps outline procedures for design of an op-amp, using an OTA. The device selected for this example is a CA3060, which is an RCA OTA in IC form. Figure 4-7 is the working schematic of a 20 dB op-amp using the CA3060. The circuit requirements are as follows:

> Closed loop voltage gain: 10 (20 dB)
>
> Offset voltage: adjustable to zero
>
> Current drain: as low as possible
>
> Supply voltage: ±6 V
>
> Maximum input voltage: ±50 mV
>
> Input resistance: 20 kΩ
>
> Load resistance: 20 kΩ

As discussed in Chapter 3, the closed-loop gain is set by the ratio of feedback resistance R_F to input resistance R_S. With R_S specified as 20 kΩ,

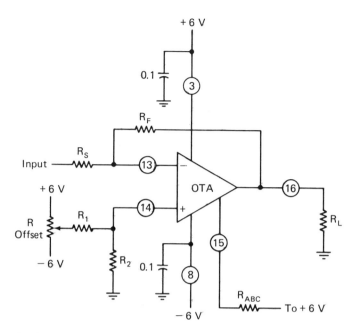

Fig. 4-7. Basic OTA system connections

and a desired closed-loop gain of 10, the value of R_F is: 10×20 kΩ or 200 kΩ.

The next step is to calculate the required transconductance g_m (also listed as g_{21} on some OTA datasheets) to produce a suitable open-loop gain. Assume that the open-loop gain A_{OL} must be at least 10 times the closed-loop gain. With a closed-loop gain of 10, the open-loop gain must be: 10×100.

Open-loop gain A_{OL} is related directly to load resistance R_L and transconductance g_m. The required transconductance is equal to A_{OL} divided by R_L. With an A_{OL} of 100 and R_L of 20 kΩ, the g_m should be: $100/20$ k$\Omega = 5$ millimhos (mmhos). However, the actual load resistance is the parallel combination of R_L and R_F, or approximately 18 kΩ, found by $(20 \times 200)/(20 + 200) \approx 18$. With an A_{OL} of 100 and an actual load R_L of 18 kΩ, the g_m should be: $100/18$ kΩ, or approximately 5.5 mmho.

The transconductance g_m is set by I_{ABC}. With a datasheet curve similar to that of Fig. 4-8, select an I_{ABC} from the *minimum value* curve to assure that the OTA will provide sufficient gain. As shown in Fig. 4-8, for a g_m (shown as g_{21}) of 5.5 mmho, the required I_{ABC} is approximately 20 μA.

Before calculating the value of R_{ABC} that will produce the desired I_{ABC}, check that the calculated I_{ABC} (20 μA) will produce the desired output swing capability. With an input of ± 50 mV, and a gain of 10, the output voltage swing is ± 0.5 V, and the 0.5 V will appear across the output load. As

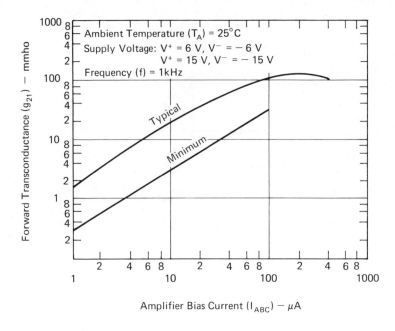

Fig. 4-8. Forward transconductance versus amplifier bias current for RCA OTA

discussed, the output load is equal to the parallel combination of R_L and R_F, or approximately 18 kΩ. With a 0.5 V swing, and an approximate load of 18 kΩ, the total amplifier current output is approximately: 0.5 V/18 kΩ, or 27.7 μA.

With a datasheet curve similar to that of Fig. 4-9, use the minimum value curve to check that an I_{ABC} of 20 μA will produce an I_{OM} of at least 27.7 μA. As shown in Fig. 4-9, for an I_{ABC} of 20 μA, the I_{OM} is approximately 40 μA, well above the desired minimum of 27.7 μA.

Once assured that the calculated I_{ABC} will meet the output swing requirements, calculate the value of R_{ABC}. As shown in Fig. 4-7, R_{ABC} is connected to the +6 V supply. (R_{ABC} can be connected to the Zener bias regulator. However, this will increase current drain on the supply.)

As shown in Fig. 4-6, with R_{ABC} connected to +V, R_{ABC} and diode D_1 are in series between the +V and −V supplies, and there is a total of 12 V across the series components. The drop across D_1, which is V_{ABC}, can be found by reference to a curve similar to Fig. 4-10. As shown in Fig. 4-10, with an I_{ABC} of 20 μA, V_{ABC} is approximately 630 mV, or 0.63 V. The drop across R_{ABC} is 12−0.63 V, or 11.37 V. For a drop of 11.37 V and I_{ABC} of 20 μA, the value of R_{ABC} is: 11.37/20 μA=568 kΩ. Use the next lowest standard resistor of 560 kΩ to assure that a minimum I_{ABC} of 20 μA will flow.

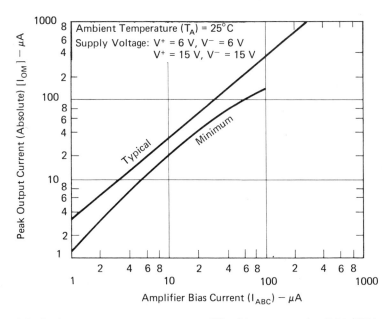

Fig. 4-9. Peak output current versus amplifier bias current for RCA OTA

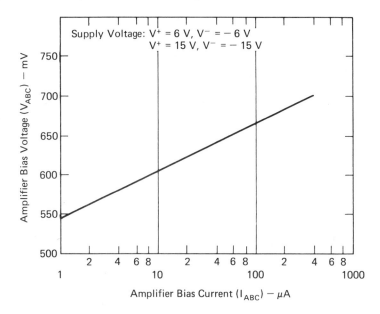

Fig. 4-10. Amplifier bias voltage versus amplifier bias current for RCA OTA

The final step is to calculate values for the input offset adjustment circuit R_{offset}, R_1 and R_2. To reduce the loading effect of the offset adjustment circuit on the power supply, the values should be selected in a manner similar to that of a conventional op-amp. For example, the value of R_2 should be approximately equal to the parallel combination of R_S and R_F. This will equalize currents between the inverting and noninverting inputs. Thus, R_2 should equal $(20 \times 200)/(20 + 200)$, or approximately 18 k$\Omega$.

With a datasheet curve similar to that of Fig. 4-11, find the input offset current. As shown in Fig. 4-11, for an I_{ABC} of 20 μA, the input offset current should be a maximum of 200 nA. With 200 nA flowing through R_2, the voltage across R_2 is: $200^{-9} \times 18^{+3} = 3600^{-6}$, or 3.6 mV. This 3.6 mV must be added to the maximum input offset voltage possible for the OTA. The OTA datasheet shows a maximum input offset voltage of 5 mV. Thus, the maximum voltage required at the noninverting input is: 5 mV + 3.6 mV or 8.6 mV.

The current necessary to provide a possible offset voltage of 8.6 mV across an 18 kΩ resistance is: $8.6^{-3}/18^{+3}$, or $0.48^{-6}(0.48 \mu$A). This current must flow through R_1.

A possible ± 6 V is available from R_{offset} to R_1. However, for a more stable circuit, assume that ± 1 V is available to R_1. With 1 V available, and a required current of 0.48 μA, the value of R_1 is: $1/0.48^{-6}$, or approximately 2 MΩ. Use the next larger standard value of 2.2 MΩ.

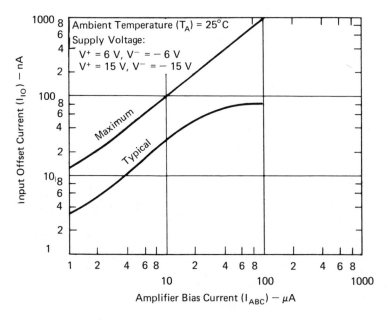

Fig. 4-11. Input offset current versus amplifier bias current for RCA OTA

The value of R_{offset} is not critical. A larger value of R_{offset} will draw less current from the supplies. As a guideline, the maximum value of R_{offset} should be less than twice the value of R_1, or less than 4.4 MΩ in this case. Use a standard value of 4 MΩ.

This completes the final step in design. Other design considerations for OTAs are essentially the same as for conventional op-amps, as discussed in Chapter 3. For example, phase compensation is accomplished by the usual open-loop or closed-loop methods. (The datasheet for the CA3060 recommends open-loop compensation by means of a resistor and capacitor across the differential input terminals.) As always, the user should follow all datasheet recommendations. If none are available, then use one of the phase compensation methods described in Chapter 3. This same recommendation applies to selection of decoupling capacitor values.

4.4.1. Effects of capacitance on OTA circuit design

One major difference between the OTA and conventional op-amps is the high output impedance of all OTA units. A typical OTA has a much higher output impedance than a comperable op-amp, even though the output impedance can be varied by adjustment of I_{ABC}. Because of this higher impedance, the effects of capacitance are much greater. For example, a 10 kΩ output load with a stray capacitance of 15 pF has an RC time constant of about 1 MHz. In designing any OTA circuit, particularly where the OTA is used as a feedback amplifier, stray capacitance must always be considered because of the adverse effect on frequency response and stability. Figure 4-12 illustrates how a 10 kΩ, 15 pF load modifies the frequency characteristics of the RCA CA3060 unit.

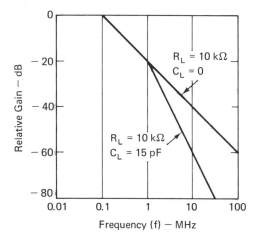

Fig. 4-12. Effect of capacitive loading on frequency response

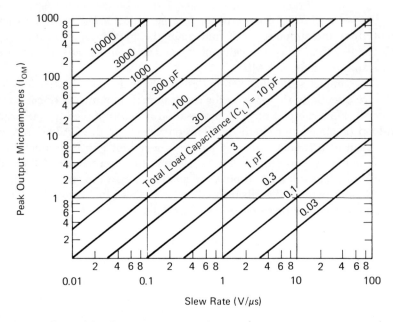

Fig. 4-13. Effect of load capacitance on slew rate

Capacitive loading also has an effect on slew rate. The maximum slew rate is limited to the maximum rate at which the capacitance can be charged by the peak output current I_{OM} (I_{OM} is set by I_{ABC} as shown in Fig. 4-9.)

As discussed in Chapter 3, slew rate is the difference in output voltage divided by the difference in time, and is expressed as dV/dt. In the case of an OTA connected as shown in Fig. 4-7, the slew rate is equal to the I_{OM} divided by C_L, where C_L is the total load capacitance, including strays.

Figure 4-13 illustrates the relationship between slew rate and total load capacitance. For example, with a given peak output current I_{OM} of 10 μA and a load capacitance of 100 pF, the slew rate is 0.1 V/μs. If the load capacitance can be decreased to 1 pF, the slew rate is increased to 10 V-μs.

4.5.   TYPICAL OTA APPLICATIONS

The following paragraphs describe how OTA units can be used. Keep in mind that an OTA can be used as a substitute for a conventional op-amp in any application, provided that the OTA has comparable characteristics (frequency response, power output, etc.). The area in which an OTA cannot be substituted for a conventional op-amp is one in which low output impedance is required. Even when OTA output impedance is reduced to the minimum by adjustment of I_{ABC}, the impedance is generally higher than for a comparable op-amp.

4.5.1. OTA multiplexer

Because an OTA has the variable bias (I_{ABC}) feature, the OTA can be gated for multiplex applications. That is, the OTA can be gated full-on or full-off by means of pulses applied to the I_{ABC} input. In a multiplex circuit, two or more OTA units are connected so that their outputs are summed together, with the inputs receiving signals from separate sources. The resultant output is the combination of the input signals.

Figure 4-14 shows a simple two-channel multiplex system using two RCA CA3080 OTA devices. In this example, positive and negative 5 V power supplies are used for the OTAs. The IC flip-flop is powered by the positive supply. If necessary to satisfy the logic supply requirements, the negative supply voltage may be increased to -15 V, with the positive supply at $+5$ V.

Outputs from the clocked flip-flop are applied through *pnp* transistors to gate the OTAs (on alternate half-cycles) at the I_{ABC} input. The transistors are connected in the grounded-base configuration to minimize capacitive feed-through (from flip-flop to OTA) via the base-collector junction of the *pnp* transistors.

There is some level shift between input and output of the multiplex system. In the case of the OTA units shown, the level shift is about 2 mV for the CA3080A and 5 mV for the CA3080. Of course, the level shift depends upon characteristics of the OTA used. In general, the level shift is due to input offset voltage of the OTA, rather than the gain. Typically, open-loop gain of the system is about 100 dB, with normal loading at the output. To further increase gain and reduce the effects of load, it is possible to add a buffer and/or amplifier stage at the OTA multiplexed output.

The circuit of Fig. 4-14 may require some phase compensation at the output. This can be done by means of a simple *RC* network as shown. The values for the *RC* phase-compensation network are dependent upon frequency characteristics of the OTA. As a general rule, let $1/(6.28\,RC)$ be approximately equal to the lowest frequency pole of the OTA. That is, let the *RC* network equal the lowest frequency at which the OTA gain begins to drop. In the example of Fig. 4-14, the frequency is approximately 2 MHz. Refer to Chapter 3 for a further discussion of phase compensation.

The datasheet for the CA3080 shows additional multiplexer circuits. Likewise, the datasheet for the CA3060 (an IC package that contains three identical OTA units) describes three-channel multiplexer systems.

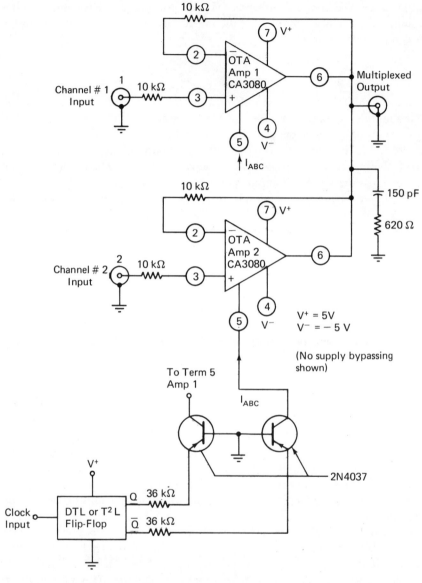

Fig. 4-14. Two-channel linear time-shared multiplex circuit using RCA OTAs

4.5.2. OTA comparator

Any OTA can be used as the amplifier in a comparator circuit. The RCA CA3060 is well suited for use in a tri-level comparator circuit since the CA3060 has three identical OTA units in one IC package. Figure 4-15 shows the functional block diagram of a tri-level comparator. The circuit has three adjustable limits. If either the upper or lower limit is exceeded, the appropriate output is activated until the input signal returns to a selected intermediate limit. Tri-level comparators are particularly suited to many industrial control applications.

As shown in Fig. 4-15, two of the three amplifiers are used to compare the input signal with the upper-limit and lower-limit reference voltages. The third amplifier is used to compare the input signal with a selected value of intermediate-limit reference voltage. By appropriate selection of resistance

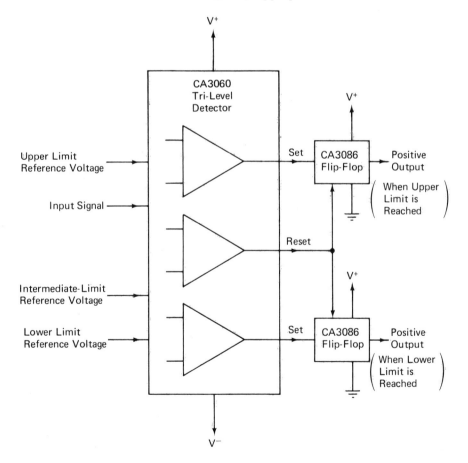

Fig. 4-15. Functional block diagram of tri-level comparator

ratios the intermediate-limit may be set to any voltage between the upper-limit and lower-limit values.

The output of the upper-limit and lower-limit comparator sets the corresponding upper- or lower-limit flip-flop. The activated flip-flop retains its state until the third comparator (intermediate limit) initiates a reset function, thereby indicating that the signal voltage has returned to the intermediate limit selected.

The full circuit diagram of the tri-level comparator is shown in Fig. 4-16. The flip-flops are made up of RCA CA3086 transistor arrays (in IC form). Discrete transistors can be used instead, with similar resistance values. Power is provided for the CA3060 by ± 6 V supplies. The built-in regulator (see Fig. 4-6) provides amplifier bias current (I_{ABC}) to the three amplifiers.

In the circuit of Fig. 4-16, upper-limit and lower-limit reference voltages are selected by appropriate adjustment of potentiometers R_1 and R_2, respectively. When resistors R_3 and R_4 are equal in value (as shown), the intermediate-limit reference voltage is automatically established at a value midway between the lower-limit and upper-limit values. Other values of intermediate-limit voltages can be selected by adjustment of R_3 and R_4. The input signal E_S is applied to the three comparators through 5.1 kΩ resistors. The comparator outputs on the upper-limit and lower-limit SET lines trigger the appropriate flip-flop whenever the input signal reaches a limit value. When the input signal returns to an intermediate value, the common flip-flop RESET line is energized. The loads shown in the circuit of Fig. 4-16 are 5 V, 25 mA lamps. Other loads can be used, within the current and voltage capabilities of the flip-flop components.

4.5.3.   OTA sample-and-hold circuits

The multiplex system described in Sec. 4.5.1. can be modified to produce sample-and-hold functions, using the "strobe" or gating characteristics of an OTA. That is, the I_{ABC} input of the OTA is pulsed to switch the OTA on and off. The output of the OTA then represents a sample of the OTA signal input (taken during the on period). The "hold" function is provided by means of a capacitor at the OTA output.

Figure 4-17 shows the basic sample-and-hold system using an RCA CA3080A as the amplifier. In this circuit, the OTA functions as a basic voltage-follower, with the phase-compensation capacitor C serving the additional function of sampled-signal storage. When the I_{ABC} input is at 0 V (SAMPLE), the OTA is on. Under these conditions, the OTA output is at the same level as the signal input, and the 300 pF phase-compensation capacitor C charges to the level of the OTA output. When the I_{ABC} input is at -15 V (HOLD), the OTA is off. However, the capacitor C remains charged.

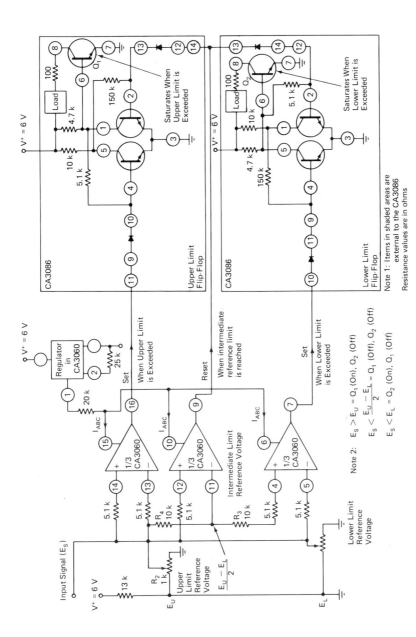

Fig. 4-16. Tri-level comparator using RCA OTAs

125

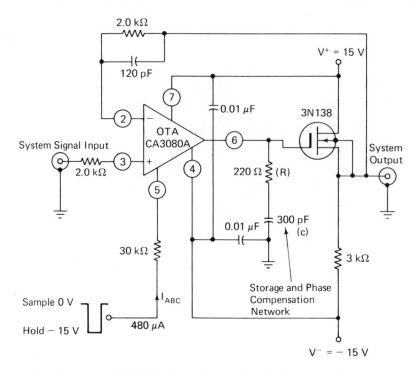

Fig. 4-17. Sample-and-hold circuit using an RCA OTA

The main problem with any sample-and-hold system using a charging capacitor is that the capacitor may discharge through leakage during the off or HOLD cycle. Such leakage can occur through the amplifier output circuit, or the 3N138 input circuit. However, since an OTA has very high output resistance, the leakage path through the OTA is practically nil. (The CA3080 has an output resistance in excess of 1000 MΩ under cut-off conditions.) This low leakage path is one of the advantages of an OTA over a conventional op-amp when used in a sample-and-hold circuit. Likewise, the gate leakage of the 3N138 is very low (typically 10 pA) since the transistor is a MOSFET (with insulated gate).

The open-loop voltage gain of the system is approximately 100 dB. The open-loop output impedance of the 3N138 is approximately 220 ohms (with a g_m of about 4600 μmhos at an operation current of 5 mA). The system closed-loop output impedance is approximately equal to Z_O (open-loop output impedance of the 3N138) divided by open-loop voltage gain, or 220 ohms/100 dB = 220 ohms/10^5 = approximately 0.0022 ohm. The output impedance is comparable to the closed-loop output impedance of a conventional op-amp.

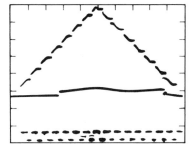

Top Trace: Sampled Signal 1 V/Div and 20 μsec/Div
Center Trace: Top Portion of Upper Signal
1 V/Div and 2 μsec/Div
Bottom Trace: Sampling Signal 20 V/Div and
20 μsec/Div

Fig. 4-18. Waveforms for sample-and-hold circuit

Figure 4-18 shows a "sampled" triangular wave using the circuit of Fig. 4-17. The lower trace is the sampling signal (pulse applied to the I_{ABC} input). When the sampling signal goes negative (to -15 V), the OTA is cut off, and the system signal is "held" on the storage capacitor C, as shown by the plateaus on the triangular waveform. The center trace of Fig. 4-18 is a time expansion of the topmost transition (in the upper trace) with a time scale of 2 μs/div.

Output tilt. As discussed, variation in the stored signal level during the hold period is of concern in any sample-and-hold system using the charging capacitor method. This variation is primarily a function of the cutoff leakage current in the OTA output (a maximum limit of 5 nA), the leakage of the storage element (capacitor C), and other extraneous paths (such as gate leakage of the 3N138). The leakage currents may be either positive or negative. Consequently, the stored signal may rise or fall during the hold interval. The term "tilt" is used to describe this condition. Figure 4-19 shows the expected pulse tilt in microvolts as a function of time for various values of the compensation/storage capacitor C. The horizontal axis shows three scales representing typical leakage currents of 50 nA, 5 nA, and 500 pA.

As an example of how Fig. 4-19 can be used, assume that the hold period is 20 μs, and that capacitor is 100 pF. If leakage can be limited to 500 pA, the pulse tilt will be 100 μV. That is, the system output level can be shifted by 100 μV during the hold period. If the leakage current increases to 5 nA, the pulse tilt increases to 1000 μV. The effects of level shift (or pulse tilt) are cumulative. For example, if 10 hold periods are required for a system signal input, the total level shift is 10 times that for each hold period.

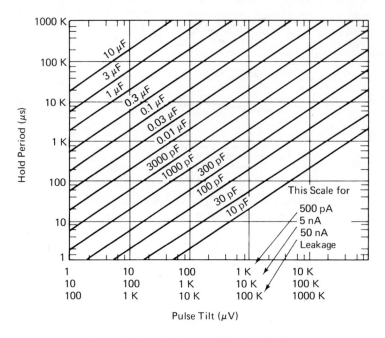

Fig. 4-19. Chart showing "tilt" in sample-and-hold potentials as a function of hold time with load capacitance as a parameter

DTL/TTL Logic levels. The circuit of Fig. 4-17 can be modified to operate with pulse levels found in DTL/TTL logic systems (refer to Chapter 8.) Typically, DTL/TTL logic pulses are at 0 V and 5 V levels. Figure 4-20 shows the required modification. The control pulse to the I_{ABC} input is applied through a *pnp* transistor (RCA 2N4037) connected in the grounded-base configuration. Grounded base is used to minimize capacitive feedthrough from the control pulse source to the I_{ABC} input. The 9.1 kΩ resistor connected to the 2N4037 emitter establishes I_{ABC} conditions similar to those used in the circuit of Fig. 4-17.

Slew rate problems. There is a trade-off on the size of capacitor C. If the value of capacitor C is increased, larger signals can be sampled and held. However, each increase in capacitor C decreases the slew rate. In general, use the largest possible capacitor C compatible with required slew rate. Figure 4-21 shows slew rate as a function of I_{ABC} and various capacitors C. The information of Fig. 4-21 applies to capacitor C in both Figs. 4-17 and 4-20. The magnitude of the current being supplied to capacitor C is equal to the I_{ABC} when the OTA is supplying its maximum output current. As in the case with the multiplexer (Sec. 4.5.1.), let the RC network equal the lowest frequency at which OTA gain begins to drop.

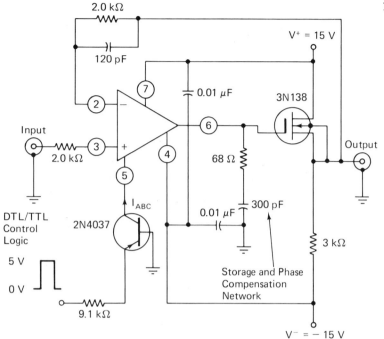

Fig. 4-20. Sample-and-hold circuit for DTL/TTL control logic using RCA OTA

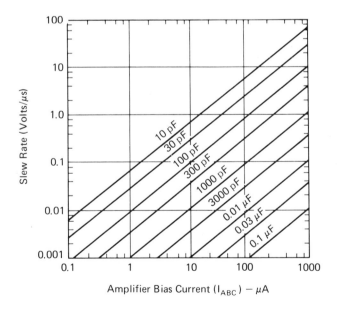

Fig. 4-21. Slew rate as a function of amplifier bias current with phase-compensation capacitance as a parameter

4.5.4. OTA gyrator

One of the problems in designing filters used at very low frequencies is the need for a very large inductance. This problem can be overcome by means of an active filter gyrator circuit. A gyrator is a circuit that appears as a variable inductance of high value (typically in the kilohenry region). However, the gyrator does not contain any inductive components.

Figure 4-22 shows a gyrator circuit composed of two OTA units (such as the CA3060). The circuit appears as an inductance of up to 10 kilohenry across terminals A and B. The circuit can be tuned to exact inductance values by means of potentiometer R_1. The setting of R_1 varies the I_{ABC} current of both OTAs simultaneously. When one I_{ABC} current is decreased, the opposite current is increased. This action varies the transconductance of both OTAs simultaneously, and serves to "tune" the circuit by changing the resistance of the OTAs.

The circuit of Fig. 4-22 makes use of the high impedance output available in OTA units. The Q of the 10 kH variable inductance is

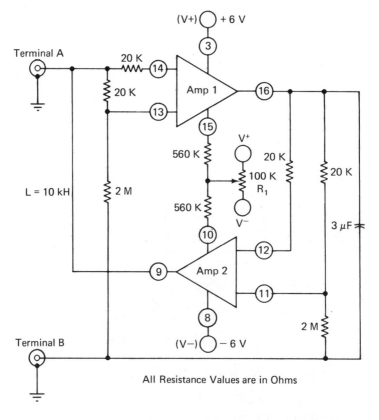

Fig. 4-22. Gyrator in an active filter circuit using two RCA OTAs

approximately 13, using the circuit values shown. The 20 kΩ to 2 MΩ attenuators in this circuit extend the dynamic range of each OTA by a factor of 100.

4.5.5. OTA gain control and modulation circuits

An obvious function of an OTA is that of a gain control element. The gain of a signal passing through an OTA can be controlled by variation of the I_{ABC}. This is because transconductance of the OTA (and thus gain of the circuit) is directly proportional to I_{ABC}.

Gain control. In the simplest form, an OTA can be connected as a conventional amplifier, but with the I_{ABC} input connected to a voltage source through a variable resistance (which acts as the gain control). For a specified value of I_{ABC} (as set by the variable resistance), the output current of the OTA is equal to the product of transconductance and the input signal magnitude. The output voltage swing is the product of output current and the load resistance.

Amplitude modulation. The gain control function can be applied to amplitude modulation of signals. Figure 4-23 shows a basic amplitude modulation system using an OTA as the modulator. In this circuit, the signal input is a voltage V_X at some carrier frequency, and the I_{ABC} input is a voltage V_M at some modulating frequency. The output signal current I_O is

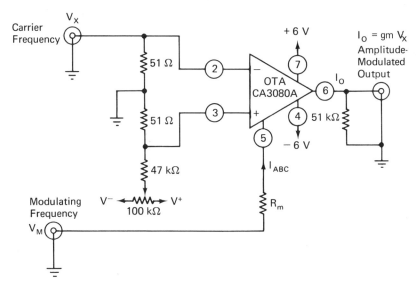

Fig. 4-23. Amplitude modulator using an RCA OTA

equal to transconductance g_m times V_X. The sign of the output signal is negative because the input signal is applied to the inverting input of the OTA.

The transconductance of the OTA is controlled by adjustment of I_{ABC} as usual. However, in this circuit the level of the unmodulated carrier output is established by a particular I_{ABC} through resistor R_M. Amplitude modulation of the carrier frequency occurs because variations of the voltage V_M force a change in the I_{ABC} supplied via resistor R_M. When V_M goes in the negative direction (toward the I_{ABC} terminal potential), the I_{ABC} decreases and reduces g_m of the OTA. When V_M goes positive, the I_{ABC} increases, resulting in an increase of the g_m. For the particular OTA shown, the g_m is approximately equal to $19.2 \times I_{ABC}$, where g_m is in mmhos and I_{ABC} is in mA.

Amplitude modulation with transistor drive. If the I_{ABC} terminal is driven by a current source (such as from the collector of a *pnp* transistor), the effect of any V_{ABC} variation is eliminated. Instead, any variation is dependent upon the *pnp* transistor base-emitter junction characteristics. Figure 4-24 shows a method of driving the I_{ABC} input using a *pnp* transistor.

If a *npn* transistor is added to the circuit of Fig. 4-24 as an emitter-follower to drive the *pnp* transistor, variations resulting from base-emitter characteristics of the *pnp* are considerably reduced because of the complementary nature of the *npn* and *pnp* base-emitter junctions. Also, the temperature coefficients of the two base-emitter junctions tend to cancel one another. Figure 4-25 shows a configuration using one transistor (in an RCA

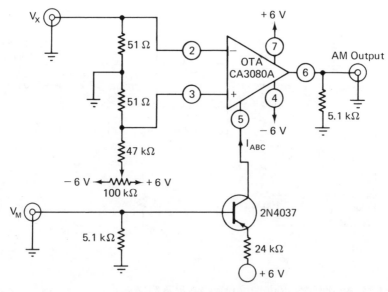

Fig. 4-24. Amplitude modulator using RCA OTA controlled by a *pnp* transistor

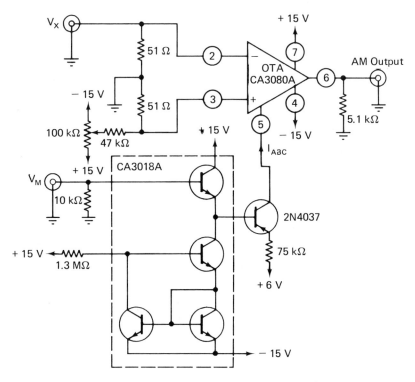

Fig. 4-25. Amplitude modulator using RCA OTA controlled by *pnp* and *npn* transistors

CA3018A *npn* transistor array) as an emitter-follower (with the three remaining transistors of the array connected as a current source for the emitter-followers).

Power supply and offset cancellation. Note that the circuits of Figs. 4-23 and 4-24 use ±6 V supplies, whereas the circuit of Fig. 4-25 requires a ±15 V supply. The 100 kΩ potentiometer shown in all three circuits is used to null the effects of amplifier input offset voltage (V_{io}). This potentiometer is used to set the output voltage symmetrically about zero (as described for conventional op-amps in Chapter 3). The OTA modulation method described here permits a range exceeding 1000-to-1 in gain, and thus provides modulation of the carrier signal input in excess of 99 percent.

4.5.6. OTA two-quadrant multiplier

Figure 4-26 shows an OTA used as a two-quadrant multiplier. Note that the circuit of Fig. 4-26a is essentially the same as for the modulator circuits described in Sec. 4.5.5. That is, when modulation is applied to the I_{ABC} input, the carrier voltage is applied to the differential input, the waveform

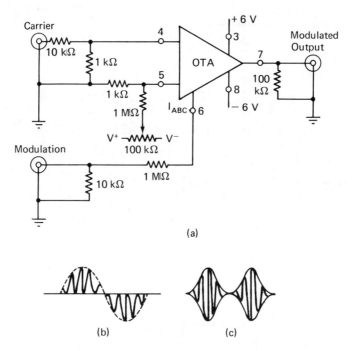

Fig. 4-26. Two-quadrant multiplier

shown in Fig. 4-26b is obtained. However, in the circuit of Fig. 4-26, the input offset control (100 kΩ potentiometer) is adjusted to balance the circuit so that no modulation can occur at the output without a carrier input. The linearity of the modulator is indicated by the solid trace of the superimposed modulating frequency, as shown in Fig. 4-26b. The maximum depth of modulation (or percentage of modulation) is determined by the ratio of the peak input modulating voltage to V^-).

The two-quadrant multiplier characteristic of the circuit is seen if modulation and carrier are reversed (modulation to differential input, carrier to I_{ABC}), as shown in Fig. 4-26c. The polarity of the output must follow that of the differential input. Thus, the output is positive only during the first (or positive) half-cycle of the modulation, and is negative only during the second half-cycle. Note that both input and output signals are referenced to ground. The output signal is zero when either the differential input or I_{ABC} is zero.

4.5.7. Four-quadrant multipliers

OTA units can be used as four-quadrant multipliers. Two circuit configurations are possible. One circuit uses three identical OTA devices. The second circuit uses a single OTA.

Four-quadrant multiplier with three OTAs. Figure 4-27 shows a schematic for a four-quadrant multiplier using the three OTA units of an RCA CA3060 package. All of the adjustment controls associated with differential input, and an adjustment for equalizing the gains of the amplifiers are included.

Adjustment of the circuit is quite simple. With both the X and Y voltages at zero, connect terminal 10 to terminal 8. This procedure disables amplifier 2. Adjust the offset voltage of amplifier 1 to zero by means of R_1. Remove the short between terminals 8 and 10. Connect terminal 15 to

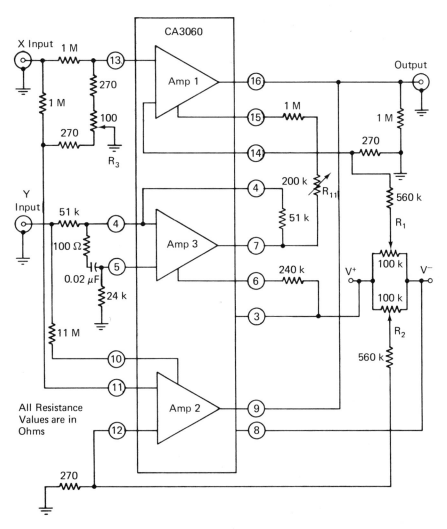

Fig. 4-27. Typical four-quadrant multiplier circuit using RCA CA3060 OTA

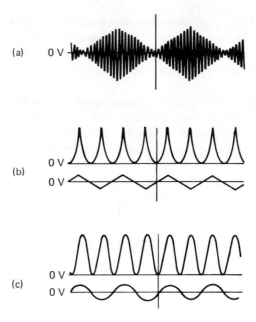

Fig. 4-28. Voltage waveforms of four-quadrant multiplier circuit

terminal 8. This disables amplifier 1. Adjust the offset voltage of amplifier 2 to zero by means of R_2.

With ac signals on both the X and Y inputs, adjust R_3 and R_{11} for symmetrical output signals. Figure 4-28 shows the output waveforms with the circuit adjusted. Figure 4-28a shows the suppressed carrier modulation of a 14 kHz carrier with a triangular wave. Figures 4-28b and c, respectively, show the squaring of a triangular wave and a sine wave. Notice that in both cases the outputs are always positive, and return to zero after each cycle.

Four-quadrant multipliers using a single OTA.   A single OTA can be used for many low-frequency, low-power, four-quadrant multiplier applications. The basic multiplier circuit shown in Fig. 4-29 using an RCA CA3080A is particularly useful in waveform generation, doubly balanced modulation, and other signal processing applications, in portable equipment where low-power consumption is essential and accuracy requirements are moderate. The multiplier circuit of Fig. 4-29 is basically an extension of the gain-control function discussed in Sec. 4.5.5. Note that the single OTA version of the four-quadrant multiplier does not provide the accuracy of the system using three OTA units (Fig. 4-27).

To obtain a four-quadrant multiplier, the first term of the modulation equation (which represents the fixed carrier) must be reduced to zero. In the circuit of Fig. 4-29, this is accomplished by placing a feedback resistor R

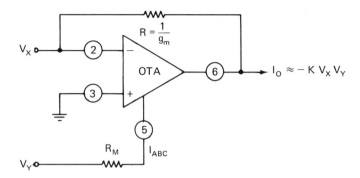

Fig. 4-29. Basic four-quadrant analog multiplier using an RCA OTA

between the output and the inverting input terminal of the OTA. The value of the feedback resistor R is equal to $1/g_m$.

Figure 4-30 shows the schematic diagram of a basic multiplier, with adjustment controls added to give the circuit an accuracy of approximately ± 7 percent of full-scale. There are only three adjustment controls: R_1 at the output compensates for slight variations in the current-transfer ratio of the current-mirrors (which would otherwise result in a symmetrical output about some current level other than zero); R_2 in the V_Y input establishes the g_m of the system equal to the value of the fixed resistor shunting the system when the Y input is zero; and R_3 compensates for an error that may arise from input offset voltage.

The following procedure is used to adjust the circuit:

1. Set R_1 to the center of its range.
2. Ground the X and Y inputs.
3. Adjust R_3 until a zero volt reading is obtained at the output.
4. Ground the Y input and apply a signal to the X input through a low-source impedance generator. It is essential that a low impedance source be used. This will minimize any change in the g_m balance for zero-point because of the Y input bias current (which is an I_{ABC} of about 50 µA).
5. Adjust R_2 until a reading of zero volts is obtained at the output. This adjustment establishes the g_m of the OTA at the proper level to cancel the output signal. The output current is diverted through the feedback resistance R_F.
6. Ground the X input and apply a signal to the Y input through a low-source impedance generator.
7. Adjust R_3 for an output of zero volts.

There will be some interaction among the adjustments and the procedure should be repeated for best circuit performance.

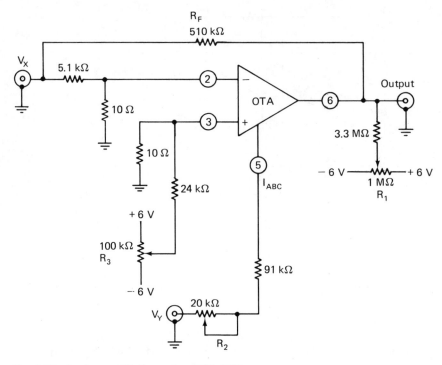

Fig. 4-30. Analog multiplier using RCA OTA

Analog multiplier with OTA controlled by a transistor.   Figure 4-31 shows the schematic of an OTA multiplier circuit with a *pnp* transistor replacing the *Y* input "current" resistor. The advantage of the Fig. 4-31 circuit is in the higher input resistance resulting from the current gain of the *pnp* transistor. The addition of another emitter-follower preceding the *pnp* transistor (similar to that shown in Fig. 4-25) will further increase the current gain, while markedly reducing the effect of the *pnp* base-emitter temperature-dependent characteristic (and possible input offset caused by base-emitter voltage).

Figure 4-32 shows output signals of the Fig. 4-31 circuit. Figures 4-32a and 4-32b show outputs when the circuit is used as a suppressed-carrier generator. Figures 4-32c and 4-32d show outputs when the circuit is used in signal squaring (that is, in squaring sinewave and triangular-wave inputs). If ± 15 V power supplies are used (as shown in Fig. 4-31), both inputs can accept ± 10 V input signals. Adjustment of the circuit in Fig. 4-31 is the same as for the circuit in Fig. 4-30.

The accuracy and stability of these multipliers are direct functions of the power supply voltage stability because the *Y* input is referenced to the

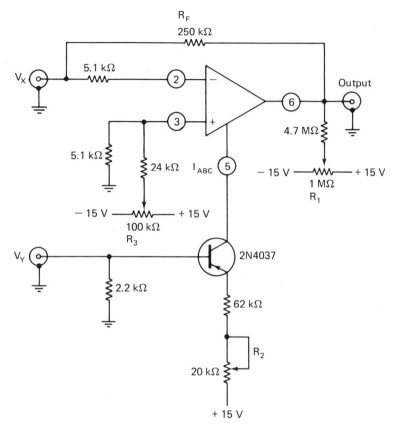

Fig. 4-31. Analog multiplier using OTA controlled by a *pnp* transistor

negative supply voltage. Tracking of the positive and negative supply is also important because the balance adjustments for both the offset voltage and output current are also referenced to these supplies.

4.5.8.   OTA decoder-multiplexer

A simple but effective system for multiplexing and decoding can be assembled using OTAs as the basic elements. The complete circuit is shown in Fig. 4-33. Although only two channels are shown, more channels can be added. Figure 4-34 shows waveforms of circuit operation.

In the multiplexer, an IC flip-flop is used to trigger two OTAs, one OTA for each channel. The decoder consists of an OTA and MOSFET used as a sample-and-hold circuit (refer to Sec. 4.5.3.) that is driven by another OTA used as a one-shot multivibrator. The OTA one-shot multivibrator

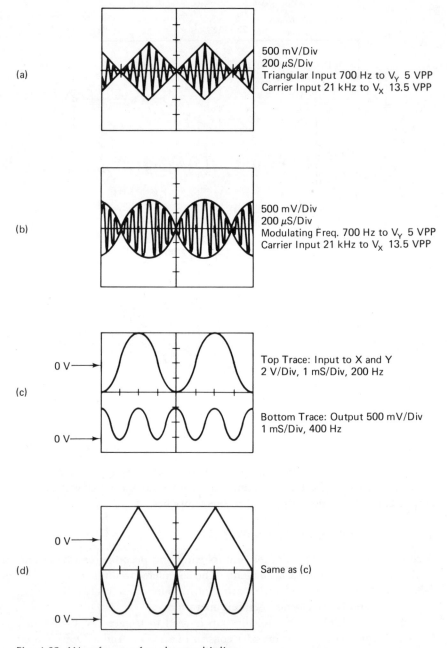

(a)

500 mV/Div
200 µS/Div
Triangular Input 700 Hz to V_Y 5 VPP
Carrier Input 21 kHz to V_X 13.5 VPP

(b)

500 mV/Div
200 µS/Div
Modulating Freq. 700 Hz to V_Y 5 VPP
Carrier Input 21 kHz to V_X 13.5 VPP

(c)

0 V

0 V

Top Trace: Input to X and Y
2 V/Div, 1 mS/Div, 200 Hz

Bottom Trace: Output 500 mV/Div
1 mS/Div, 400 Hz

(d)

0 V

0 V

Same as (c)

Fig. 4-32. Waveforms of analog multiplier

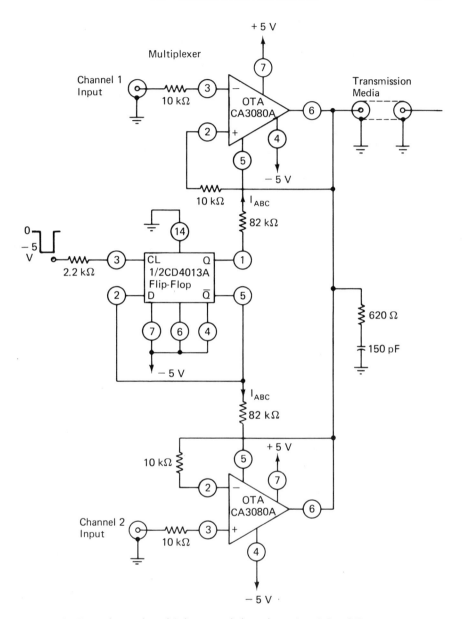

Fig. 4-33. Two-channel multiplexer and decoder using RCA OTAs

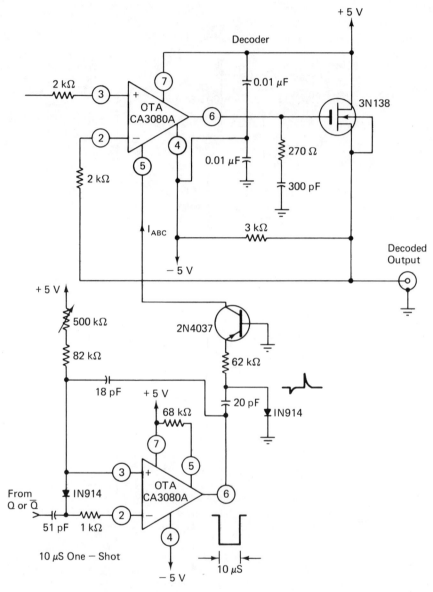

Fig. 4-33 (*continued*)

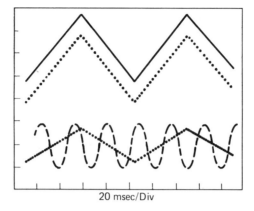

20 msec/Div

Top Trace: Input Signal (1 Volt/Div)
Center Trace: Recovered Output (1 Volt/Div)
Bottom Trace: Multiplexed Signals (2 Volts/Div)

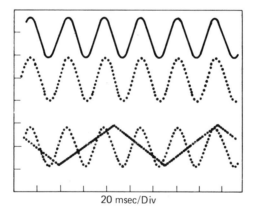

20 msec/Div

Top Trace: Input Signal (1 Volt/Div)
Center Trace: Recovered Output (1 Volt/Div)
Bottom Trace: Multiplexed Signals (1 Volt/Div)

Fig. 4-34. Waveforms showing operation of a linear multiplexer/sample-and-hold diode circuitry

introduces a 10 μs delay in the decoder to ensure that the sample-and-hold circuit can sample only after the input has settled. Thus, the trailing edge of the OTA one-shot output signal is used to sample the input at the sample-and-hold circuit for approximately 1 μs. Note that either the Q or $\bar{Q}$ output from the flip-flop may be used to trigger the 10 μs one-shot to decode a signal.

144

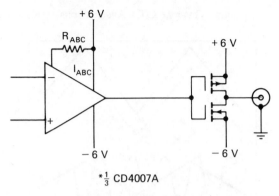

+ 6 V

R_{ABC}

I_{ABC}

+ 6 V

− 6 V

− 6 V

$*\frac{1}{3}$ CD4007A

*Additional current output can be
obtained when remaining two
amplifiers of the CD4007A are
connected in parallel with the
single stage shown.

Fig. 4-35. OTA combined with MOS to form open-loop amplifier circuit with
high current output

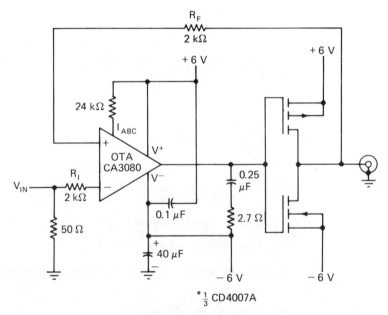

R_F

2 kΩ

+ 6 V

+ 6 V

24 kΩ

I_{ABC}

OTA
CA3080

V^+

V^-

R_I

V_{IN}

2 kΩ

0.25
µF

0.1 µF

2.7 Ω

50 Ω

+
40 µF

− 6 V

− 6 V

$*\frac{1}{3}$ CD4007A

*Additional current output can be
obtained when remaining two
amplifiers of the CD4007A are
connected in parallel with the
single stage shown.

Fig. 4-36. OTA combined with MOS to form unity gain amplifier with increased current capacity

4.5.9   OTA with high current output stages

It is possible to combine and OTA with a MOS device to produce high-gain, high-current circuits. For example, the sample-and-hold circuit of Fig. 4-17 (Sec. 4.5.3) combines an OTA with a MOSFET. The resultant voltage gain is about 100 dB. The actual voltage gain of the overall circuit is equal to the product of OTA g_m and output resistance (this product is typically 142,000 or 103 dB). Thus, the overall gain is set by the OTA characteristics. However, the output voltage swing and current swing are set by the MOSFET characteristics (and the source-terminal load).

Figure 4-35 shows an OTA combined with a MOS inverter/amplifier to form a simple, open-loop amplifier circuit. The MOS device shown is one-third of an RCA CD4007A COS/MOS inverter. (The term COS/MOS is the RCA designation for complementary-symmetry MOS devices. Such devices are discussed in Chapter 8.) Each of the three inverter/amplifiers in the CD4007A has a typical voltage gain of 30 dB. This gain, combined with the typical 100 dB gain of the OTA, results in a total voltage gain of about 130 dB. (Note that OTA gain is affected by I_{ABC} which, in turn, is set by R_{ABC}. The rules for selecting R_{ABC} values are the same as described in Sec. 4.4.)

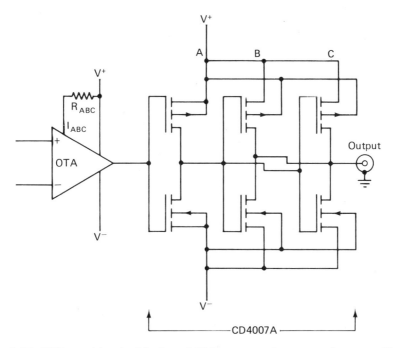

Fig. 4-37. OTA combined with three MOS stages to form open-loop amplifier circuit with high current output and increased gain

There are several circuit configurations available when an OTA is combined with MOS inverter/amplifiers. Some of these configurations provide additional gain, while others provide additional current capacity.

The circuit shown in Fig. 4-36 provides unity gain (with the same resistance values for R_F and R_I), and increased current capacity. The MOS inverter/amplifier (one-third of a CD4007A) can source or sink a current of about 6 mA.

The circuit shown in Fig. 4-37 provides a voltage gain of about 160 dB (10^8), and a source or sink current capacity of about 12 mA. The voltage gain results from open-loop operation. The OTA provides about 100 dB, MOS inverter/amplifier A provides about 30 dB, and the parallel-connected MOS inverter/amplifiers B and C provide the remaining 30 dB. Since the outputs of B and C are connected in parallel, the normal source/sink current capacity of 6 mA is doubled to 12 mA.

The circuit of Fig. 4-38 provides unity gain (with the same resistance values for R_F and R_I) but with increased current capacity. The parallel-connected B and C outputs provide approximately 12 mA of sink or source current capacity.

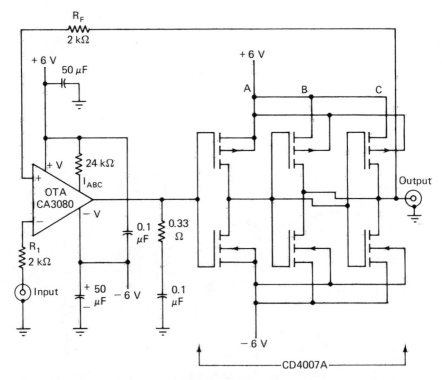

Fig. 4-38. OTA combined with three MOS stages to form unity-gain amplifier with increased current capacity

4.5.10.   OTA multistable circuits

An OTA can be set to draw very little standby current by proper adjustment of I_{ABC}. Likewise, MOS inverter/amplifiers draw very little current when used in switching applications. A MOS inverter/amplifier switch draws current when changing states, but very little current when the output signal voltage swings either positive or negative. The low standby power capabilities of the OTA, when combined with the characteristics of the MOS inverter/amplifier, are ideally suited for use in precision multistable circuits.

Free-running multivibrator.   Figure 4-39 shows an OTA combined with a MOS inverter/amplifier to form an astable (free-running) multi-vibrator. As shown by the equations, the frequency of operation is set by feedback resistance R and capacitor C, as well as resistors R_1 and R_2. Since resistors R_1 and R_2 also set output impedance of the circuit, a trade-off may be necessary in selecting values of R and C. The value of R_{ABC} is set by the I_{ABC} requirements, as described in Sec. 4.4. For a multistable circuit, I_{ABC} is usually set so that the OTA draws minimum power, but enough so that the MOS inverter/amplifier is properly driven to provide the desired output voltage swing. Using the values shown, the output frequency is about 7.7 kHz, with an output voltage swing equal to $V+$ and $V-$.

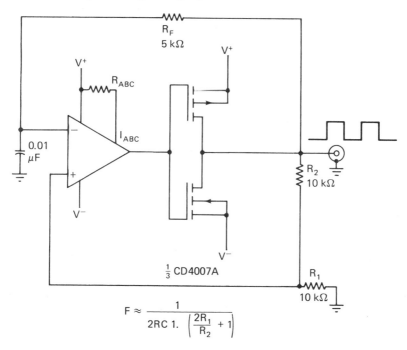

$$F \approx \frac{1}{2RC\,1.\left(\dfrac{2R_1}{R_2} + 1\right)}$$

Fig. 4-39. OTA combined with a MOS to form an astable (free-running) multivibrator

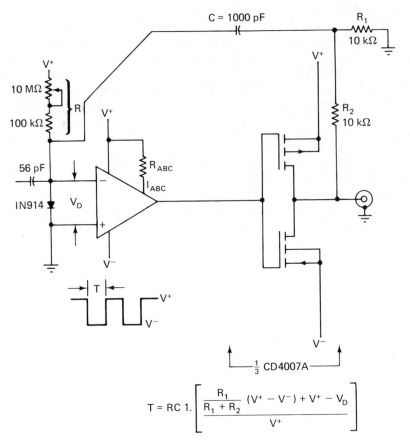

$$T = RC \cdot 1. \left[\frac{\dfrac{R_1}{R_1 + R_2}(V^+ - V^-) + V^+ - V_D}{V^+} \right]$$

Fig. 4-40. OTA combined with MOS to form a monostable (one-shot) multivibrator

One-shot multivibrator.   Figure 4-40 shows an OTA combined with a MOS inverter/amplifier to form a monostable (one-shot) multivibrator. Generally, monostable multivibrators are used to produce output pulses of some specific time duration (T), regardless of trigger input pulse duration and frequency. As shown by the equations, the time duration T of output pulses is set by feedback capacitance C and resistance R, as well as resistors R_1 and R_2.

The relationship of $V+$, $V-$ and V_D also affects output pulse duration T. Note that V_D, the voltage across the input diode, is typically 0.5 V. Again, R_{ABC} is adjusted or selected so that the OTA provides just enough output to drive the MOS inverter/amplifier for the desired output voltage swing. Also, it may be necessary to trade-off the values of R_1, R_2, R, and C, since R_1 and R_2 set the output impedance of the circuit.

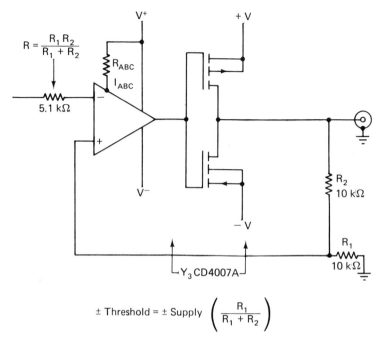

$$\pm\text{ Threshold} = \pm\text{ Supply} \left(\frac{R_1}{R_1 + R_2} \right)$$

Fig. 4-41. OTA combined with MOS to form a threshold detector

Threshold detector. Figure 4-41 shows an OTA combined with a MOS inverter/amplifier to form a threshold detector. The threshold voltage point is set by R_1 and R_2, as well as the supply voltage. Standby power consumption is set by I_{ABC}, which, in turn, is set by R_{ABC}.

The standby power consumption of the circuits shown in Figs. 4-39 through 4-41 is typically 6 mW. However, the standby power can be made to operate in the micropower region by changes in the value of R_{ABC}. Also, for greater current output from any of the circuits shown in Figs. 4-39 through 4-41, the remaining MOS amplifier-inverters in the RCA CD4007A can be connected in parallel with the single stage. Each of the three elements in the CD4007A will sink or source about 6 mA. Thus, with all three elements in parallel, the circuit should be able to sink or source about 18 mA.

4.5.11. OTA micropower comparator

Figure 4-42 shows an OTA combined with two MOS inverter/amplifiers to form a micropower comparator. Circuit output is proportional to the differential signal at the OTA inputs. If both inputs are at the same level, there will be no output. Either of the inputs (inverting or noninverting) can be adjusted to some reference level by means of a voltage divider network (such

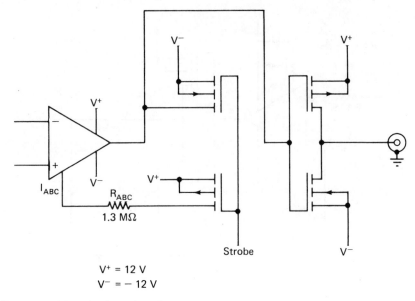

Fig. 4-42. OTA combined with two MOS stages to form a micropower comparator

as shown in Fig. 4-16) if desired. Under these conditions, the output is proportional to the difference between the signal input level and the reference level.

The circuit is on only when a strobe or gate pulse is applied. The standby power consumption of the circuit is about 10 μW. When the circuit is strobed, the OTA consumes about 420 μW. Under these conditions the circuit response to a differential input signal is about 6 μs. By decreasing the value of R_{ABC} the circuit response time can be decreased to about 150 ns. However, the standby power consumption will rise to about 20 mW.

The differential input common-mode range of the circuit is approximately −1 V to +10.5 V. Voltage gain of the circuit is about 130 dB.

5

LINEAR
APPLICATIONS
FOR OP-AMPS

In this chapter, we shall discuss linear applications for op-amps. These linear applications include those cases in which input and output are essentially sinewaves, even though modified by the op-amp or the associated circuit. Nonlinear applications are discussed in Chapter 6. For the reader's convenience, the same format is used for each application (where practical).

First, a working schematic is presented for the circuit, together with a brief description of its function. Where practical, the working schematic also includes the operational characteristics of the circuit (in equation form), as well as rule-of-thumb relationships of circuit values (also in equation form).

Next, design considerations, such as desired performance, use with external circuits, amplification, operating frequency, and so forth, are covered. This is followed by reference to equations (on the working schematic) and procedures for determining external component values that will produce the desired results.

Finally, a specific design problem is stated and a design example is given. The value of each external circuit component is calculated in step-by-step procedures, using guidelines established in the design considerations and/or working schematic equations.

The reader will note that the power supply and phase compensation connections are omitted from the schematics related to applications described here, except in Sec. 5.1. In all of the remaining applications it is assumed that the op-amp is connected to a power source as described in Sec. 2.7. Likewise, it is assumed that a suitable compensation scheme has been selected for the op-amp, as discussed in Sec. 3.2. Unless otherwise stated, all of the design considerations for the basic op-amp described in Chapter 3 apply to each application covered here.

5.1. BASIC OP-AMP SYSTEM DESIGN

Figure 5-1 is the working schematic of a closed-loop op-amp system, complete with external circuit components. The design considerations discussed in Sec. 2.7 and Chapter 3 apply to the circuit of Fig. 5-1. The following paragraphs provide a specific design example for the circuit.

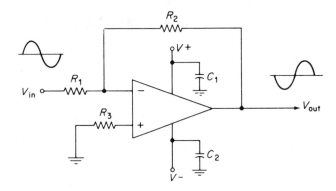

$$\text{Voltage gain} = \frac{V_{out}}{V_{in}} = \frac{R_2}{R_1} \qquad C_1 = C_2 = 0.1 - 0.001\,\mu F$$

$$R_3 = \frac{R_1 R_2}{R_1 + R_2}$$

Fig. 5-1. Basic op-amp system connections

5.1.1. Op-amp characteristics

Supply voltage: $+15$ and -15 V nominal, ± 19 V maximum
Total device dissipation: 750 mW, derating 8 mW/°C
Temperature range: 0 °C to $+70$ °C
Input offset voltage: 3 mV typical
Input offset current: 10 nA nominal, 30 nA maximum
Input bias current: 100 nA nominal, 200 nA maximum
Input offset voltage sensitivity: 0.2 mV/V
Device dissipation: 300 mW maximum
Open-loop voltage gain: as shown in Fig. 3-15

Slew rate: 4 V/μs at a gain of 1, 6 V/μs at a gain of 10, 33 V/μs at a gain of 100.
Open-loop bandwidth: as shown in Fig. 3-15
Common-mode rejection: 94 dB
Output voltage swing: 23 V (*p–p*) typical
Input impedance: 1 megohm
Output impedance: 300 ohms
Input voltage range: -13 V, $+10$ V
Output power: 250 mW typical

5.1.2. Design example

Assume that the circuit of Fig. 5-1 is to provide a voltage gain of 100 (40 dB), the input signal is 80 mV (RMS), the input source impedance is not specified, the output load impedance is 500 ohms, the ambient temperature is 25 °C, the frequency range is dc up to 300 kHz, and the power supply is subject to 10 percent variation.

Frequency/gain relationship.   Before attempting to calculate any circuit values, make certain that the op-amp can produce the desired voltage gain at the maximum frequency. This can be done by reference to a graph similar to Fig. 3-15. Note that the maximum frequency of 300 kHz intersects the phase curve at about 135°. This allows a phase margin of 45°. Thus, the op-amp should be stable over the desired frequency range.

Note also that the maximum frequency of 300 kHz intersects several capacitance curves above the 40 dB open-loop voltage gain level. Thus, the op-amp should be able to produce more than 40 dB of open-loop gain.

Supply voltage.   The positive and negative supply voltages should both be 15 V since this is the nominal value listed. Most op-amp datasheets will list certain characteristics as "maximum" (temperature range, total dissipation, maximum supply voltage, maximum input signal, etc., and then list the remaining characteristics as "typical" with a given "nominal" supply voltage.

In no event can the supply voltage exceed the 19 V maximum. Since the available supply voltage is subject to a 20 percent variation, or 18 V maximum (15 V$\times$20%$=3$ V; 15 V $+3$ V$=18$ V), the supply is within the 19 V limit.

Decoupling or bypass capacitors.   The values of C_1 and C_2 should be found on the datasheet. In the absence of a value, use 0.1 μF for any frequency up to 10 MHz. If this value produces a response problem at any frequency (high or low), try a value between 0.001 and 0.1 μF.

Closed-loop resistances. The value of R_2 should be 100 times the value of R_1 to obtain the desired gain of 100. The value of R_1 should be selected so that the voltage drop across R_1 (with the nominal input bias current) is comparable to the input signal (never larger than the input signal).

A 50 ohm value for R_1 will produce a 10 μV drop with the maximum 200 nA input bias current. Such a 10 μV drop is less than 10 percent of the 80 mV input signal. Thus, the fixed drop across R_1 should have no appreciable effect on the input signal. With a 50 ohm value for R_1, the value of R_2 must be 5000 (50×100 gain $= 5000$).

Offset minimizing resistance. The value of R_3 can be found using the equation of Fig. 5-1 once the values of R_1 and R_2 have been established. Note that the value of R_3 works out to about 49 ohms, using the Fig. 5-1 equation:

$$\frac{R_1 R_2}{R_1 + R_2} = \frac{50 \times 5000}{50 + 5000} \approx 49 \text{ ohms}$$

Thus, a simple trial value for R_3 is always slightly less than the R_1 value. The final value of R_3 should be such that the no-signal voltages at each input are equal.

Comparison of circuit characteristics. Once the values of the external circuit components have been selected, the characteristics of the op-amp and the closed-loop circuit should be checked against the requirements of the design example. The following is a summary of the comparison.

Gain versus phase compensation. The closed-loop gain should always be less than the open-loop gain. As a guideline, the open-loop gain should be at least 20 dB greater than the closed-loop gain. Figure 3-15 shows that open-loop gain up to about 66 dB is possible with a proper phase compensation capacitor. Figure 3-15 also shows that a capacitance of 0.001 μF (1000 pF) will produce an open-loop gain of slightly less than 60 dB at 300 kHz, whereas a 300 pF capacitance will produce a 66 dB open-loop gain. To assure an open-loop gain of 60 dB, use a capacitance of about 700 pF.

With a 60 dB open-loop gain, and values of 50 to 5000 ohms, respectively, for R_1 and R_2, the closed-loop circuit should have a flat frequency response of 40 dB (gain of 100) from zero up to 300 kHz. A rolloff will start at frequencies above 300. Thus, the closed-loop gain is well within tolerance.

Input voltage. The peak input voltage must not exceed the rated maximum input signal. In this case, the rated maximum is $+10$ V and -13 V, whereas the input signal is 80 mV (RMS), or approximately 112 mV peak (80×1.4). This is well below the $+10$ V maximum limit.

When the rated maximum input signal is an uneven positive and negative value, always use the lowest value for total swing of the input signal. In this case, the input swings from $+112$ mV to -112 mV, far below $+10$ V. An input signal that started from zero could swing as much as $+10$ V and -10 V, without damaging the op-amp. An input signal that started from -2 V could swing as much as ± 11 V.

Output voltage. The peak-to-peak output voltage must not exceed the rated maximum output voltage swing (with the required input signal and selected amount of gain).

In this case, the rated output voltage swing is 23 V (peak-to-peak), whereas the actual output is approximately 22.4 V (80 mV RMS $\times$ a gain of $100 = 8000$ mV output; 8000 mV $\times 2.8 = 22.4$ V peak-to-peak). Thus, the anticipated output is within the rated maximum.

However, the actual output voltage is dependent upon slew rate, which in turn depends upon compensation capacitance. As shown in the characteristics, the slew rate is given as 33 V/μs when gain is 100. The datasheet does not show a relationship between slew rate and compensation capacitance. However, since slew rate is always maximum with the lowest value of compensation capacitance, it can be assumed that the slew rate will be 33 V/μs with a capacitance of 700 pF (which is near the lowest recommended value of 300 pF, Fig. 3-15).

Using the equations of Sec. 3-3.3, it is possible to calculate the output voltage capability of the op-amp. With a maximum operating frequency of 300 kHz, and an assumed slew rate of 33 V/μs, the peak output voltage capability is:

$$\frac{33,000,000}{6.28 \times 300,000} \approx 17 \text{ V}$$

Thus, the peak-to-peak output voltage capability is 34 V, well above the anticipated 22.4 V. A quick approximation of the output voltage capability can also be found by reference to Fig. 3-20.

Output power. The output power of an op-amp is usually computed on the basis of RMS output voltage (rather than peak or peak-to-peak) and output load.

In this example, the output voltage is 8 V RMS (80 mV $\times 100$ gain $= 8000$ mV $= 8$ V). The load resistance or impedance is 500 ohms as stated in the design assumptions. Thus, the output power is:

$$\frac{(8)^2}{500} = 0.128 \text{ W} = 128 \text{ mW}$$

A 128 mW output is well below the 250 mW typical output power of the op-amp. Also, 128 mW output plus a device dissipation of 300 mW is 428 mW, well below the rated 750 mW total device dissipation. Thus, the op-amp should be capable of delivering full power output to the load.

Note that power output ratings usually are at some given temperature; 25 °C in this case. Assume that the temperature is raised to 50 °C. The total device dissipation must be derated by 8 mW/ °C, or a total of 200 mW for the 25 °C increase in temperature. This derates or reduces the 750 mW total device dissipation to 550 mW. However, the 550 mW is still well above the anticipated 428 mW.

Output impedance.   Ideally, the closed-loop output impedance should be as low as possible and always less than the load impedance. The closed-loop output impedance can be found using the equations of Fig. 3-9. In this example, the approximate output impedance is:

$$\frac{300}{1 + 1000 \times \dfrac{50}{50 + 5000}} \approx 30 \text{ ohms}$$

5.2. ZERO OFFSET SUPPRESSION

Figure 5-2 is the working schematic of an op-amp used to provide amplification of a small signal riding on a large, fixed direct current level. For example, assume that the input signal varies between 3 and 8 mV, and that the signal source never drops below 7 V. That is the source is $+7$ V with no signal, and $+7.003$ to $+7.008$ V with signal. Now assume that the output is to vary between 300 and 800 mV.

The obvious solution is to apply a fixed $+7$ V to the noninverting input. This will offset the $+7$ V at the inverting input, and result in a 0 V output (under no-signal conditions). This solution ignores the fact that the op-amp probably has some input offset, or assumes that the op-amp has some provisions for neutralizing the offset. The solution also assumes that the signal is rating on exactly $+7$ V.

If input offset cannot be ignored (say because it is large in relation to the signal), or if the fixed dc voltage is subject to possible change, the alternate offset circuit of Fig. 5-2 can be used. The circuit in Fig. 5-2 is a simple resistance network that makes use of existing V_{CC} and V_{EE} voltages. In use, potentiometer R_4 is adjusted to provide zero output from the op-amp under no-signal conditions. That is, the 3 to 8 mV signal is removed, but the $+7$ V remains at the inverting input, while R_4 is adjusted for zero at the op-amp output.

The values for the offset network are not critical. However, the values should be selected so that a minimum of current is drawn from the V_{CC} V_{EE}

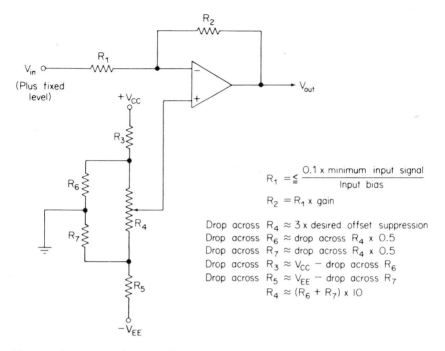

$$R_1 = \underset{\leq}{=} \frac{0.1 \times \text{minimum input signal}}{\text{Input bias}}$$

$$R_2 = R_1 \times \text{gain}$$

Drop across $R_4 \approx 3 \times$ desired offset suppression
Drop across $R_6 \approx$ drop across $R_4 \times 0.5$
Drop across $R_7 \approx$ drop across $R_4 \times 0.5$
Drop across $R_3 \approx V_{CC} -$ drop across R_6
Drop across $R_5 \approx V_{EE} -$ drop across R_7
$R_4 \approx (R_6 + R_7) \times 10$

Fig. 5-2. Op-amp with zero offset suppression to provide amplification of small signal riding on large, fixed level

supplies, and a minimum of current should flow through R_4. This is discussed further in the design example.

The technique shown in Fig. 5-2 can be applied to most of the op-amp applications described in this chapter and in Chapter 6. That is, any zero offset (at the op-amp input and output) can be suppressed by applying a fixed (or adjustable) voltage of correct polarity and amplitude to the opposite input.

For example, in the basic circuit in Fig. 5-1 (Sec. 5.1), the offset can be suppressed by application of a positive voltage at the noninverting input, in place of R_3 (or in addition to R_3). The offset suppression voltage can be fixed, or adjusted, using the circuits of Fig. 5-2 as applicable. Of course, the circuit of Fig. 5-2 is not necessary if the op-amp has provisions for zero offset suppression (such as the op-amp in Fig. 3-5).

5.2.1. Design example

Assume that the circuit of Fig. 5-2 is used to monitor a 3 to 8 mV signal, riding on a $+7$ V level, and to produce a 300 mV to 800 mV output (a gain of 100). The op-amp does not have provisions for input offset neutralization or null. It is important that the no-signal output be exactly 0 V. The op-amp has an input bias of 200 nA, and V_{CC} and V_{EE} are 15 V. The voltage drop across R_1 caused by the input bias current should be no greater than 10 percent of the lowest input signal. With a low signal of 3 mV, and a 200 nA bias current, the value of R_1 is 1500 ohms (3 mV×0.1=0.3 mV; 0.3 mV/200 nA=1500). With R_1 at 1500, and a required gain of 100, the value of R_2 is 150 kΩ.

With V_{CC} and V_{EE} at +15 V and −15 V, respectively, the total drop across the offset adjustment network is 30 V. Allowing an arbitrary 1 mA current through the network, the total resistance should be about 30 kΩ. Since the desired offset suppression is approximately 7 V (to offset the +7 V level), the drop across R_4 should be approximately 21 V. This results in a drop of about 10.5 V each across R_6 and R_7. In turn, the drops across R_3 and R_5 (each) should be about 4.5 V(15 V − 10.5 V). With a desired 4.5 V drop, and approximately 1 mA current flow, the values of R_3 and R_5 should be 4500 ohms each. Likewise, the values of R_6 and R_7 should be 10.5 kΩ each. With R_6 and R_7 at 10.5 kΩ each, the value of R_4 should be 210 kΩ ($R_6 + R_7 = 21$ kΩ; 22 kΩ×10=221 kΩ).

5.3. VOLTAGE FOLLOWER (SOURCE FOLLOWER)

Figure 5-3 is the working schematic of an op-amp used as a voltage follower (also known as a source follower). The circuit is essentially a *unity-gain amplifier*. There is no feedback or input resistance in the circuit. Instead, the output is fed back directly to the inverting input. Signal input is applied directly to the noninverting input. With this arrangement, the output voltage equals the input voltage (or may be slightly less than the input voltage). However, the input impedance is very high, with the output impedance very low (as shown by the equations). In effect, the input

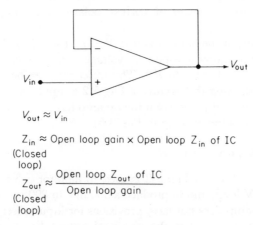

$$V_{out} \approx V_{in}$$

$Z_{in} \approx$ Open loop gain × Open loop Z_{in} of IC
(Closed loop)

$Z_{out} \approx \dfrac{\text{Open loop } Z_{out} \text{ of IC}}{\text{Open loop gain}}$
(Closed loop)

Fig. 5-3. Voltage follower (unity gain) using an op-amp

impedance of the op-amp is multiplied by the open-loop gain, whereas the output impedance is divided by the open-loop gain. Keep in mind that open-loop gain varies with frequency. Thus, the closed-loop impedances are frequency dependent.

Another consideration that is sometimes overlooked in this application is the necessity of supplying the input bias current to the op-amp. In a conventional circuit, bias is supplied through the input resistances. In the circuit of Fig. 5-3, the input bias must be supplied through the signal source. This may alter input impedance.

One more problem with the circuit in Fig. 5-3 is that the total input voltage is a common-mode voltage. That is, a voltage equal to the total input voltage appears across the two inputs. Should the input signal consist of a large dc value, plus a large signal variation, the common-mode input voltage range may be exceeded. One solution for the problem is to capacitively couple the input signal to the noninverting input. This eliminates any dc voltage, and only the signal appears at the inputs. This will solve the common-mode problem, but will require the use of a resistor from the noninverting input to ground. The resistance provides a path for the input bias current. However, the resistance also sets the input impedance of the op-amp. If the resistance is large, the input bias current will produce a large offset voltage (both input and output) across the resistance.

Some manufacturers recommend that resistances (of equal value) be used in the feedback loop (from output to inverting input) and at the noninverting input. This will still result in unity gain, and the circuit will perform as a voltage or source follower. Such a circuit is shown in Fig. 5-4.

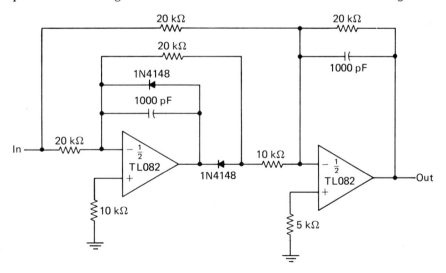

Fig. 5-4. Unity gain absolute value amplifier using Texas Instruments dual BiFET op-amp

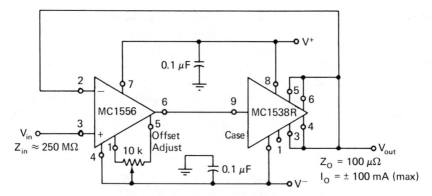

Fig. 5-5. High input impedance, high output current voltage follower using Motorola ICs

Note that two op-amps (actually two halves of a dual IC op-amp) are used. The second op-amp restores the gain to unity or 1 so that the output is equal to the absolute value of the input. Also, in the circuit of Fig. 5-4, the input and output impedances are set (primarily) by the resistance values rather than the op-amp characteristics, as is the case with the circuit of Fig. 5-3.

Another recommendation is that the circuit of Fig. 5-3 be followed by a power amplifier capable of high current output. This is shown in Fig. 5-5. Here, the op-amp output is fed back to the power amplifier input. The output from the power amplifier is fed back to the op-amp inverting input. The voltage gain remains at approximately 1, but the available current is set by the power amplifier capability. The input and output impedances of the overall circuit are increased by the gains of both amplifiers. For example, if the op-amp and power amplifier open-loop gains are both 100, the input impedance is multiplied by approximately 100 and the output impedance is divided by 100.

5.3.1.   Design example

Assume that the circuit of Fig. 5-3 is to provide unity gain with high input impedance and low output impedance. Also assume that the op-amp has an open-loop gain of 100 (40 dB) at a specific frequency, and output impedance of 300 ohms, and an input impedance of 1 megohm. With these characteristics, the closed-loop input impedance is: 100×1 megohm ≈ 100 megohms. The closed-loop output impedance is: $300 \div 100 \approx 3$ ohms.

5.4.   UNITY GAIN WITH FAST RESPONSE

One of the problems of a unity gain amplifier is that the slew rate is very poor. That is, the response time is very slow, and the power bandwidth is

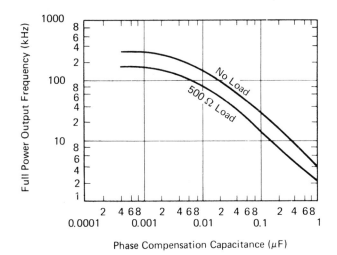

Fig. 5-6. Frequency for full power output as a function of phase-compensating capacitance

decreased. The reason for poor bandwidth with unity gain is that most op-amp datasheets recommend a large-value compensating capacitor for unity gain.

As an example, assume that the op-amp has the characteristics as shown in Fig. 3-15, and the desired operating frequency is 200 kHz (the unity-gain op-amp must have a full power bandwidth up to 200 kHz). The recommended compensation for 60 dB gain is 0.001 μF, whereas the unity-gain compensation is 1 μF. Now assume that the op-amp has the characteristics of Fig. 5-6, and that the load is 100 ohms. With a compensation of 0.001 μF, full power can be delivered at frequencies well above 200 kHz. However, if the compensating capacitor is 1 μF, the maximum frequency at which full power can be delivered is below 4 kHz.

Several methods are used to provide fast response time (high slew rate) and good power bandwidth with unity gain. Two such methods are described next.

5.4.1. Using op-amp datasheet phase compensation

The circuit of Fig. 5-7 shows a method of connecting an op-amp for unity gain, but with high slew rate (fast response and good power bandwidth). With this circuit, the phase compensation recommended on the datasheet is used, but with modifications. Instead of using the unity-gain compensation, use the datasheet phase compensation recommended for a gain of 100. Then select values of R_1 and R_3 to provide unity gain ($R_1 = R_3$). As shown by the

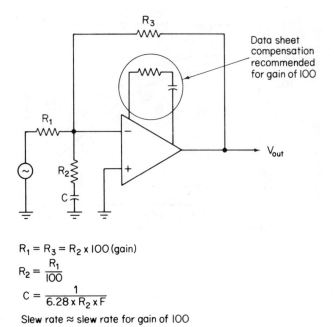

R₃

Data sheet
compensation
recommended
for gain of 100

R_1

R_2

C

V_out

$R_1 = R_3 = R_2 \times 100 \text{ (gain)}$

$R_2 = \dfrac{R_1}{100}$

$C = \dfrac{1}{6.28 \times R_2 \times F}$

Slew rate ≈ slew rate for gain of 100

Fig. 5-7. Unity gain op-amp with fast response (good slew rate) using datasheet
compensation

equations, the values of R_1 and R_3 must be approximately 100 times the
value of R_2. Thus, R_1 and R_3 must be fairly high values for practical design.

Assume that the circuit of Fig. 5-7 provides unity gain, but with a slew
rate approximately equal to that which results when a gain of 100 is used.
Also assume that the input/output signal is 8 V and that the op-amp has an
input bias current of 200 nA and compensation characteristics similar to
those of Fig. 3-15, but with higher gain.

The compensation capacitance recommended for a gain of 100 is 0.001
µF. With this established, select the values of R_1, R_2 and R_3. The value of R_1
(and consequently that of R_3) is selected so that the voltage drop with
nominal input bias current is comparable (preferably 10 percent or less) to
the input signal. Ten percent of the 8 V input is 0.8 V. Using the 0.8 V
value and the 200 nA input bias, the resistance of R_1 is 4 megohms. With R_1
at 4 megohms, R_3 must also be 4 megohms, and R_2 must be 40 kΩ (4
megohms/100 = 40 kΩ).

The value of C_1 is found using the equations of Fig. 5-7 once the value
of R_2 and the open-loop rolloff point are established. Assume that the
op-amp has characteristics similar to those shown in Fig. 3-12, where the 6
dB/octave rolloff starts at about 200 kHz. With these figures, the value of C

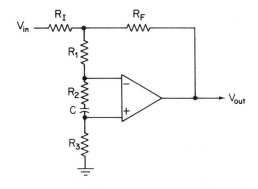

$$R_1, R_2, R_3, C = \text{see text}$$
$$R_I = R_F = 0.25 \times R_1$$

Fig. 5-8. Unity gain op-amp with fast response (good slew rate) using input phase compensation

is:

$$C \approx \frac{1}{6.28 \times 40 \ \mathrm{k\Omega} \times 200 \ \mathrm{kHz}} \approx 19 \ \mathrm{pF}$$

5.4.2. Using input phase compensation

The circuit of Fig. 5-8 shows a method of connecting an op-amp for unity gain, but with high slew rate, using input compensation. With this circuit, the phase compensation recommended on the op-amp datasheet is not used. Instead, the input phase compensation system in Sec. 3.2.6 is used.

The first step is to compensate the op-amp by modifying the open-loop input impedance, as described in Sec. 3.2.6 and Fig. 3-13. The recommended values are $R_1 = R_3 = 5 \ \mathrm{k\Omega}$, $R_2 = 2100$, $C = 0.0004 \ \mu\mathrm{F}$.

Next, select values of R_I and R_F to provide unity gain (both R_I and R_F must be the same value). The values of R_I and R_F are not critical, but they must be identical. Using the equations of Fig. 5-8, the values of R_I and R_F should be 0.25 times the value of R_1, or $0.25 \times 5 \ \mathrm{k\Omega} = 1250 \ \Omega$.

5.5. ACTIVE FILTER CIRCUITS
USING OP-AMPS

Because an op-amp has considerable gain and is provided with both positive and negative feedback inputs, an op-amp is well suited for use in active filter circuits. The use of an active filter eliminates the signal loss normally associated with passive filters (either RC or LC). The following paragraphs describe some typical circuits.

5.5.1. Narrow bandpass filter (tuned peaking)

Figure 5-9 is the working schematic of an op-amp used as a narrow bandpass (or tuned peaking) filter. Note that the circuit is essentially a narrow bandpass amplifier.

Circuit gain is determined by the ratio of R_1 and R_F in the usual manner. However, the frequency at which maximum gain occurs (or the narrow bandpeak) is the resonant frequency of the $L_1 C_1$ circuit. Capacitor C_1 and inductance L_1 form a parallel-resonant circuit that rejects the

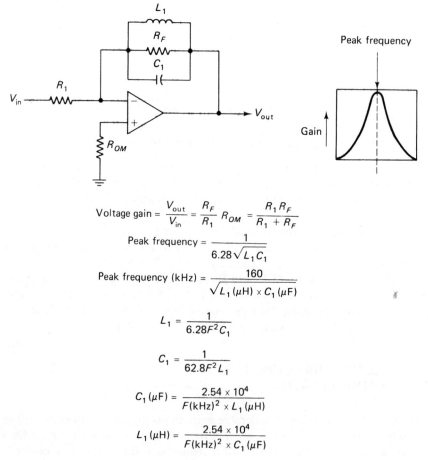

$$\text{Voltage gain} = \frac{V_{out}}{V_{in}} = \frac{R_F}{R_1} \qquad R_{OM} = \frac{R_1 R_F}{R_1 + R_F}$$

$$\text{Peak frequency} = \frac{1}{6.28\sqrt{L_1 C_1}}$$

$$\text{Peak frequency (kHz)} = \frac{160}{\sqrt{L_1\,(\mu H) \times C_1\,(\mu F)}}$$

$$L_1 = \frac{1}{6.28 F^2 C_1}$$

$$C_1 = \frac{1}{62.8 F^2 L_1}$$

$$C_1\,(\mu F) = \frac{2.54 \times 10^4}{F(kHz)^2 \times L_1\,(\mu H)}$$

$$L_1\,(\mu H) = \frac{2.54 \times 10^4}{F(kHz)^2 \times C_1\,(\mu F)}$$

Fig. 5-9. Narrow bandpass amplifier (tuned peaking)

resonant frequency. Thus, there is minimum feedback (and maximum gain) at the resonant frequency. Signals at frequencies on either side of the peak frequency are reduced in gain, or completely filtered out.

Design example. Assume that the circuit of Fig. 5-9 is to provide 40 dB gain at a peak frequency of 800 kHz. Use the datasheet phase compensation recommended for 40 dB, or the nearest gain value.

Select a value of R_1 on the basis of input bias current and voltage drop, as described in Sec. 5.1. Assume an arbitrary value of 1 kΩ for R_1 to simplify the example. The value of R_{OM} is then the same, or slightly less. With a value of 1 kΩ for R_1, and value of R_F is 100 k (or $R_1 \times 100$) for a 40 dB gain.

Any combination of L_1 and C_1 can be used, provided the resonant frequency is 800 kHz. For frequencies below 1 MHz, the value of C_1 should be between 0.001 and 0.01 μF. Assume an arbitrary 0.002 μF for C_1. Using the equations of Fig. 5-9, the value of L_1 is:

$$L_1 = \frac{2.54 \times 10^4}{(800)^2 \times 0.002} \approx 20 \ \mu\text{H}$$

5.5.2. Wide bandpass filter

Figure 5-10 is the working schematic of an op-amp used as a wide bandpass filter. Note that the circuit is essentially a wide bandpass amplifier.

Maximum circuit gain is determined by the ratio of R_R and R_F. The gain of the passband or flat portion of the response curve is set by R_F/R_R. Minimum circuit gain is determined by the ratio of R_1 and R_F.

The ratio of relationships of R_F, R_R and R_N also determines the relationships of the passband frequencies. For example, if the value of R_F is increased in relationship to R_R, and all other factors remain the same, the frequency F_1 will be decreased, but with F_2 unchanged. Thus, there will be a greater frequency spread between F_1 and F_2.

The ratios given in the equations of Fig. 5-10 are selected to provide a differential of about 10 dB between minimum and maximum gain. That is, if minimum gain is 20 dB, maximum gain is about 30 dB, and so on. In turn, minimum gain can be set by the ratio of R_1 to R_F.

There is also a direct relationship between the values of the capacitors, and between the capacitors and resistors. For example, if the value of C_N is increased, and all other factors remain the same, the frequencies F_3 and F_4 will be decreased, with F_1 and F_2 unchanged. Thus, there will be a narrower passband. The same conditions occur for R_N (an increase in R_N decreases the passband).

It is obvious from this analysis of the equations that the shape of the passband is determined by capacitance and resistance ratios. Also, there

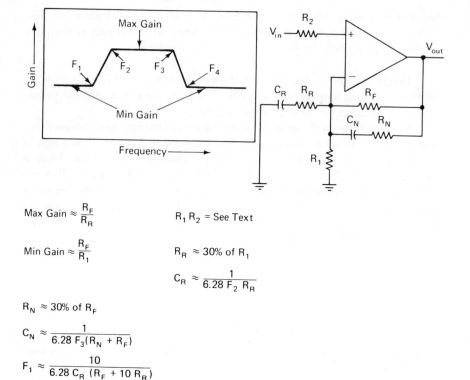

$$\text{Max Gain} \approx \frac{R_F}{R_R}$$

$$\text{Min Gain} \approx \frac{R_F}{R_1}$$

$$R_1 \, R_2 = \text{See Text}$$

$$R_R \approx 30\% \text{ of } R_1$$

$$C_R \approx \frac{1}{6.28 \, F_2 \, R_R}$$

$$R_N \approx 30\% \text{ of } R_F$$

$$C_N \approx \frac{1}{6.28 \, F_3 (R_N + R_F)}$$

$$F_1 \approx \frac{10}{6.28 \, C_R \, (R_F + 10 \, R_R)}$$

$$F_2 \approx \frac{1}{6.28 \, C_R \, R_R}$$

$$F_3 \approx \frac{1}{6.28 \, C_N \, (R_N + R_F)}$$

$$F_4 \approx \frac{40}{6.28 \, C_N \, (40 \, R_N + R_F)}$$

Fig. 5-10. Wide bandpass amplifier

must be a trade-off between gain and frequency relationships. If the ratios given in the equations of Fig. 5-10 do not provide the desired gain-frequency trade-off, alter the ratios as needed.

Design example. Assume that the circuit of Fig. 5-10 is to provide approximately 100 dB minimum gain at all frequencies, and approximately 110 dB gain at the passband. Frequency F_2 is to be about 200 kHz, with frequency F_3 at about 400 kHz. Use the phase compensation recommended for the 100 dB minimum gain, and not the passband gain of 110 dB.

Select a value of R_1 on the basis of input bias current and voltage drop, as described for the closed loop resistances in Sec. 5-1. Assume an arbitrary value of 1 kΩ for R_1 to simplify this example. The value of R_2 is then the same, or slightly less.

With a value of 1 kΩ for R_1, a value of 100 kΩ is used for R_F. With a value of 100 kΩ for R_F, the value of R_N is about 30 kΩ. Also with R_1 at 1 kΩ, R_R is 300 ohms. These relationships produce a minimum gain of 100 dB and a maximum gain of 110 dB.

With a value of 300 ohms for R_R, and F_2 of 200 kHz, the value of C_R is:

$$C_R \approx \frac{1}{6.28 \times 200 \text{ kHz} \times 300} \approx 0.003 \text{ } \mu\text{F}$$

With a value of 30 kΩ for R_N, 100 kΩ for R_F and an F_3 of 400 kHz, the value of C_N is:

$$C_N \approx \frac{1}{6.28 \times 400 \text{ kHz} \times (30 \text{ k} + 100 \text{ k})} \approx 3 \text{ pF}$$

With a value of 0.003 μF for C_R, an R_F of 100 kΩ, and an R_R of 300 ohms, the frequency of F_1 is:

$$F_1 \approx \frac{10}{6.28 \times 0.003 \text{ } \mu\text{F} \times \left[100 \text{ kΩ} + (10 \times 300) \right]} \approx 5 \text{ kHz}$$

With a value of 3 pF for C_N, an R_F of 100 kΩ, and an R_N of 30 kΩ, the frequency of F_4 is:

$$F_4 \approx \frac{40}{6.28 \times 3 \text{ pF} \times \left[(40 \times 30 \text{ kΩ}) + 100 \text{ kΩ} \right]} \approx 1.6 \text{ MHz}$$

5.5.3. Active low-pass (high-cut) filter

Figure 5-11 is the working schematic of an op-amp used as an active low-pass (high-cut) filter. Figure 5-11 also shows the basic equations for determining circuit values, together with the corresponding characteristic curves for several sets of component values. Note that these values are approximate and will usually require trimming to get an exact curve. The typical voltage gain is slightly less than 1 (unity) since the op-amp is operated as a voltage follower (Sec. 5.3).

The amount of gain and the shape of the curves are set by the amount of feedback in relation to signal (which in turn is set by component values). Note that the feedback is positive, and thus adds to the signal. However, the feedback amplitude (across the entire frequency range) is just below the point necessary for oscillation.

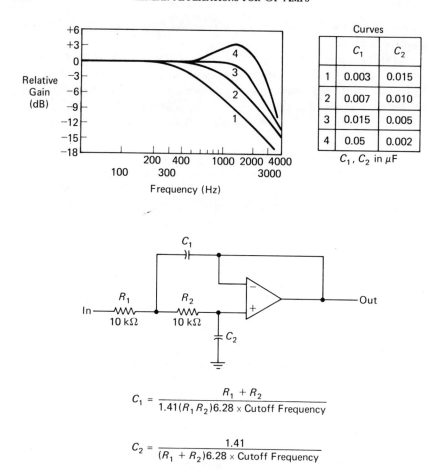

$$C_1 = \frac{R_1 + R_2}{1.41(R_1 R_2)6.28 \times \text{Cutoff Frequency}}$$

$$C_2 = \frac{1.41}{(R_1 + R_2)6.28 \times \text{Cutoff Frequency}}$$

Fig. 5-11. Active low-pass (high-cut) filter using an IC op-amp

5.5.4. Active high-pass (low-cut) filter

Figure 5-12 is the working schematic of an op-amp used as an active high-pass (low-cut) filter. Figure 5-12 also shows the basic equations for determining circuit values, together with the corresponding characteristic curves for several sets of component values. Note that these values are approximate and will usually require trimming to get an exact curve. The typical voltage gain is slightly less than 1 (unity) since the op-amp is operated as a voltage follower (Sec. 5.3).

    The circuit of Fig. 5-12 is the inverse of the Fig. 5-11 circuit. That is, the Fig. 5-12 circuit uses capacitors in series with the base, with feedback obtained through R_1 rather than C_1. The gain and shape of the curves are set by the amount of feedback (determined by circuit values).

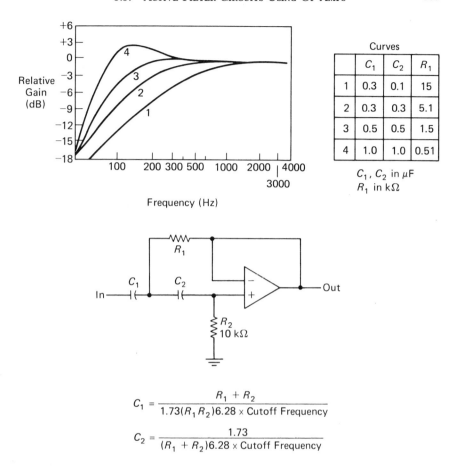

	C_1	C_2	R_1
1	0.3	0.1	15
2	0.3	0.3	5.1
3	0.5	0.5	1.5
4	1.0	1.0	0.51

C_1, C_2 in μF
R_1 in kΩ

$$C_1 = \frac{R_1 + R_2}{1.73(R_1 R_2)6.28 \times \text{Cutoff Frequency}}$$

$$C_2 = \frac{1.73}{(R_1 + R_2)6.28 \times \text{Cutoff Frequency}}$$

Fig. 5-12. Active high-pass (low-cut) filter using an IC op-amp

5.5.5. Active bandpass filter

The circuits of Figs. 5-11 and 5-12 can be cascaded to provide a bandpass filter. Any of the curves can be used. However, curve 3 is the most satisfactory because it has the sharpest break at cutoff. Curves 1 and 2 have considerable slope with no sharp break, and curve 4 produces some peaking at the breakpoints.

If the circuits are cascaded, the low-pass filter (Fig. 5-11) should follow the high-pass filter (Fig. 5-12), for best results.

5.5.6. High-Q bandpass filter with gain

Figure 5-13 is the working schematic of four op-amps used as a high-Q bandpass filter with gain. The four op-amps shown are contained in a single

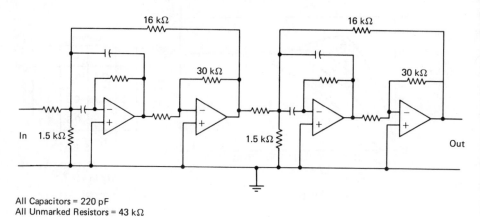

All Capacitors = 220 pF
All Unmarked Resistors = 43 kΩ
All ICS = Texas Instruments TL084

Fig. 5-13. High-Q bandpass filter with gain

package (the Texas Instruments TL084 which is an internally compensated BiFET op-amp).

With the values shown, the bandpass center frequency is 100 kHz, the Q is approximately 69, and the gain is 16. With a Q of 69, the bandwidth is approximately 1400 to 1500 Hz about the center frequency of 100 kHz. If desired, the center or peak frequency can be changed by as much as three decades when *all four capacitors are changed by a common factor*. However, the input resistance values may require some slight trimming.

5.5.7. High-Q notch filter

Figure 5-14 is the working schematic of an op-amp used as a high-Q notch filter without gain. The op-amp is a Texas Instruments TL081 (which is an internally compensated BiFET op-amp). The typical voltage gain is slightly less than 1 (unity).

With the values shown, the bandpass center frequency is 1 kHz. Other frequencies can be selected using the correct values of R and C, as shown by the equations of Fig. 5-14. Any combinations of R and C can be used to produce the desired frequency. That is, the capacitors can be made very large or very small, with corresponding resistor values. Generally, larger capacitor values and smaller resistor values produce a higher Q (narrower bandwidth), all other factors being the same. Of course, there are practical limits since the input signal must pass through R_1 and R_2. Extremely large R values can seriously attenuate the signal. Using the circuit values of Fig. 5-14, the center or peak frequency can be changed by as much as three decades when all three capacitors are changed by a common factor. This results in a minimum of trimming for the three resistor values.

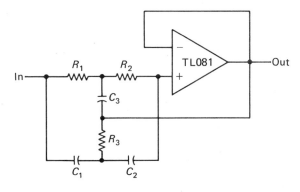

Bandpass
Center $= \dfrac{1}{6.28 \, R_1 C_1}$
Frequency

$\left. \begin{array}{l} R_1 = R_2 = 2R_3 = 1.5 \text{ M}\Omega \\[2mm] C_1 = C_2 = \dfrac{C_3}{2} = 110 \text{ pF} \end{array} \right\}$ 1 kHz Center Frequency

Fig. 5-14. High-Q notch filter

5.6. AUDIO CIRCUITS USING OP-AMPS

Although there are a number of special-purpose ICs designed specifically for audio work (Chapter 7), op-amps can also be used to form simple, effective audio circuits. The following paragraphs describe two typical circuits.

5.6.1. Audio distribution amplifier using op-amps

Figure 5-15 is the working schematic of four op-amps used as an audio distribution amplifier. The four op-amps are contained in a single package (Texas Instruments TL084). A single input is distributed to three outputs (A, B, and C). The three output ICs are operated as voltage followers and provide unity gain. The input IC provides a gain of about 10 across the audio range (down to a few Hz). Thus, the overall circuit gain is about 10.

5.6.2. Audio preamplifier using op-amps

Figure 5-16 is the working schematic of two op-amps used as an audio preamplifier. The two op-amps are contained in a single package (Texas Instruments TL080, which is an externally compensated BiFET op-amp). Frequency compensation is provided by the 10 pF capacitors. The circuit

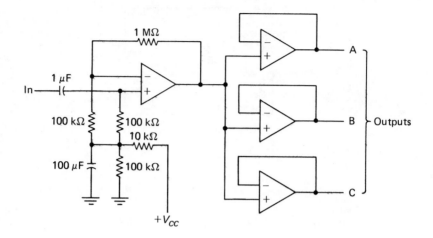

All ICS = Texas Instruments TL084

Fig. 5-15. Audio distribution amplifier using op-amps

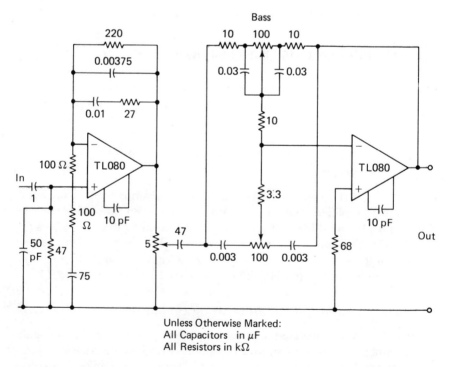

Unless Otherwise Marked:
All Capacitors in μF
All Resistors in kΩ

Fig. 5-16. Audio preamplifier using op-amps

includes the conventional treble, bass, and gain controls normally associated with preamplifiers, and provides sufficient gain to raise an input signal (from a microphone, phono pickup, etc.) to a level suitable for a typical power amplifier. The shape of the audio frequency response curve can be altered to meet specific requirements by means of potentiometer R_1. The setting of R_1 determines the amount of feedback in the input IC at various audio frequencies.

5.7.   OTHER OP-AMP LINEAR APPLICATIONS

The linear applications for IC op-amps described thus far represent only a portion of the possible applications. An IC op-amp can be used in any amplifier application, provided the gain and frequency characteristics are compatible. The IC op-amp has the advantage over other amplifiers in that the gain and frequency characteristics of IC op-amps can be programmed to meet specific requirements. Some other typical linear applications for IC op-amps include: summing amplifiers (analog adders), scaling amplifiers (weighted adders), difference amplifiers (analog subtractors), high input impedance amplifiers, feed-forward amplifiers (where high frequencies are routed around the first stages through a booster to extend bandwidth), voltage-to-current converters (transadmittance amplifiers), voltage-to-voltage converters, bridge circuit amplifiers, differential input/differential output amplifiers, and angle generators. These applications, and more, are described in the author's *Manual for Operational Amplifier Users* (Reston Publishing Company, Reston, Virginia, 22090, 1976).

6

NONLINEAR
APPLICATIONS
FOR OP-AMPS

In this chapter, we shall discuss nonlinear applications for op-amps. These nonlinear applications include those cases in which inputs and outputs are essentially non-sinewave, or where the output is drastically modified by the op-amp. Also included are those circuits in which the op-amp is used to generate nonlinear signals (such as squarewaves, pulses, integrated waves, etc.).

The format used here is essentially the same as found in Chapter 5. Unless otherwise stated, all of the design considerations for the basic op-amp described in Chapter 3 apply to each application covered here. Particular attention must be given to the information regarding op-amp power supplies and phase compensation.

6.1. PEAK DETECTORS

Op-amps can be used effectively as the active elements in peak detector circuits. The following paragraphs describe two typical peak detector circuits. One circuit uses a conventional capacitor-diode detector. The other circuit uses the base-collector junction of a transistor as the detecting diode.

6.1.1. Peak detector with capacitor-diode detector

Figure 6-1 is the working schematic of three op-amps used as a capacitor-diode type peak detector. The three op-amps shown are contained in a single package, the Texas Instruments TL084 (the fourth op-amp in the package is unused).

In operation, capacitor C_1 is charged to the peak value of V_{in} through diode D_1. The input IC operates as a voltage follower, producing a peak-to-peak output voltage equal to V_{in}. The negative peaks are clipped off by D_1 (which conducts only on the positive peaks), thus permitting C_1 to charge to the peak value. The output IC also operates as a voltage follower, producing an output voltage V_o equal to the peak voltage charge on C_1. The circuit is reset to zero when the RESET switch is closed. This applies a positive voltage to the noninverting input of the third (or reset) IC. The output of the reset IC is applied to the base of *npn* transistor Q_1, driving Q_1 into saturation, and permitting C_1 to discharge. The circuit will accept input voltages (V_{in}) up to the limit of the IC (± 15 V, in this case). V_{CC} must be larger than the anticipated peak voltage.

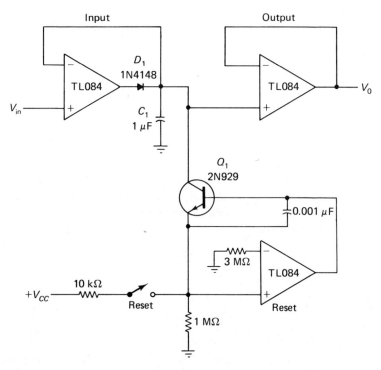

Fig. 6-1. Peak detector with capacitor-diode detector

6.1.2. Peak detector with transistor junction detector

Figure 6-2 is the working schematic of an op-amp used as the major active element in a peak detector system. Where accuracy is required for peak detection, the conventional capacitor-diode detector is often inadequate because of changes in the diode's forward voltage drop (caused by variations in the charging current and temperature). Ideally, if the forward drop of the diode can be made negligible, the peak value is the absolute value of the peak input, and will not be diminished by the diode-voltage drop (typically 0.5 to 0.7 V for a silicon diode). This can be accomplished by means of an op-amp.

The circuit of Fig. 6-2 uses the base-collector junction of a transistor as the detecting diode. The transistor (acting as a diode) is contained within the feedback loop (between the op-amp output and the circuit output). This reduces the effective forward voltage drop of the diode by an amount equal to the loop gain. For example, assume that the transistor has a base-emitter drop of 0.5 V, that the open-loop gain of the op-amp is 1000, and the

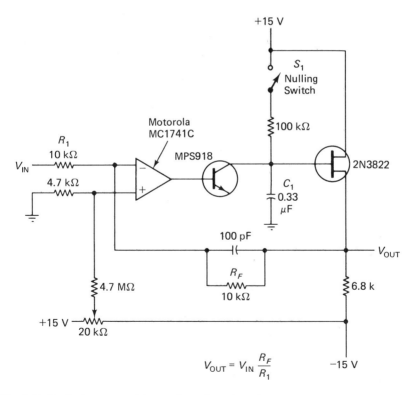

Fig. 6-2. Peak detector with transistor junction detector

closed-loop gain is 1 (unity). Under these conditions, the loop gain is 1000, and the effective forward voltage drop is 0.0005 V. Any transistor can be used, provided the leakage is low (preferably 10 nA, or less, with ± 15 V at the collector).

The storage time of the circuit in Fig. 6-2 is dependent upon leakage of the diode (transistor) and the FET, as well as the value of C_1. A larger value of C_1 will increase storage time.

Note that the values shown in Fig. 6-2 apply to a typical op-amp that requires a $V_{CC} - V_{EE}$ of 15 V (such as a Motorola MC1741, or other general purpose op-amp). These values can be used as a starting point for design. In use, the output is adjusted to zero offset (zero volts output with no signal input) by closing the nulling switch S_1 and adjusting offset potentiometer R_1. Switch S_1 also serves as a reset switch.

6.2. OSCILLATORS USING OP-AMPS

Since an op-amp has both positive and negative inputs, as well as gain, it is possible to use positive feedback from the output to sustain oscillation. Op-amp oscillators are particularly effective at very low frequencies where the high values of inductance (and capacitance) required make it impractical to use conventional LC oscillators. The following paragraphs describe some typical op-amp oscillator circuits.

6.2.1. Low-frequency sinewave generator

Figure 6-3 is the working schematic of an op-amp used as a low-frequency sinewave generator. The op-amp can be general purpose, such as the Motorola MC1741C. This circuit is a parallel-T oscillator. Feedback to the noninverting input becomes positive at the frequency indicated in the equation. Positive feedback is applied at all times. The amount of positive feedback (set by the ratio of R_1 to R_2) is sufficient to cause the op-amp to oscillate. In combination with the feedback to the noninverting input, feedback to the inverting input can be used to stabilize the amplitude of oscillation.

The value of R_1 is approximately 10 times the value of R_2. The ratio of R_1 and R_2, as set by the adjustment of R_2, controls the amount of positive feedback. Thus, the setting of R_2 determines the stability of oscillation.

The amplitude of oscillation is set by the peak-to-peak output capability of the op-amp, and the values of Zener diodes CR_1 and CR_2. As shown by the equations, the Zener voltage should be approximately 1.5 times the desired peak-to-peak output voltage. The nonlinear resistance of the back-to-back Zener diodes is used to limit the output amplitude and maintain good linearity.

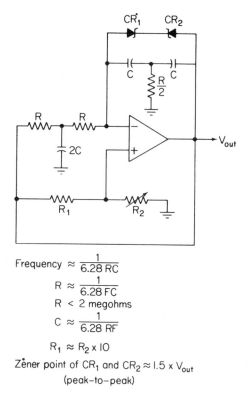

$$\text{Frequency} \approx \frac{1}{6.28\,RC}$$

$$R \approx \frac{1}{6.28\,FC}$$

$$R < 2 \text{ megohms}$$

$$C \approx \frac{1}{6.28\,RF}$$

$$R_1 \approx R_2 \times 10$$

Zener point of CR_1 and $CR_2 \approx 1.5 \times V_{out}$
(peak-to-peak)

Fig. 6-3. Low-frequency sinewave generator

The frequency of oscillation is determined by the values of C and R. The upper-frequency limit is approximately equal to the bandwidth of the basic op-amp. That is, if the open-loop gain drops 3 dB at 100 kHz, the oscillator should provide full voltage output up to about 100 kHz.

Design example. Assume that the circuit in Fig. 6-3 is to provide 6 V sinewave signals at 8 Hz.

Since R_2 is variable, the exact value is not critical. Assume a maximum value of 10 kΩ for R_2. With R_2 at 10 kΩ, R_1 is 100 kΩ.

With a required 6 V peak-to-peak output, the values (Zener voltage) of CR_1 and CR_2 should be 9 V. It is assumed that the basic op-amp is capable of 9 V peak-to-peak output.

The values of R and C are related to the desired frequency of 8 Hz. Any combination of R and C can be used, provided that the combination works out to a frequency of 8 Hz. For practical design, the value of R should not exceed about 2 M. Assume a value of 1 M for simplicity. With R and 1

M, and a desired frequency of 8 Hz, the value of C is:

$$C \approx \frac{1}{6.28 \times 8 \times 1^{+6}} \approx 0.02 \ \mu F$$

6.2.2. Low-frequency squarewave generator

Figure 6-4 is the working schematic of an op-amp used as a low-frequency squarewave generator. The circuit is similar to that shown in Fig. 6-3. However, the function of the Fig. 6-4 circuit is more like a blocking oscillator. The frequency of the circuit is set by the values of C and R_F. The oscillator can be made to operate at any frequency up to about 100 Hz simply by changing the value of C, so that the values of C and R_F work out to the desired frequency. Above 100 Hz, it may be necessary to trim the value of R_F to maintain oscillation.

Circuit stability is controlled by the R_1/R_2 ratio. If the circuit appears to be unstable with changes in temperature and/or power supply, trim the value of R_2. Keep in mind that the oscillator frequency should remain stable over wide temperature and power supply variations. The same is true of output amplitude. However, the maximum output amplitude is set by the

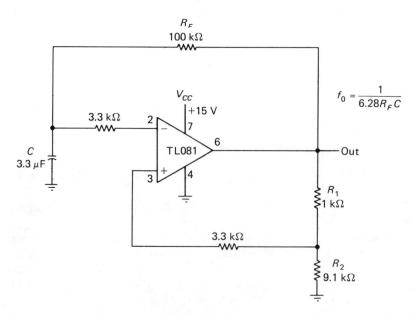

Fig. 6-4. Low-Frequency square wave generator

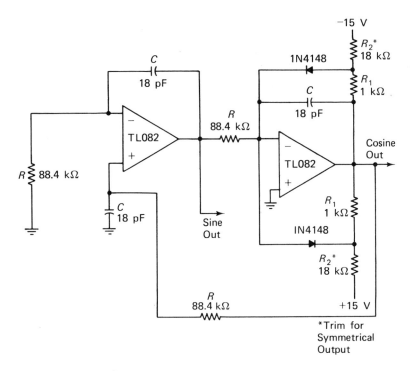

Fig. 6-5. Quadrature oscillator

power supply value. Using the 15 V shown in Fig. 6-4, the peak-to-peak amplitude of the squarewave output will be slightly less than 15 V.

6.2.3. Quadrature oscillator

Figure 6-5 is the working schematic of two op-amps used as a quadrature oscillator. The two op-amps are contained on a single package (the Texas Instruments TL082). The purpose of a quadrature oscillator is to produce two output signals of identical frequency, but shifted in phase by a fixed amount. As shown in Fig. 6-5, one output represents the sine of the signal, whereas the other output represents the cosine of the same signal.

Using the values shown, the signal frequency is 100 kHz. This can be lowered by as much as three decades when the values of all three capacitors (C) are changed simultaneously. However, it may be necessary to trim the values of resistors (R) to obtain an exact frequency. If so, always change the values of all three resistors (R) simultaneously. Also, it may be necessary to trim the values of resistors R_1 and R_2 to produce symmetrical outputs, as shown by the note on Fig. 6-5.

6.3. TEMPERATURE SENSORS USING OP-AMPS

Op-amps can be used as the major active element in a temperature sensor circuit. Temperature sensing with thermistors is popular because thermistors are inexpensive and easy to use. However, thermistors are nonlinear and can supply only very small output signals (generally a few microwatts). These problems can be overcome by means of an op-amp. The following paragraphs describe two typical op-amp temperature sensor circuits.

6.3.1. Basic temperature sensor circuit

In the circuit of Fig. 6-6, temperature is sensed by a thermistor in the normal manner and is read out on a voltmeter. The circuit can then be calibrated so that a given voltage indicates a given temperature (in F or C, as desired). As an alternate, the voltmeter scale can be altered to indicate temperature directly.

The output voltage is relatively linear over the temperature range of the op-amp. The output is exactly linear near the temperature at which the thermistor resistance equals the resistance of R_1. Thus, R_1 should be equal to R_T (the thermistor resistance) at the *center* of the desired temperature range.

The reference voltage V_{ref} is obtained from the Zener diode CR_1 and the voltage divider. The upper limit is a value determined by the power rating (P_T) of the thermistor, as shown by the equations. With the values

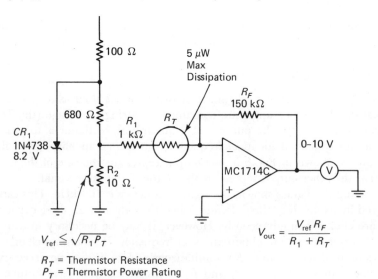

$$V_{ref} \leq \sqrt{R_1 P_T}$$

$$V_{out} = \frac{V_{ref} R_F}{R_1 + R_T}$$

R_T = Thermistor Resistance
P_T = Thermistor Power Rating
R_1 = R_T at Center of Temperature Range

Fig. 6-6. Basic temperature sensor circuit

shown, R_2 is adjusted so that V_{ref} is -0.067 V. Under these conditions, the maximum power dissipated by the thermistor is about 2.5 μW, well under the 5 μW rating.

6.3.2.   Freezing temperature sensor

The circuit of Fig. 6-7 uses a thermistor to sense temperature and turns on a lamp when the temperature drops to 32 °F (0 °C) or below. A typical use for such a circuit is an icy road warning indicator. In this circuit, the read out lamp is an LED (light emitting diode) which is flashed on and off when the temperature drops to 0 °C (or to another temperature as selected by adjustment of R_1).

All three op-amps shown in Fig. 6-7 are contained on a single IC package (Texas Instruments TL084). The input op-amp receives an adjustable reference voltage (set by R_1) at the noninverting input, and a variable voltage (determined by the temperature) at the inverting input. The output is applied to the noninverting input of the LED-drive op-amp. The inverting input of the LED-drive op-amp receives a pulse signal from the oscillator

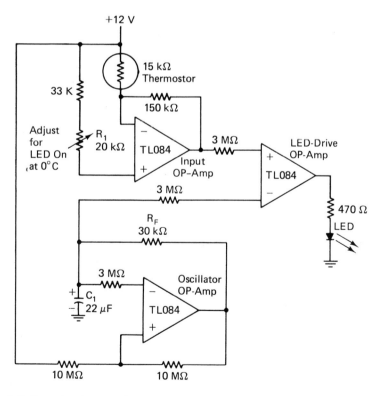

Fig. 6-7. Freezing temperature sensor

op-amp. (Note that the oscillator op-amp circuit is similar to that of the squarewave generator in Fig. 6-4.) When the thermistor resistance (and temperature) equal 0 °C (or what ever temperature is selected by adjustment of R_1) the oscillator pulses are passed to the LED, and serve to flash the LED on and off at a frequency determined by C_1 and R_F.

6.4. LINEAR STAIRCASE
AND RAMP GENERATORS

Op-amps can be used as the active elements in linear ramp generator and linear staircase generator circuits. The staircase circuit (using an op-amp) is the more useful of the two since there are many other circuits capable of producing a linear ramp (and which do not require the expense of an op-amp). However, design of a linear staircase generator using an op-amp is nearly identical to that of the ramp generator. For that reason, we shall discuss both types of circuits.

Figure 6-8 shows a linear ramp generator in which the noninverting input of an op-amp is grounded, and switch S_1 returns the output to the inverting input. When S_1 is closed, the op-amp is in the unity gain configuration, and the output is at ground (less any input offset voltage). When switch S_1 is opened, the output moves in the positive direction when the reference voltage V_{ref} is negative, or in the negative direction when V_{ref} is positive.

Because the output under closed-loop conditions (feedback through capacitor C) tries to maintain the input terminal at zero volts, the charging

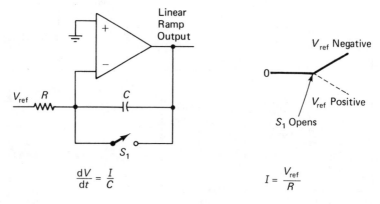

Fig. 6-8. Basic linear ramp generator using an op-amp

current to the capacitor C is constant at a rate of $dV/dt = I/C$, where:

> dV/dt is the increase (or decrease) of the ramp voltage for a given amount of time (such as one volt increase for each second after S_1 opens); C is the value of capacitor C (if C is expressed in μF and I is in amperes, then dV/dt is in volts per μs); I is V_{ref}/R (with V_{ref} in volts and R in ohms).

As an example, assume that V_{ref} is $+10$ V, R is 100 ohms, and C is 1 μF. Under these conditions, $I = 0.1$ A ($+10$ V/100 ohms) and the ramp decrease at a rate of one volt per 0.1 μs (0.1 A/1 μF). (The ramp decreases since V_{ref} is given as a positive voltage.)

Note that this equation for the linear ramp generator is accurate only if the charging current is *much greater* than the input bias current of the op-amp. As a guideline, make the charging current at least 100 times that of the input bias current. If the bias is not small in relation to the charging current, the bias will offset the charging current (either add to or subtract from). This will tend to make the ramp nonlinear, particularly at the start or end of the ramp. In any event, the dV/dt equation will no longer be accurate.

The *maximum possible* height or amplitude of the linear ramp depends upon the supply voltages for the op-amps. That is, the maximum possible positive ramp is set by V_{CC}, whereas V_{EE} sets the maximum negative ramp. Assuming a given dV/dt (set by V_{ref}, R and C), the *actual height* of the ramp is set by the amount of time switch S_1 is open. For example, if dV/dt is one volt per μs, and S_1 is open for 10 μs, the ramp can increase from zero to 10 V, provided the supply voltage is 10 V or greater. In practical applications, the maximum ramp height is slightly less than the supply voltage.

Figure 6-9 shows the circuit for a *linear staircase generator using an op-amp*. Note that the circuit is similar to the ramp generator, except for the input. In the Fig. 6-9 circuit, charging resistance R is replaced by two diodes and capacitor C_1. Likewise, the fixed V_{ref} is replaced by pulses. In the circuit of Fig. 6-9, a pulse of amplitude E couples a charge Q to the op-amp input. The charge Q is equal to $C_1 (E - 2 V_{ak})$, where $2 V_{ak}$ is the forward voltage drop across the two diodes. Typically, $2 V_{ak}$ is equal to about 1 V, since the drop across each diode is about 0.5 V.

When switch S_1 is open, capacitor C_2 is charged (by each pulse) in steps equal to $(E - 2 V_{ak}) (C_1/C_2)$. For example, assume that the pulses are 11 V, C_1 is 500 pF and C_2 is 1 μF. Under these conditions, each step will be about 5 mV. This is shown as follows:

$$(11 - 1)(0.0005/1) = 0.005 \text{ V} = 5 \text{ mV}$$

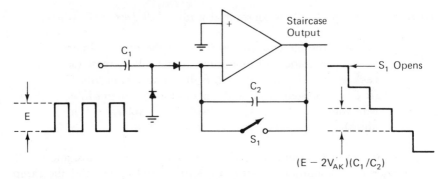

Fig. 6-9. Basic linear staircase generator using an op-amp

Note that the equations for the linear staircase generator are accurate only if the charging current is much greater than the input bias current of the op-amp. Thus, op-amps with minimum input bias are recommended for staircase generator circuits. Of more importance, the pulse amplitude must be much greater than the voltage/temperature variations of the diodes. If the diode voltage variations are not small in relation to the charging pulses, the voltage/temperature variations will add to (or subtract from) the pulse amplitudes, and produce uneven staircase steps.

The maximum possible height of the staircase is set by the op-amp supply voltages. The actual height of the staircase is set by the amount of time switch S_1 is open (assuming an infinite number of pulses and steps).

6.4.1. Practical staircase generator circuit

In practical ramp and staircase circuits, the switch S_1 is replaced with an electronic switch. Fig. 6-10 shows a staircase generator circuit controlled by a FET switch. This circuit is used as part of a digital voltmeter. Clock pulses are applied to the FET gates to control operation of the staircase circuit. When the clock pulses are "on," the drain-source resistance of the FET is reduced to zero, and capacitor C_3 is shorted, or "closed." When the clock pulses are "off," the drain-source resistance rises to several hundred megohms, removing the "short" from C_3. With C_3 in the feedback circuit, the staircase generator produces steps in response to each signal pulse applied through C_1 and C_2. The generator will continue to produce steps until C_3 is again shorted by the FET in response to "off" clock pulses.

Note that capacitor C_1 is adjustable so that the staircase output can be set to exactly 5 mV steps. The signal pulse at the input are fixed at approximately 11 V. This is reduced to about 10 V by the drop across diodes D_1 and D_2. In theory, when C_1 is adjusted so that the parallel capacitance of C_1 and C_2 is 500 pF, the ratio of input capacitance to output capacitance is

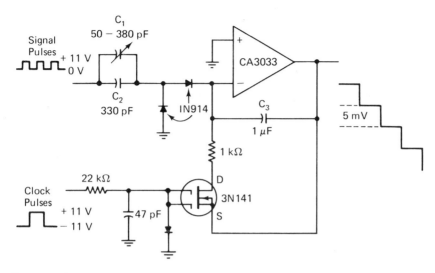

Fig. 6-10. Staircase generator circuit controlled by a FET electronic switch

0.0005 to 1, and the 10 V input produces 5 mV steps at the output. However, because the diodes do not always have the same voltage drop, and the op-amp input bias is not always the same (because of temperature variations, etc.), the value of $C_1 - C_2$ is not always precisely 500 pF. Instead, C_1 is simply adjusted until the output steps are precisely 5 mV.

6.5. OTHER OP-AMP NONLINEAR APPLICATIONS

The nonlinear applications for IC op-amps described thus far represent only a portion of the possible applications. Some other typical nonlinear applications for IC op-amps include: multiplexers, multivibrators, comparators, log and antilog amplifiers, envelope detectors, track and hold amplifiers, integration amplifiers, differentiation amplifiers, T-filters, and voltage regulators. These applications, and more, are described in the author's *Manual for Operational Amplifier Users* (Reston Publishing Company, Reston, Virginia, 22090, 1976).

7

LINEAR
IC PACKAGES
AND ARRAYS

In this chapter, we discuss linear ICs that are not op-amps or OTAs. There are two basic types of such ICs. First, there are the general-purpose ICs. These include diode arrays, transistor arrays, Darlington-pair arrays, amplifier arrays, and similar groups of solid-state elements fabricated on a single semiconductor chip. Second, there are the special purpose ICs that can replace all of the active elements in a solid-state circuit. These include regulators, comparators, differential amplifiers, video, IF and RF amplifiers, audio amplifiers, DC amplifiers, balanced modulators, and so on. Both types of ICs are discussed from the design standpoint.

Unlike Chapters 4, 5, and 6 where the applications or uses apply to many types of op-amps and OTAs, the values and design examples in this chapter are based on the characteristics of the particular IC discussed. Although these examples are "typical," they do not necessarily apply to all ICs of a given type. Thus, the datasheet must be consulted to determine the IC characteristics and values. However, the values given here can be used as a starting point for design.

7.1. IC DUAL DIFFERENTIAL COMPARATOR

The dual differential comparator performs the same function as the op-amp comparator described in Sec. 4.5.2. However, no OTAs are needed when an

IC differential comparator is used. In effect, the circuit is complete, with the possible exception of a few resistors and external power.

The IC differential comparator is similar to an op-amp. This is shown in Fig. 7-1 which is the schematic of a Motorola MC1514 comparator. In general, op-amps are designed for analog applications, and are not easily adapted to digital use. For example, the output voltage swing is not usually compatible with the various logic families. Likewise, there is a long propagation time and slow recovery time in most op-amps, making them unsuited for high-speed digital use.

The IC differential comparator provides a circuit that is easily interfaced with digital logic (in regard to propagation time, recovery time, and

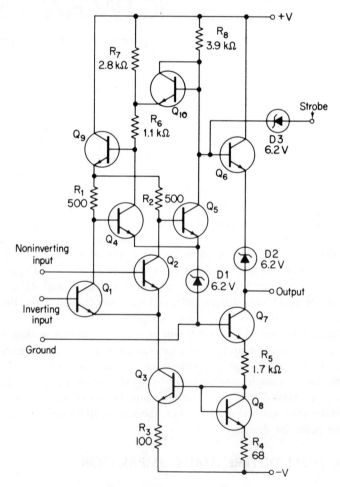

Fig. 7-1. One-half of a Motorola MC1514 comparator

output swing; refer to Chapter 8), but still retains some op-amp characteristics. For example, as shown in Fig. 7-1, the comparator has two inputs (inverting and noninverting), a single-ended output, dual power supplies, plus a strobe to enable the output. The strobe feature is essential in digital use where data must be held until called for at a specific time.

7.1.1. Basic circuit features

Basically, the comparator consists of two amplifier stages, output level shifting, and strobe capability, as shown in Fig. 7-1. The combined voltage gain of the two amplifier stages is typically 1700 (open-loop). The single-ended output of the second stage is referenced several volts positive with respect to ground. In order to make the output voltage swing compatible with the popular TTL (Chapter 8) logic levels, the output level is shifted down by Q_6 and the 6.2 V Zener diode D_2. When using the recommended power supply voltages of $+12$ V and -6 V, the typical output voltage swing is from -0.5 V to $+3$ V. The output has sufficient current sink capability to drive one TTL load.

The strobe circuit clamps the output one base-emitter voltage drop below the strobe input potential. The strobe can be driven with standard TTL logic. The strobe adds considerable flexibility to the comparator, especially when used in noisy environments. Typically, minimum strobe width is 30 ns, with a strobe load of less than 1 mA. Delay time between strobe input and comparator output is 4 to 12 ns, depending on temperature.

7.1.2. Application as level detector

Figure 7-2 shows an IC comparator connected as a basic level detector. A reference voltage between ± 5 V can be applied to one input, and the signal to be detected is applied to the opposite input. When the input signal exceeds the reference voltage, the comparator switches state. The voltage transfer function for a positive reference voltage is shown in Fig. 7-3. If a negative reference voltage is applied to the inverting input, then the transfer curve is inverted about the vertical axis.

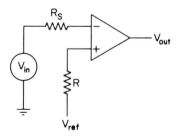

Fig. 7-2. Basic level detector using IC comparator

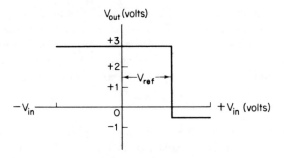

Fig. 7-3. Level detector transfer characteristics

In order to minimize the turn-on error voltage of the circuit in Fig. 7-2, the dc resistance of the reference voltage source R should equal the source resistance R_S of the signal source. Minimum input offset voltage is generated when the two resistances are equal, because the input bias currents to the comparator produce nearly equal voltages across each source resistance (as is the case with an op-amp). Also, a low source resistance will minimize the effects of the input offset current. From a practical standpoint, R and R_S should be selected for minimum resistance, and to match any differences between source and reference resistances.

The circuit of Fig. 7-2 can be used for a variety of threshold detector applications. Other applications include a voltage comparator in analog-to-digital converters, and a high noise immunity buffer circuit.

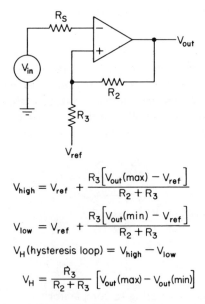

$$V_{high} = V_{ref} + \frac{R_3\left[V_{out}(max) - V_{ref}\right]}{R_2 + R_3}$$

$$V_{low} = V_{ref} + \frac{R_3\left[V_{out}(min) - V_{ref}\right]}{R_2 + R_3}$$

$$V_H \text{ (hysteresis loop)} = V_{high} - V_{low}$$

$$V_H = \frac{R_3}{R_2 + R_3}\left[V_{out}(max) - V_{out}(min)\right]$$

Fig. 7-4. Basic level detector with hysteresis

7.1.3. Application as level detector with hysteresis

Figure 7-4 shows an IC comparator connected as a level detector with hysteresis. In some slow-speed switching applications, noise can cause oscillation during the time the input signal is passing through the transition region of the comparator. This problem can be greatly reduced or eliminated by using hysteresis in the form of external feedback as shown. The turn-on and turn-off points of the circuit are then controlled by the ratio of the feedback resistances (all other factors being equal). This is shown by the equations of Fig. 7-4 and the graph of Fig. 7-5.

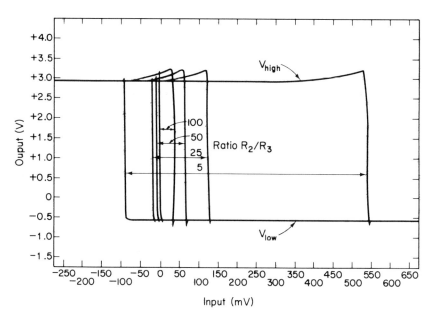

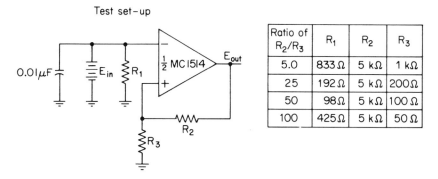

Test set-up

Ratio of R₂/R₃	R₁	R₂	R₃
5.0	833 Ω	5 kΩ	1 kΩ
25	192 Ω	5 kΩ	200 Ω
50	98 Ω	5 kΩ	100 Ω
100	425 Ω	5 kΩ	50 Ω

Fig. 7-5. Motorola MC1514 comparator hysteresis curves

If possible, the source resistance should be equal to the parallel resistance of R_2 and R_3. This will minimize input offset voltage. Hysteresis curves for various values of feedback resistors are given in Fig. 7-5. There is one precaution that should be observed with the feedback resistors. Oscillation can occur if the ratio of the feedback resistors is greater than the small-signal gain of the comparator. Note that the circuit of Fig. 7-4 can also be used as a variable-threshold Schmitt trigger.

7.1.4.   Application as a double-ended limit detector

Figure 7-6 shows two IC comparators connected as a double-ended limit detector. The circuit switches the output of the comparators to the high state whenever the input voltage is outside the bounds of the reference voltages. This circuit can be used in test equipment to test for some set limits. The input-output transfer function is also given in Fig. 7-6. Note that the strobe input can be used to disable the circuit when not in use.

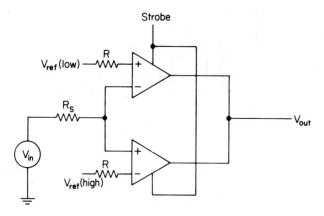

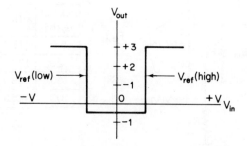

Fig. 7-6. Double-ended limit detector using IC comparator

7.1.5. Application as a monostable multivibrator

Figure 7-7 shows two IC comparators connected as a monostable multivibrator with a fixed threshold. Figure 7-8 shows the connections for an adjustable threshold monostable MV. In either circuit, the monostable multivibrator action is accomplished with the two cross-coupled comparators. Both circuits are switched from the stable state to the quasi-stable state with a positive-going signal.

In the circuit of Fig. 7-7, the threshold or triggering point is determined solely by the comparator output in the high state (V_{OH}). In the circuit of Fig. 7-8, the trigger point is determined by V_{OH} and the ratio of R_1 and R_2, as shown by the equations.

In either circuit, the amount of time the circuit remains in the quasi-stable state (duration of the monostable or one-shot pulse) is determined by the RC time constant, and the ratio of ΔV_{out} and V_{ref}, as shown by the equations. For example, assume an RC time constant of 1000 ns (say an R of 1 kΩ and a C of 1000 pF), and a ratio of 3 (ΔV_{out} of 3 and a V_{ref} of 1). This will produce an approximate 1300 ns duration of the monostable pulse (1000 ns $\times$ 1.3 = 1300).

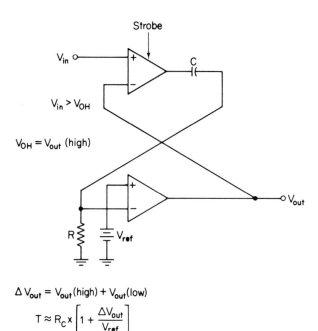

$$\Delta V_{out} = V_{out}(high) + V_{out}(low)$$

$$T \approx R_C \times \left[1 + \frac{\Delta V_{out}}{V_{ref}} \right]$$

T = Duration of monostable output pulse

Fig. 7-7. Monostable multivibrator with fixed reference using IC comparator

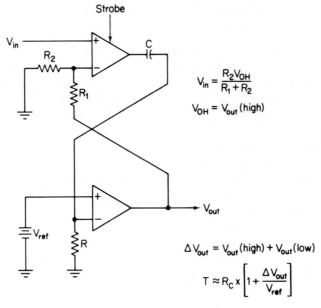

$$V_{in} = \frac{R_2 V_{OH}}{R_1 + R_2}$$

$$V_{OH} = V_{out}\,(high)$$

$$\Delta V_{out} = V_{out}\,(high) + V_{out}\,(low)$$

$$T \approx R_C \times \left[1 + \frac{\Delta V_{out}}{V_{ref}}\right]$$

T = Duration of monostable output pulse

Fig. 7-8. Monostable multivibrator with adjustable reference using IC comparator

A positive reference voltage is required to hold the comparator output V_{out} high during the time the circuit is in the stable state (no input pulse). Also, the reference voltage can be used to vary the output pulse. For example, using the previous values, if the reference voltage is increased to 1.5 V, the ratio is 2 (ΔV_{out} of 3 and V_{ref} of 1.5 = 2). This will produce an approximate 1200 ns duration of the pulse.

The strobe feature of the comparator can provide additional flexibility. For example, the strobe can be used to gate output pulses of different time duration. This can be accomplished since the strobe and input signals are essentially an AND function. If desired, the strobe can be used to inhibit selected input pulses.

In the circuit of Fig. 7-8, the input voltage need not be greater than V_{OH}, as is the case with the Fig. 7-7 circuit. As shown, the threshold (trigger point) of the circuit is set by the values of R_1 and R_2, for a given value of V_{OH}. For example, assume that V_{OH} is 3 V, R_1 is 500 Ω and R_2 is 50 Ω. Under these conditions, the circuit should trigger with an approximate 0.27 V input pulse:

$$V_{in} = \frac{50 \times 3}{500 + 50} = \frac{150}{550} = 0.27 \text{ V}$$

7.1.6. Application as an astable multivibrator

Figure 7-9 shows an IC comparator connected to an astable multivibrator. As shown by the equations, the period of the output pulse waveform (and hence the frequency) is determined by the RC time constant, the ratio of feedback resistors R_2 and R_3, the ratio of high- and low-output voltages, and the level of the negative power supply voltage.

Resistors R, R_1, R_2, and R_3 bias the comparator in the active region to ensure self-starting. In practical applications, R_1 can be made variable to ensure turn-on of the circuit, and to adjust the frequency (within limits set by the RC values).

The capacitor C produces a net positive feedback at high frequencies which insures oscillation. The resistance value of R should be limited from

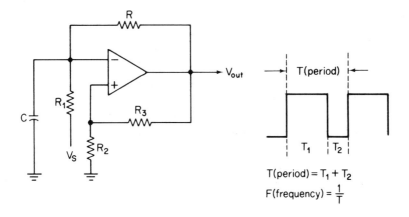

$$T(period) = T_1 + T_2$$
$$F(frequency) = \frac{1}{T}$$

$$T_1 \approx \frac{R_1 RC}{R+R_1} \ln \left[\frac{R_1 V_{OH} - R V_s + \dfrac{R_2}{R_2+R_3}(R_1+R)V_{OL}}{R_1 V_{OH} - R V_s - \dfrac{R_2}{R_2+R_3}(R_1+R)V_{OH}} \right]$$

$$T_2 \approx \frac{R_1 RC}{R+R_1} \ln \left[\frac{R_1 V_{OL} + R V_s + \dfrac{R_2}{R_2+R_3}(R_1+R)V_{OH}}{R_1 V_{OL} + R V_s - \dfrac{R_2}{R_2+R_3}(R_1+R)V_{OL}} \right]$$

ΔV_{out} = the total output voltage swing
V_{OH} = total voltage in the high state
V_{OL} = total voltage in the low state
V_S = negative power supply voltage

Fig. 7-9. Astable multivibrator using IC comparator

1kΩ to 50 kΩ. If R is less than 1 kΩ, the current capability of the comparator may not be sufficient to allow the output to switch fully to the low state. For values of R larger than 20 kΩ, the input offset current can affect the output pulse width.

For high frequency operation, the value of C has to be small. If the value of the C calculated to generate a desired frequency approaches the stray capacitance of the physical layout, the frequency generated may vary considerably from the calculated value. To minimize this effect, a minimum C of 100 pF is recommended. With these values of R and C, operation is limited to frequencies below about 5 MHz.

In order to minimize the effects of offset current and voltage, the parallel combination of R_2 and R_3 should be equal to the parallel combination of R_1 and R.

7.1.7.   Application as a zero-crossing pulse generator

Figure 7-10 shows two IC comparators connected as a zero-crossing pulse generator. This circuit is a special application of a double-ended limit detector, and is designed to generate a digital output pulse at the zero crossover points. This is shown in Fig. 7-11.

Two comparators are needed. One comparator is set to detect a minus voltage, and the other comparator is set to detect a positive voltage. The

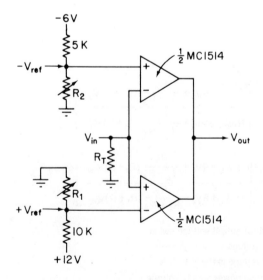

Fig. 7-10. Zero-crossing pulse generator using IC comparator

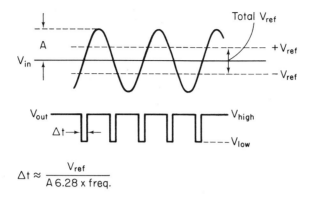

$$\Delta t \approx \frac{V_{ref}}{A\ 6.28 \times freq.}$$

Fig. 7-11. Input-output waveforms for zero-crossing pulse generator

output of the two wired-OR comparators (Chapter 8) is low only during the time the input signal is between $- V_{ref}$ and $+ V_{ref}$. Otherwise, the comparator output is high as shown in Fig. 7-11. The circuit is most accurate when the reference voltages are as close to the zero as is possible.

As shown by the equations, the duration of the output pulse is determined by the total reference voltage, the amplitude (RMS) of the sinewave, and the sinewave frequency. For example, assume a total reference voltage of 20 mV ($+ 10$ mV and $- 10$ mV), a frequency of 10 kHz, and a sinewave of 2.5 V. The duration of the output is approximately:

$$\frac{20\ mV}{2.5 \times 6.28 \times 10\ kHz} \approx 127\ ns$$

The value of R_T should be chosen to minimize input offset voltage.

7.1.8. Application as a peak voltage detector

Figure 7-12 shows an IC comparator connected as a peak voltage detector. With the RC network connected at the comparator input, the voltage at the noninverting input will always lag V_{in} (at the inverting input) by an amount dependent upon frequency. This lag in voltage results in an input voltage polarity reversal every time the input signal changes direction. The reversal in input voltage is sufficient to cause the comparator output to switch states. Capacitor C_1 is used to filter out possible oscillation during switching. Typical values and waveforms are included on Fig. 7-12.

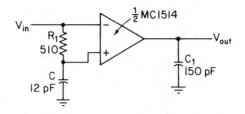

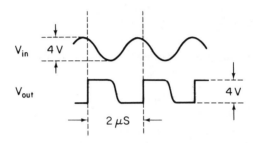

Fig. 7-12. Peak voltage detector using IC comparator

7.1.9.   Application as a core memory sense amplifier

Figure 7-13 shows two IC comparators connected as a core memory sense amplifier. This was one of the original applications for the comparator. A sense amplifier is essentially an interface between the memory elements and the logic elements of a digital computer. The signals generated by the tiny ferrite cores of the memory are of insufficient amplitude to drive the logic circuits directly. The function of the sense amplifier is to convert the relatively small signals from the core memory into logic levels such as TTL. Not only must the sense amplifier detect small signals, but it also has to distinguish between signals which differ by only a few millivolts in amplitude. The problem is further complicated by the fact that noise and relatively large common-mode signals are present during most of the computer's memory cycle.

For example, a typical core memory produces outputs of less than 50 mV, while a typical logic level is +5 V. This requires a gain of 100. To make the problem more complex, a logic 1 output from a core could be represented by a 15 to 50 mV signal, whereas a logic 0 output might be a 0 to 10 mV signal. The problem can be solved by setting the comparator threshold so that the output (to the computer logic) goes to a logic 1 (+5 V) whenever the input (from the core memory) is above about 20 mV. The comparator output switches to a logic 0 (0 V) with inputs below 20 mV.

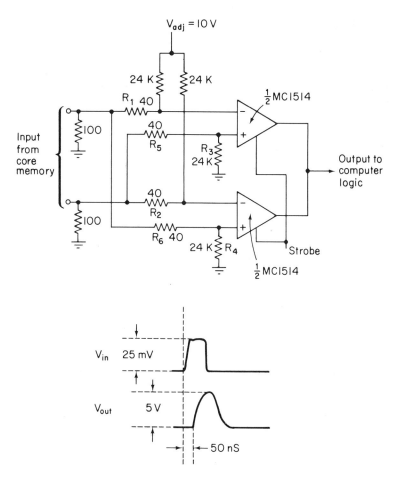

Fig. 7-13. Core memory sense amplifier using IC comparator

In the circuit of Fig. 7-13, the threshold is set by the resistor divider network. The two comparators with their outputs wire-ORed are required to respond to both positive and negative input signals. The threshold is essentially equal to the voltage across resistors R_1 and R_2. The 100 Ω input resistors provide sense line (input from core memory) termination. Resistances R_3 through R_6 are used to increase common-mode rejection by balancing the input threshold circuit. With V_{adj} tied to the $+12$ V line, the input threshold of the resistor divider is set at 20 mV. However, the input threshold can be varied by changing the V_{adj} voltage. This feature allows the memory designer to tailor the sense amplifier to each memory plane.

In practical use, the resistor values must be closely matched to keep the uncertainty region about the threshold level (20 mV) as small as possible.

Considerable care should be used in laying out the resistor network since any stray capacitance can result in poor high-frequency common-mode rejection. Since the input threshold voltage is a function of the V_{adj} voltage, it must also be tightly controlled once set. If the strobe input to the comparator is grounded, the comparator output will be clamped to about -0.5 V. This will prevent undesired noise that may pass through the comparator from entering the computer logic circuits.

All other factors being equal, the response time of the comparator is dependent upon the amount of overdrive from the core memory. The more the threshold level is exceeded by the input voltage, the faster the comparator will respond. As in the case of any logic element, the faster (or shorter) the response time, the greater the system speed. In the circuit of Fig. 7-13, with a 25 mV input and a 20 mV threshold (5 mV overdrive), the response time of the comparator is about 50 ns.

7.1.10. Miscellaneous applications

In addition to the applications described thus far, a comparator can be used whenever it is necessary to produce pulsed or uniform level-changing output (suitable for digital circuits) in response to various input levels and waveforms. The comparator switches output states in a uniform manner (1 or 0, high or low, etc.) in response to a differential input signal. This input can be pulsed, sinewave, or simply a difference in dc levels (or a combination). The comparator is an important building block in most analog/digital conversion circuits. No matter what input is used, the output is a pulse or level change that can be adapted to most of the existing logic families. In this regard, the comparator functions as a pulse restorer.

As an example, the comparator can be used as a *line receiver* where digital pulse data must be transmitted through long lines. Assume that the digital logic levels are 0 V and $+5$ V (representing 0 and 1), and that the pulses are attenuated to about 3 V by the transmission line. Also assume that there is a 1 V noise level on the line. If the line is fed directly to the digital circuit, the 3 V pulse might not be sufficient to trigger the digital circuit elements, unless gain is introduced. On the other hand, a high gain could cause the 1 V noise signals to trigger the digital circuits. This problem can be solved by a comparator where the input threshold is set about 1 V, but below 3 V (say 2 V), and the output is set for 0 V and $+5$ V.

7.2. IC DIFFERENTIAL AMPLIFIER

There are many applications where an amplifier with differential input and differential output is required. A laboratory differential voltmeter is a classic example. Such meters must amplify very small differential signals in the

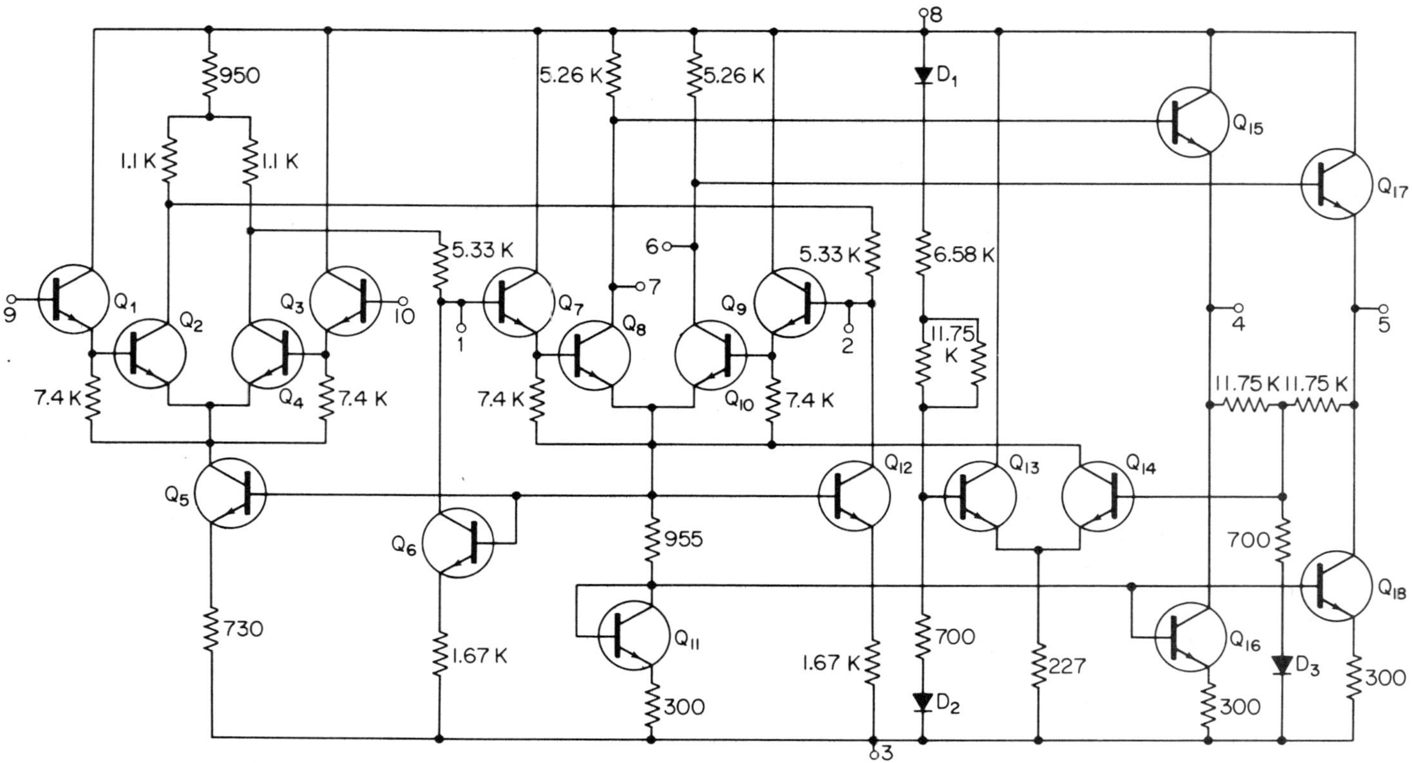

Fig. 7-14. Motorola MC1520 differential output op-amp

presence of large common-mode signals. The output of the meter amplifier is fed to a meter movement where both the amplitude and polarity of the differential signal is indicated, but common-mode signals are rejected.

A differential output amplifier can be made up using two op-amps. To be really effective, both op-amps must be identical in characteristics (gain, temperature drift, etc.). This is best accomplished using a dual-channel op-amp. Also available are differential amplifiers in IC package form. These differential amplifiers are similar to dual-channel op-amps, except that both channels are already interconnected on the same semiconductor chip. Such an amplifier is shown in Fig. 7-14 which is the schematic of a Motorola MC1520 differential amplifier.

7.2.1. Typical connections and applications

The IC shown in Fig. 7-14 can be used wherever a high-gain, wide frequency range differential amplifier is required. Figure 7-15 shows the connections for both balanced-input and unbalanced-input operation. The recommended values for the external resistors and capacitors, together with corresponding gain and frequency ranges, are given on the datasheets. The IC can also be used as a single-ended op-amp (both inverting and noninverting) if desired.

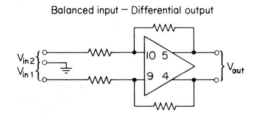

Fig. 7-15. Typical connections for differential output op-amp

7.3. IC HIGH-FREQUENCY (VIDEO) AMPLIFIERS

There are a number of high-frequency amplifiers in IC package form. Often, such amplifiers are referred to as video amplifiers since they have the wideband/high gain/fast rise time characteristics necessary for the video sections of television receivers, radar circuits, active filters, etc., which must pass pulses as well as high-frequency sinewaves. Some of these IC amps are special purpose (low distortion, fixed gain option, adjustable gain, etc.) whereas others are general purpose (typically differential input and differential output, with or without AGC).

There is an almost infinite variety of applications for these amplifiers. It is impractical to discuss all of the ICs and their applications. (Such discussions are available in the manufacturer's datasheets and application notes.) However, the following paragraphs describe a typical high-frequency video amplifier IC package and its applications. Where practical, both the external circuit values and the performance characteristics are given. Needless to say, an ingenious experimenter can often adapt the application circuits to other ICs of similar design.

Figure 7-16 is the schematic of a Motorola MC1545, which is a gated video amplifier. The IC is designed as a wideband video amplifier. However, because of the gated, two-channel, differential design, a number of other applications are possible.

7.3.1. Basic circuit features

As shown in Fig. 7-16, the circuit consists of a constant current source transistor Q_7 and a switching differential amplifier Q_5-Q_6 that splits the constant current between two differential amplifiers or channels, composed to transistor pairs Q_1-Q_2 and Q_3-Q_4 depending on the voltage applied at pin 1, the gating pin. The collectors of both channels are tied together and connected to a common load resistor (1 kΩ). The amount of current flowing through each of the 1 kΩ load resistors is constant and independent of whichever channel, Q_1-Q_2 or Q_3-Q_4, is conducting. As a result, there is essentially no dc level shift at the output when one channel is turned off and the other channel is turned on. The steady-state change measured in the differential output voltage when switching from one channel to the other is typically 15 mV. The amplified signal that appears at the collectors of the input channels is transferred to the output via Darlington emitter followers for a low output impedance and, at the same time, buffers the input differential amplifiers from any capacitive loading that would tend to lessen the frequency response.

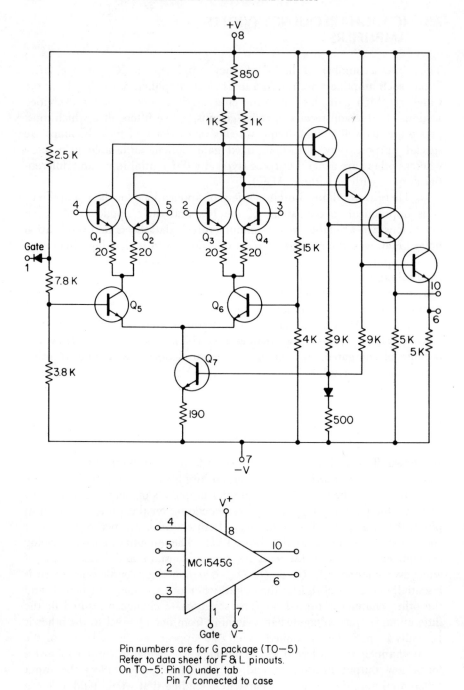

Pin numbers are for G package (TO–5)
Refer to data sheet for F & L pinouts.
On TO–5: Pin 10 under tab
          Pin 7 connected to case

Fig. 7-16. Motorola MC1545 amplifier

Common-mode feedback is provided from the emitter of the first emitter followers back to constant current source Q_7. This stabilizes the dc operation point of the circuit and provides common-mode rejection. The circuit is biased such that with a sufficiently positive voltage applied to the gating pin, or with the gating pin left open, the voltage at the base of Q_6 is more negative than the voltage at the base of Q_5. As a result Q_5 is on and Q_6 is off. Under this condition, all of the constant current that is established in the collector of Q_7 passes through Q_5, and establishes a bias current in the Q_3–Q_4 differential amplifier. Thus, any signal applied to Q_3–Q_4 is amplified and appears in differential form at the output.

If the gating pin is connected to ground or some negative value, the voltage at the base of Q_5 becomes more negative than the voltage at the base of Q_6, which causes constant current flow through Q_6, and establishes the bias current in the Q_1–Q_2 differential amplifier. In this state, any signal applied to Q_1–Q_2 appears at the output, whereas the signal applied to Q_3–Q_4 is gated off. The voltage required to perform this gating function at pin 1 is compatible with standard forms of saturated logic (TTL, etc.). Either of two signals can be gated through the amplifier, depending upon the application of a logic signal to the gating pin.

The amount of attenuation given to an input signal when the amplifier is gated off is primarily a function of the dc voltage at the gating pin. Figure 7-17 is a curve showing the attenuation versus gate voltage. Note that a variable voltage at the gating pin provides an automatic gain control (AGC) feature for the amplifier.

The amount of signal attenuation is also a function of input frequency, as shown in Fig. 7-18. This curve indicates that above 30 kHz, the amplifier begins to show a certain amount of capacitive feedthrough, primarily because of the physical closeness of the pins, and can be improved by the use

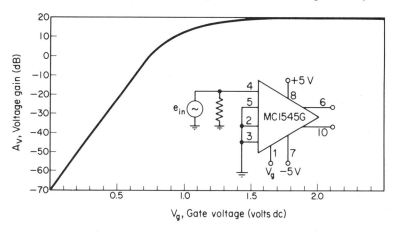

Fig. 7-17. Video amplifier with AGC

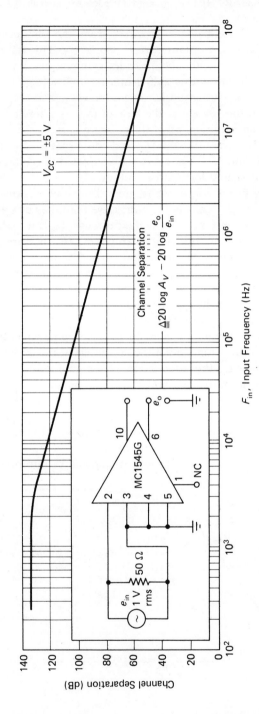

Fig. 7-18. Channel separation versus input frequency

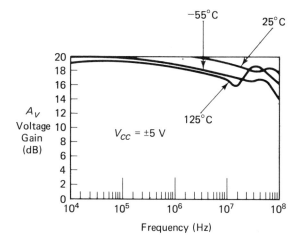

Fig. 7-19. Single-ended voltage gain versus frequency

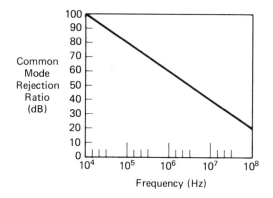

Fig. 7-20. Common-mode rejection versus frequency

of proper shielding between pins. However, even at an input frequency of 10 MHz, a channel separation of better than 60 dB can be achieved.

Figures 7-19 and 7-20 show voltage gain and common mode rejection of the IC, respectively. Rise time, fall time, and propagation delay are typically 6 ns for the IC.

7.3.1.   Application as a video switch

Figure 7-21 shows the IC connected as a gated analog switch. Only one external resistor is required. When used as a switch, a signal (analog or digital) is applied at pin 4. When a logic 1 (positive voltage) is applied to the gate, the input signal is amplified and passed through the amplifier. The amplifier is switched off when a logic 0 is applied to the gate. If the design

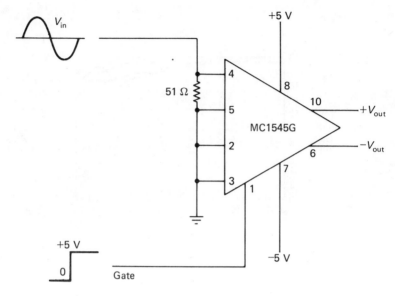

Fig. 7-21. Basic video switch

requirements are such that the conditions must be reversed (a logic 1 blocks the signal), the input signal can be applied to pins 2 or 3, with pins 4 and 5 grounded, eliminating the need for inverters.

The channel select time (time delay from 50% point of the gate pulse to the 50% point of the full output swing) is approximately 20 ns. When the gate logic is at 0, the IC must sink about 2.5 mA maximum. With the gate at 1, the IC must source only the leakage current of a reverse biased diode (about 2 μA maximum). These requirements are similar to the input requirements of a standard TTL logic gate.

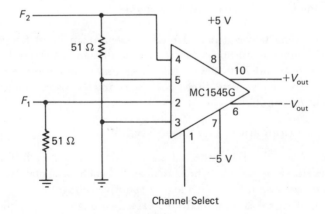

Fig. 7-22. Frequency shift keyer (FSK) (multiplexer)

7.3.2. Application as a frequency shift keyer (FSK)

Figure 7-22 shows the IC connected as a frequency shift keyer. This is similar to the video switch, except that a second frequency is applied to input pins 2 and 3. Either of the two inputs (but not both) is selected by a control voltage at the gate. Frequency F_2 is passed when the voltage at pin 1 is greater than +1.5 V. Frequency F_1 is passed when the pin 1 voltage is approximately zero.

7.3.3. Application as a data selector

Figure 7-23 shows three ICs connected as a data selector. As shown in the truth table of Fig. 7-23, one and only one input signal passes to the output with appropriate control voltages at the gates of the ICs. Any number of ICs can be connected in parallel and cascade to handle the required number of input signals.

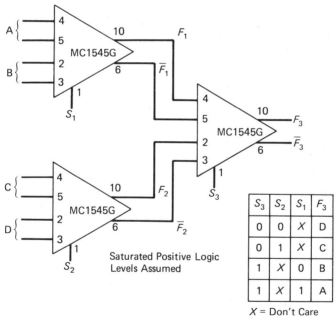

S_3	S_2	S_1	F_3
0	0	X	D
0	1	X	C
1	X	0	B
1	X	1	A

X = Don't Care

Fig. 7-23. One-out-of-four data selector

7.3.4. Application with a temperature-compensated gate

The amount of gate voltage required to switch the IC varies with temperature. This presents no problem with digital logic circuits where the gate signal is a pulse, and the gate is either "full on" or "full off." Typically, the

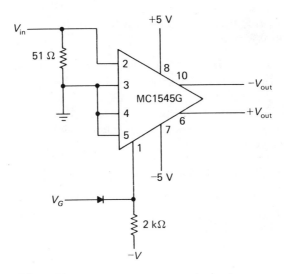

Fig. 7-24. Amplifier with temperature-compensated gate

required gate switching voltage varies about 2 mV/°C. A digital logic pulse of one or two volts can easily overcome this variation. However, when the IC is used with an analog voltage in linear applications, some form of temperature compensation may be required.

A quick method of temperature compensation is to use another silicon diode in series with the gate input diode, as shown in Fig. 7-24. Note that the external series diode is connected in reverse polarity to the IC gate diode. This provides forward-biased, back-to-back diode compensation so that the temperature drift of the two diodes will (ideally) cancel. Typically, the temperature variation is reduced to about 0.2 mV/°C with the two-diode scheme. Of course, the gate characteristics are shifted higher by about 0.6 to 0.7 V (or the equivalent of the normal voltage drop across a silicon diode).

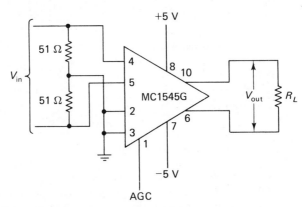

Fig. 7-25. Wideband amplifier with AGC

7.3.5. Application as wideband differential amplifier with AGC

The gate characteristics of the IC, as shown in Fig. 7-17, make the IC useful as an AGC amplifier. With a dc voltage applied to the gate pin, as much as 100 dB of AGC can be obtained. Since there is essentially no dc level shift with changes in AGC, the output waveform will collapse symmetrically about zero with little or no distortion. Figure 7-25 shows the IC connected as a wideband amplifier with AGC.

7.3.6. Application as an amplitude modulator

Figure 7-26 shows the IC connected as an amplitude modulator. The carrier signal is introduced into the input of one channel (pin 4). The remaining channels are disabled by connecting the inputs to ground. The modulation signal is introduced into the gate through a coupling capacitor. The gate is also connected to ground through a 5 kΩ potentiometer which sets a bias level on the gate. By using a potentiometer to bias the gate at the midpoint of the channel linear region (with a typical gate voltage of 1.2 to 1.4 V), a symmetrically modulated result is obtained. Of course, modulation voltage must not exceed the value that would operate the amplifier out of its linear range. Typically, the input carrier voltage is limited to 150 mV, while the input modulation voltage is approximately 400 mV (peak-to-peak).

Figure 7-27 shows the measured gate characteristics, with gain plotted on a linear scale. By biasing the gate at point B and impressing an audio

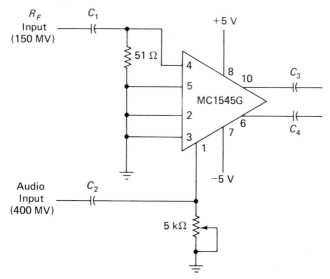

Fig. 7-26. Amplitude modulator

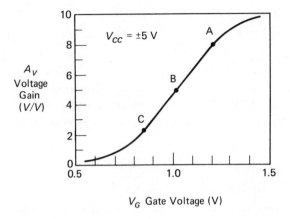

Fig. 7-27. Voltage gain versus gate voltage

signal on the bias, the gain of the channel varies quite linearly between the points A and C on the curve of Fig. 7-27, giving very little distortion on the output. Using the gate characteristics in Fig. 7-27, the up-modulation (M_U) and down-modulation (M_D) may be calculated. Referring to Fig. 7-28, the up and down modulation factors are defined as:

$$M_U = \frac{E_{max} - E}{E} \text{(upward modulation)}$$

$$M_D = \frac{E - E_{min}}{E} \text{(down modulation)}$$

where: E = peak amplitude of the unmodulated carrier

         E_{max} = maximum amplitude attained by the modulated carrier envelope

         E_{min} = minimum amplitude of the modulated carrier envelope

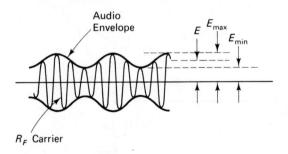

Fig. 7-28. Amplitude modulated waveforms

By making the gate voltage vary about point B (Fig. 7-27), with a maximum occurring at point A and a minimum at point B, and letting the RF carrier input be e_{in}, then:

$$e_o = e_{in}A_{V1}$$

Thus, the values of E, E_{max} and E_{min} are:

$$E = e_{in}(A_{V1})_B$$
$$E_{max} = e_{in}(A_{V1})_A$$
$$E_{min} = e_{in}(A_{V1})_C$$

and, by using the M_U and M_D equations, obtain

$$M_U = \frac{(A_{V1})_A - (A_{V1})_B}{(A_{V1})_B}$$

$$M_D = \frac{(A_{V1})_B - (A_{V1})_C}{(A_{V1})_B}$$

Substituting the values of $(A_{V1})_A$, $(A_{V1})_B$, and $(A_{V1})_C$ from Fig. 7-27 into the equations, we find:

$$M_U = 0.58$$
$$M_D = 0.54$$

These are the values of up and down modulation which can be expected without appreciable distortion.

7.3.7. Application as a balanced modulator

Figure 7-29 shows the IC connected as a balanced modulator. Note that the circuit of Fig. 7-29 is similar to that of Fig. 7-27, except for connections at the input. In the balanced circuit of Fig. 7-29, the internal differential amplifiers are connected in a manner that cross-couples the collectors.

Assuming that the carrier level is adequate to switch the cross-coupled pair of differential amplifiers, the modulating signal at the gate is switched between collector loads at the carrier rate. This results in multiplying the modulation by a symmetrical switching function. Care must be exercised not to overdrive the modulation input (typically modulation is limited to 200 mV peak-to-peak). This will ensure good carrier suppression.

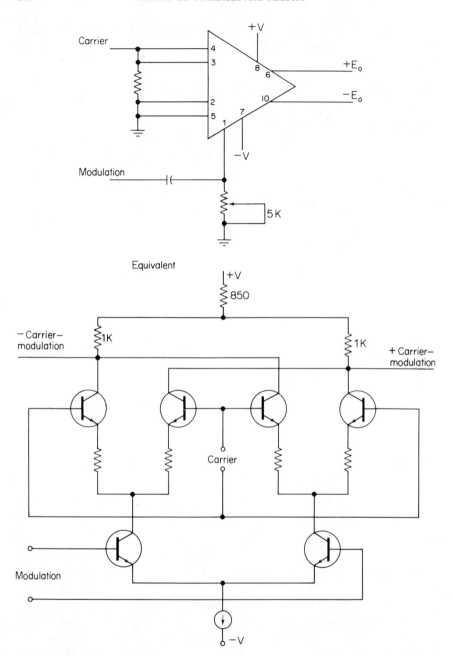

Fig. 7-29. Balanced modulator

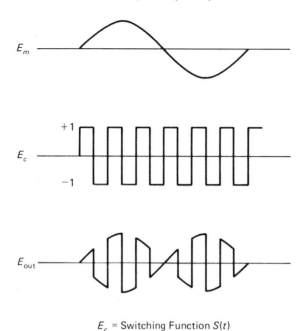

E_c = Switching Function $S(t)$

Fig. 7-30. Modulator waveforms

To balance the IC when acting as a modulator, equal gain must be given to the two differential channels. This results in an output similar to that shown in Fig. 7-30. In effect, the output is composed of the sum and difference frequencies (sidebands), and the carrier is suppressed. Equal gain for the two channels is produced by setting the potentiometer for a proper balance. In theory, the potentiometer is adjusted to set both channels at the exact center of their linear operating range. However, in practice, the center point may not be exactly the same for both channels, so the potentiometer must be set to some point of equal gain near the center.

7.3.8. Application as a gated astable multivibrator

Figure 7-31 shows two ICs connected as a gated astable multivibrator. The output is a squarewave of approximately 2 V peak-to-peak (assuming power supply voltages of 5 or 6 V). As shown by the equations, the output frequency is useful up to about 2 MHz, and is set by the values of R, C, R_1 and R_2. Rise times and fall times of the squarewave can be improved by means of the "speed-up" capacitor C_S across R_1, as shown. The circuit of Fig. 7-31 is free-running (or self-generating) so long as the gates are open. Thus, the circuit can be turned on or off by means of a bias at the gates.

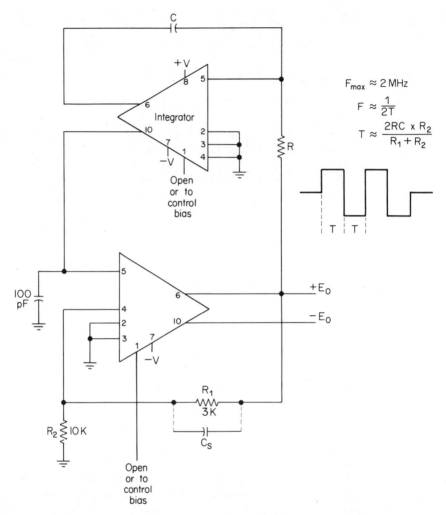

Fig. 7-31. Astable multivibrator

7.3.9. Application as a gated oscillator with level control

Figure 7-32 shows the IC used as a gated oscillator with an adjustable level control. The output is a sinewave of approximately 2 V peak-to-peak (with power supply voltages of 5 or 6 V). The useful frequency range is 1 kHz to 10 MHz. As shown by the equations, the frequency is set by values of R and C (both R values and both C values must be the same). The output level is set by the potentiometer.

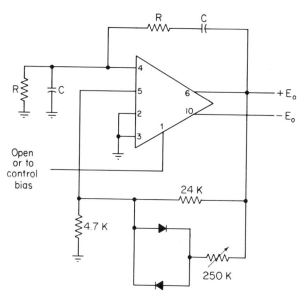

$F_{max} \approx 10 \text{ MHz}$

$F \approx \dfrac{1}{6.28\,RC}$

Fig. 7-32. Gated oscillator with level control

Rise times of the circuit are dependent upon the RC network, and may vary from 1 to several cycles of the oscillator frequency. At 160 kHz, a typical rise time is 30 μs. At 8 MHz, the rise time decreases to something near 100 ns. The circuit of Fig. 7-32 is normally free-running, and can be turned on or off by means of a bias at the gate.

7.3.10.   Application as a pulse width modulator

Figure 7-33 shows the IC used as a pulse-width modulator. Note that the gate is left open, and the IC is used as a differential amplifier.

A pulse-width modulator is essentially a comparator (Sec. 7.1). When the amplitude of the modulation signal is greater than that of the carrier signal, the output is driven to one extreme (full on or full off), assuming that amplifier gain is sufficient to produce saturation with the given input level.

A sawtooth signal is used as the carrier in the circuit of Fig. 7-33. One edge of the output pulse is always at the trailing edge of the sawtooth, with the other edge of the output derived from the modulation signal. Thus, the rise time of the output pulses is approximately the time it takes the input

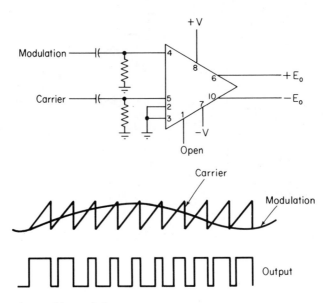

Fig. 7-33. Pulse-width modulator

signals (carrier and modulation) to give a differential voltage large enough to drive the output to full on.

Assuming that the carrier frequency is made much higher than the modulation frequency, the modulation signal level is approximately constant during the switching time (that is, approximately constant during each sawtooth cycle). The rise and fall times of the output will then depend on the carrier frequency, carrier amplitude, and amplifier gain (assuming the rise and fall times do not exceed the capability of the IC). For maximum rise time, it is desirable to have a large signal for the carrier (limited by the common mode input of the IC, or ± 2.5 V in this case).

As an example, assume that the carrier frequency is 250 kHz (a 4 μs period), the carrier signal is maximum at ± 2.5 V (or 5 V peak-to-peak), that the fall time of the sawtooth is negligible in relation to the rise time (which is normally the case), that amplifier voltage gain is 10, and that the desired output is 2 V peak-to-peak.

The slope of the sawtooth is then:

$$\frac{5\text{ V}}{4\ \mu\text{s}} = 1.25 \text{ V}/\mu\text{s}$$

With a gain of 10, the input change to give full output is 200 mV. The approximate rise time of the output is then: 1 μs/1.25 V$=0.8$ μs per volt; 0.8 μs$\times 0.2$ V$=0.16$ μs (or 160 ns).

In addition to rise and fall times, another characteristic that indicates the quality of a pulse-width modulator is the ratio of largest output pulse width to the smallest pulse width. A number of factors or variables are involved in this ratio, including pulse rise times, amplifier gain, nonlinearity, input differential mode breakdown, and input loading considerations. No matter what characteristics are involved, it is obvious that a large ratio will permit more information to be transmitted by means of pulse width than a small ratio, all other factors being equal. That is, pulse width modulation with a ratio of 10 is capable of passing more information than an identical modualtion system with a ratio of 5.

From a practical standpoint, using the circuit of Fig. 7-33, the modulation ratio can be increased by decreasing the carrier amplitude, while maintaining all other factors (modulation signal, supply voltage, etc.) constant. Of course, a decrease in carrier amplitude will increase rise and fall times of the output pulses.

7.3.11.   Application as a pulse amplifier

Figure 7-34 shows the IC used as a pulse amplifier. Such a configuration can be used in pulse radar IF circuits, and telemetry pulse-width or pulse-amplitude modulation systems. An IC of this type has several advantages as a pulse amplifier. First, the IC has a large bandwidth, as is shown in Fig. 7-19. Next, the IC uses direct coupling which provides low frequency response. Also, since both the input and output are differential, common-mode signals (such as noise) are greatly reduced, compared to a single-ended amplifier of the same gain and bandwidth.

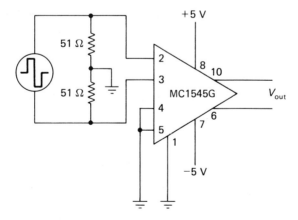

Fig. 7-34. Pulse amplifier

7.4. IC AUDIO AMPLIFIERS

Many of the linear amplifiers discussed in this chapter can be used at audio frequencies. Also, there are ICs specifically designed as audio amplifiers. In the past, and for the most part today, IC audio amplifiers are low power (1 or 2 W). These ICs are used as microphone amplifiers, drivers, preamplifiers, etc., to drive discrete component power amplifiers. However, the future trend is toward medium- and high-power IC audio amplifier packages. Of course, such ICs require a large heat sink capability. The following discussion is limited to a low-power IC which is typical of present-day units.

Figure 7-35 is the schematic of a Motorola MC1554, which is described by the manufacturer as a low frequency power amplifier. The IC provides one watt of power with either a single or split power supply. Typical voltage gains are 10, 18 or 36 with 0.4% THD (total harmonic distortion).

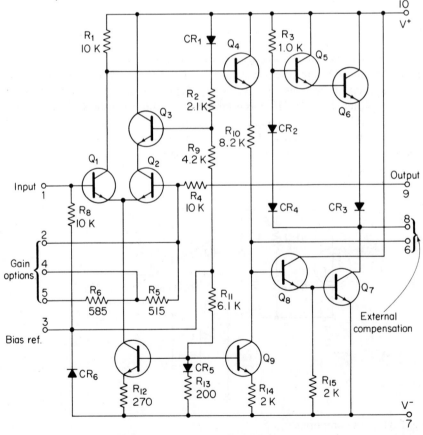

Fig. 7-35. Motorola MC1554 audio power amplifier

7.4.1. Basic circuit features

As shown in Fig. 7-35, the output stage is quasi-class B, obtained by use of diodes CR_2, CR_3 and CR_4. Diodes CR_2 and CR_4 are conducting for most of the complete cycle. This fixes the voltage across the input of Q_5, Q_6 and the diode CR_3 at two diode drops (about 1 V). Thus, Q_6 and CR_3 cannot be conducting simultaneously. This prevents excessive current flows from one power source to another through the output circuit elements. The resulting dead-band or crossover distortion is easily removed from the output signal by the action of the overall feedback loop.

Operation of the output stage can be seen by noting that V_{out} depends upon the current flow in Q_7. A rising output voltage is obtained by decreasing current in Q_7 which reduces the voltage drop across R_3, and makes the base of Q_5 rise toward $+V_{CC}$. This action turns both Q_5 and Q_6 on to supply current to the load for the positive output swing. For the negative swing Q_7 is driven to a larger value of collector current which increases the drop across R_3 and turns both Q_5 and Q_6 off.

7.4.2. Application as a noninverting power amplifier

Figure 7-36 shows the IC used as a basic noninverting power amplifier. Both single-supply and dual-supply connections are shown. Note that the output must be capacitively coupled to the load when the single supply is used. A dual-supply (or split-supply as it is sometimes called) eliminates the need for capacitive coupling.

Dissipation. With either supply system, the IC package dissipates about 1.5 W (with about 1 W output). If the IC case is mounted directly against a metal chassis, a heat sink is usually not necessary. If the IC is mounted on a typical composition printed-circuit board, a heat sink is required. The manufacturer recommends a $2 \times 2 \times \frac{1}{8}$ inch aluminum plate with a center hole drilled and reamed such that the heat sink fits snugly over the top of the TO-5 style package, using some type of thermal conducting grease.

Dual-supply operation. In the dual-supply circuit, note that a 10 μF electrolytic capacitor C_2 is connected to $-V_{EE}$. This keeps a fixed bias on the capacitor. Also, an extra bypass capacitor C_6 is required for dual-supply operation. However, the need for this extra capacitor is usually offset by the improved low-frequency response. With the values shown, typical low-frequency response of the dual-supply connection is 40 Hz, whereas the single-supply low end is 200 Hz. Another advantage of the dual-supply is elimination of the "thump" caused in a loudspeaker by the turn-on charging

Single Supply Operation

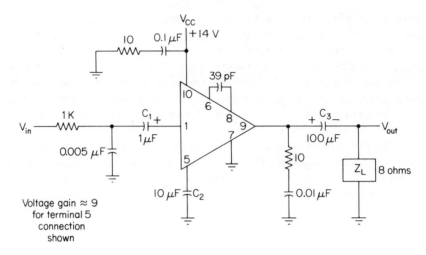

Voltage gain ≈ 9
for terminal 5
connection
shown

Split Supply (Dual-supply) Operation

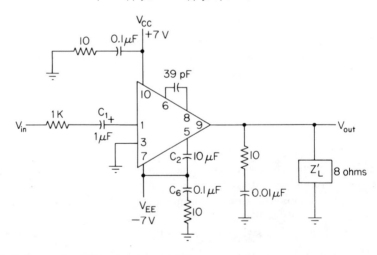

Fig. 7-36. Basic noninverting power amplifier

transient of C_3. On the other hand, direct coupling to the dual-supply circuit input produces an offset at the output. This may be objectionable. For example, if the output load is a loudspeaker, an offset voltage at the output will apply a constant current through the loudspeaker winding.

High-frequency limits. The high-frequency limits of both circuits is about 22 kHz (well beyond the normal audio range). This limit is set by the input RC filter network (1 kΩ resistor and 0.005 μF capacitor). The RC filter can be eliminated, thus extending the high frequency limit. The IC will provide 20 dB gain up to about 2 or 3 MHz, and over 30 dB gain up to about 500 kHz. However, at higher frequencies, the harmonic distortion increases. At frequencies below 22 kHz, and with 20 dB gain, the total harmonic distortion is less than 1% (typically 0.75%).

Low-frequency limits. The low-frequency limit of the dual-supply circuit is set by the value of C_1 and the input impedance of the IC (typically 10 kΩ), as well as the feedback capacitor C_2. The single-supply low-frequency limit is also affected by the output capacitor C_3 and the impedance of the load. As in the case with discrete-component circuits, the coupling capacitors and their related impedances form RC low-pass filters.

As a guideline, to find the low-frequency cutoff limit (where the signal is about 3 dB down) of any RC filter, the equation is: $1/(6.28\ RC)$. For example, assume that the load impedance is 8 Ω and the value of C_3 is 100 μF, as shown. Then the approximate low-frequency limit is: $1/(6.28\times8\times 100\ \mu F)\approx200$ Hz. Thus, if the capacitor value is doubled to 200 μF, the low-frequency limit is reduced to about 100 Hz.

Since C_3 is eliminated in the dual-supply circuit, the value of C_1 and the IC input impedance set the low-frequency limit. Using the same equation, the approximate low-frequency limit for dual-supply operation is: $1/(6.28\times1\ \mu F\times10\ k\Omega)\approx16$ Hz. The fact that this is lower than the actual low limit of about 40 Hz is caused by the effect of feedback capacitor C_2 on low-frequency operation.

Gain. The connection of capacitor C_2 sets the amount of feedback from output to input, and thus fixes the IC gain. Capacitor C_2 can be connected to pins 2, 4, or 5 to produce fixed gains of approximately 32, 25, and 20 dB, respectively. The opposite side of C_2 is connected to ground for the single-supply, or the $-V_{EE}$ for the dual-supply. Capacitor C_2 thus forms a low-pass filter (for the feedback voltage) in conjunction with internal IC resistors.

Oscillation suppression. Note that an RC bypass network (10 Ω and 0.01 μF) is shown at the output of both circuits. This network is to prevent oscillation at radio frequencies. The network is not always required, but should be included as a precaution in any power IC where high-frequency operation is possible. For example, the IC here is designed to provide some gain up to about 10 MHz, but will operate at frequencies up to 50–150 MHz. If there is any feedback (say harmonics) because of stray lead inductance, stray capacitance, etc., the IC can oscillate at these high frequencies. This will pass unnoticed in normal audio operation. Likewise, the bandwidth of most oscilloscopes will not pass such oscillations. But these undetected oscillations can cause the IC to overheat, and possibly burn out.

Offset problems.   With single-supply operation, there is a small output offset (in the order of 20 mV) caused by an unbalance in the input differential stage. However, since single-supply operation requires capacitor coupling at the output, the offset should have little effect. No current is drawn through the load because of this offset, except the initial charging current of the coupling capacitor.

With dual-supply operation, where no coupling capacitor is used and the output must have no offset, terminal 3 of the IC is grounded. This removes the base resistance unblanace and establishes a zero-volt reference at the input. Thus, the no-signal level of the output is zero volts.

7.4.3. Application as an inverting power amplifier

Figure 7-37 shows the IC used as an inverting power amplifier. The dual-supply connections are used. The output and supply connections are essentially the same as for the noninverting circuit. However, the input connections are entirely different. The input signal is introduced at the opposite (inverting) side of the differential amplifier at pin 4. The noninverting input, pin 1, is returned to ground through a 0.1 μF capacitor. Terminals 2 and 5 are shorted together. This arrangement provides the full gain of 32 dB. However, the input impedance (at pin 4) is reduced to about 250 Ω. As a result, a large value input coupling capacitor is needed to pass low frequencies. For example, using the equation and the 40 Hz input described in Sec. 7.4.2, the value of input capacitor C is about 16 μF.

The input can be direct coupled from the source. However, the output will then be offset by any no-signal dc that appears at the input, and by any drop across the approximate 250 Ω input impedance. This offset can be

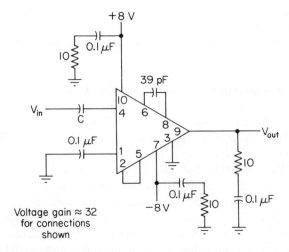

Fig. 7-37. Inverting power amplifier

compensated for by properly biasing pin 1, or terminating pin 1 with approximately 250 Ω.

With the values shown, the IC delivers about 1 W output, depending upon the value of the load. The available peak-to-peak output voltage swing is about 12 V into a 12 Ω load. If the load resistance is lower than 12 Ω, the peak-to-peak voltage swing must be reduced (by reducing the input signal) to comply with the absolute maximum peak current rating of 500 mA.

7.4.4. Application as a pulse power amplifier

Figure 7-38 shows the IC used as a pulse power amplifier. The single-supply connections are used, with the gain set for approximately 25 dB (pin 4 connected to ground through the 10 μF capacitor). Because the output is a pulse, and not a continuous linear signal, the normal power dissipation of about 1.5 W can be increased so as to produce a peak power output of about 3 W. This is based on the assumption that the pulse duty cycle is 50 to 75%.

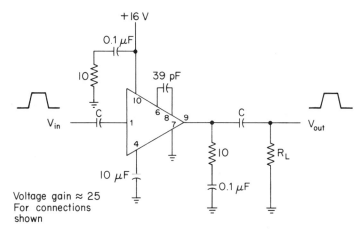

Fig. 7-38. Pulse power amplifier

In addition to not exceeding the peak power, care must be taken not to exceed peak output current (which in this case is 500 mA). For example, assume that the load must be 10 Ω. Then the peak input voltage must be limited to 5 V ($E = IR$; $0.5 \times 10 = 5$). Since the gain is set for approximately 25 dB, or a voltage gain of about 20, the input pulse must be limited to a peak of 0.25 V.

7.4.5. Application as a differential output power amplifier

Figure 7-39 shows two ICs connected as a differential output power amplifier. The single-supply connections are used for both ICs. Both amplifiers are connected for a 20-dB gain. However, the top amplifier is noninverting,

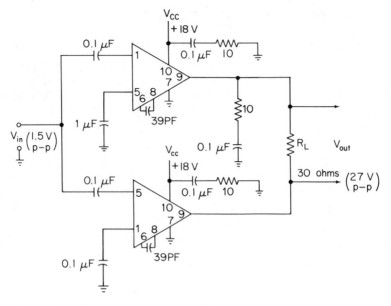

Fig. 7-39. Differential output power amplifier

while the bottom amplifier is inverting. This results in an effective overall voltage gain of twice that of either IC operating alone. That is, if the voltage gain of each IC is 10, the effective voltage gain of the circuit is 20.

The input impedance of the upper amplifier is 10 kΩ, while that of the lower amplifier is 1 kΩ. This unequal input impedance requires unequal capacitance values for the input coupling capacitors (to produce the same low-frequency cutoff point).

Because of the differential output connections, no output coupling capacitor is required, even though the single-supply connections are used. Also, the peak-to-peak output voltage swing capabilities of the amplifier can and do exceed the supply voltage. As a guideline, the maximum possible peak-to-peak output voltage is about 75% for the sum of the two supply voltages (in this case $18 + 18 = 36$; $36 \times 0.75 = 27$ V).

7.5. IC IF AMPLIFIERS FOR AM/FM AND FM RADIOS

ICs are used extensively in AM/FM and FM radio systems. The IC is particularly useful in stereo systems since two channels of amplification can be fabricated on one chip. Thus, both channels should have identical characteristics.

Generally, ICs are used to replace the IF strip in AM/FM and FM radios, as pre-amplifiers in stereo audio sections, and as complete FM decoder systems. In the case of audio pre-amps and FM decoders, the ICs are usually designed to be used with specific external components (gain, bass, treble controls, and networks). The values for the external components are given on the datasheets, and will not be duplicated here. Instead, we concentrate on the design of complete IF sections (including filters, tuning coils, etc.) for AM/FM and FM radios.

The combination of high-gain ICs and block filters provides a fairly straightforward approach to AM/FM and FM IF design. Three specific designs are given here:

A high performance FM using two Motorola MC1355 limiting gain blocks, and an external ratio detector

A quadrature detector FM using a Motorola MC3310P gain block with the Motorola MC1357 detector/gain block

A composite AM/FM using the MC1350 AGC/gain block and the Motorola MC1355 limiting gain block with an external ratio detector.

7.5.1.   High performance FM IF strip

Some of the desirable characteristics for a high quality FM IF amplifier system are: Sensitivity of less than 15 μV (assuming 10.7 MHz input signal with a frequency deviation of ± 75 kHz and a modulating frequency of 1 kHz, requred for 3% THD in detected audio), total harmonic distortion of less than 0.5%, and AM rejection greater than 45 dB. The following system meets all of these requirements.

Block diagram.   The block diagram of the overall system is shown in Fig. 7-40. Each filter block consists of a TRW linear phase filter with matching impedances of 470 ohms. Required terminations for the filter are shown in Fig. 7-41. With the filter terminated as shown in Fig. 7-41, both input and output impedances are 235 ohms.

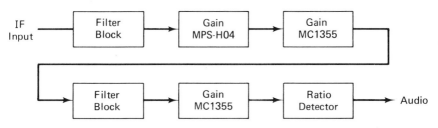

Fig. 7-40. High performance FM IF block diagram

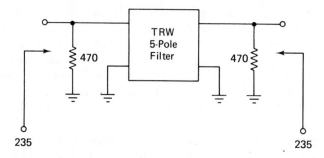

Fig. 7-41. Filter terminations

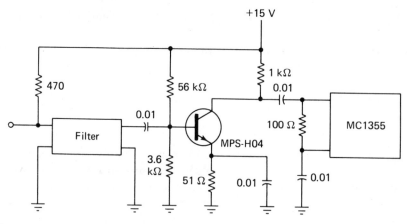

Fig. 7-42. Transistor amplifier

Design of blocks. The transistor stage in Fig. 7-42 is designed to provide a voltage gain of approximately 10. The input impedance of the MPS-HO4 is approximately 500 ohms which satisfactorily terminates the filter. A 100 ohm load is required to lower the source impedance of the first MC1355. Gain requirements of the overall board can be determined from Fig. 7-43 by noting the limiting voltage of the second MC1355 (250 mV) and the desired input limiting level of 1 μV. Assuming a voltage loss of one third through each filter, the voltage levels are as shown in Fig. 7-43.

For a theoretical -3 dB of limiting at approximately 1 μV, the voltage gain of the first MC1355 and the value of R_2 must be determined. Figure 7-44 illustrates the required calculations. The value of the load impedance on the first MC1355 is approximately 100 ohms. The load consists of R_1, R_2, and R_3 and is sufficient to provide stability and the required gain.

The gain of the second MC1355 is set by the output impedance of the filter and the primary impedance of the ratio detector used. Impedance levels are given in Fig. 7-45. With the 235 ohms source impedance and 1500

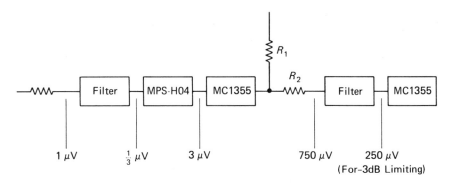

Fig. 7-43. Gain considerations

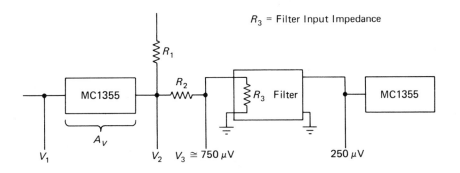

R_3 = Filter Input Impedance

Given: $V_3 = 750\ \mu V$, $R_3 = 470\ \Omega$, $V_1 = 3\ \mu V$, $y_{21} = g_m \approx 5$ mhos
For filter termination: $R_1 + R_2 = R_3$
The MC1355 load is:

$$R_L = \frac{R_1(R_2 + R_3)}{R_1 + R_2 + R_3}$$

and

$$V_3 = V_2 \frac{R_3}{R_2 + R_3} \qquad \frac{V_2}{V_1} = A_V = y_{21} R_L$$

Substituting:

$$V_2 = y_{21} V_1 \frac{R_1(R_2 + R_3)}{R_1 + R_2 + R_3} \qquad V_3 = \frac{R_3}{R_2 + R_3}\left[\frac{y_{21} V_1 R_1(R_2 + R_3)}{R_1 + R_2 + R_3}\right]$$

$$V_3 = \frac{y_{21} V_1 R_1 R_3}{R_1 + R_2 + R_3}$$

Substituting filter termination equation:

$$V_3 = \frac{y_{21} V_1 R_1 R_3}{2R_3} \qquad V_3 = \frac{y_{21} V_1 R_1}{2} \qquad R_1 = \frac{2V_3}{y_{21} V_1} \qquad R_1 = \frac{2(750)\ 10^{-6}}{5(3)\ 10^{-6}} = 100\ \Omega$$

$R_2 = R_3 - R_1 = 470 - 100 = 370\ \Omega$ Value used: $R_2 = 360\ \Omega$

Fig. 7-44. Gain requirements and calculations

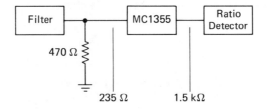

Stern stability factor (k) =
$$\frac{2(g_{11} + G_S)(g_{22} + G_L)}{|y_{12}y_{21}| + R_e(y_{12}y_2)}$$

where: G_S = real part of the source admittance
G_L = real part of the load admittance

Stern k factors for MC1355 at 10.7 MHz are:

k	R_L kΩ	R_S Ω
2	1.8	235
3	1.5	190
4	1.3	165
5	1.1	145

Fig. 7-45. Terminations of second MC1355

ohms load impedance, the second MC1355 will have a Stern stability factor (k) of approximately 2 (if k is less than 1, instability results). Some stability considerations and calculations are included on Fig. 7-45. A thorough discussion of the Stern k factor and its effect on amplifier design is given in the author's *Manual for MOS Users* (Reston Publishing Company, Reston, Virginia, 22090, 1975).

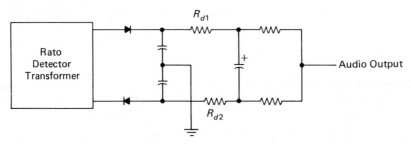

Fig. 7-46. Ratio detector circuitry

The ratio detector external circuitry can be adjusted to provide minimum distortion by varying R_{d1} and/or R_{d2}, shown in Fig. 7-46. This adjustment compensates for any unbalance in the detector diodes.

Figure 7-47 shows the emitter follower and the calculations necessary for the component values. An emitter follower is used after the ratio detector to provide a low output impedance and a simple de-emphasis network. The MPS6571 used in the emitter follower has high $h_{FE}(500)$, thus providing a high input impedance and allowing for a smaller coupling capacitor.

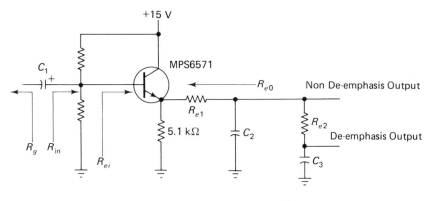

Biasing the MPS6571 for an emitter current of 1 mA gives:

$r_e \approx \dfrac{26 \text{ mV}}{1 \text{ mA}} = 26 \ \Omega$

MPS6571 input impedance is: $R_{ei} \approx h_{fe}$ (5.1 kΩ)

Given: $h_{fe} \approx 500$; $R_{ei} = 500(5.1 \text{ k}\Omega) \approx 2.5 \text{ M}\Omega$

Parallel combination of bias resistors: $R_{bias} = \dfrac{470 \text{ k}\Omega(330 \text{ k}\Omega)}{800 \text{ k}\Omega} = 195 \text{ k}\Omega$

Input impedance including the bias resistors is:

$R_{in} = \dfrac{R_{bias} R_{ei}}{R_{bias} + R_{ei}} = \dfrac{195 \text{ k}\Omega(2.5 \text{ M}\Omega)}{2.695} \approx 180 \text{ k}\Omega$

For low frequency response of 5 Hz, make $R_{in} C_1$ break at 1 Hz:

$R_{in} = 1/(6.28FC_1)$; $C_1 = 1/(R_{in} \ 6.28F)$

$= \dfrac{1}{(1.8 \times 10^5)(6.28)(1)} = 0.885 \ \mu\text{F}$; value used $C_1 = 1 \ \mu\text{F}$

High frequency rolloff is due to $(R_{eo} + R_{ei})C_2$ combination.

Assuming $R_g \approx 10 \text{ k}\Omega$ (looking into ratio detector) and noting that:

$R_{eo} \approx r_e + (R_g/h_{fe})$; substituting $R_{eo} \approx 26 + (10 \text{ k}\Omega/500)$

R_{eo} 26 + 20 = 46 Ω; choosing $R_{ei} \approx 100$; results in $R_{eo} + R_{ei} \approx 150 \ \Omega$

For a 700 kHz break point: $C_2 = \dfrac{1}{(R_{eo} + R_{ei})6.28 \ F}$

Substituting $C_2 = \dfrac{1}{1.5 \times 10^2 (6.28)(7 \times 10^5)} = C_2 = 1.52 \times 10^9 \ C_2 = 1500 \text{ pF}$

$R_{e2} = 8.2 \text{ k}\Omega$; $C_3 = 0.01 \ \mu\text{F}$ for rolloff at 2 kHz.

Fig. 7-47. Emitter follower circuit and calculations

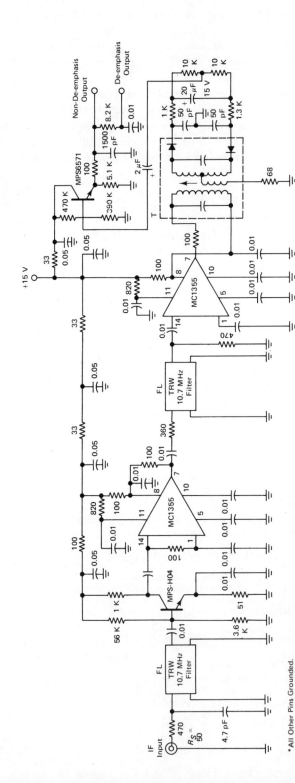

Fig. 7-48. High-performance FM IF schematic diagram

*All Other Pins Grounded.
T—Ratio Detector (Input Impedance ≈ 1.5 K) G.I. #36231 or equivalent.
FL—TRW Linear Phase 5-Pole Filter

234

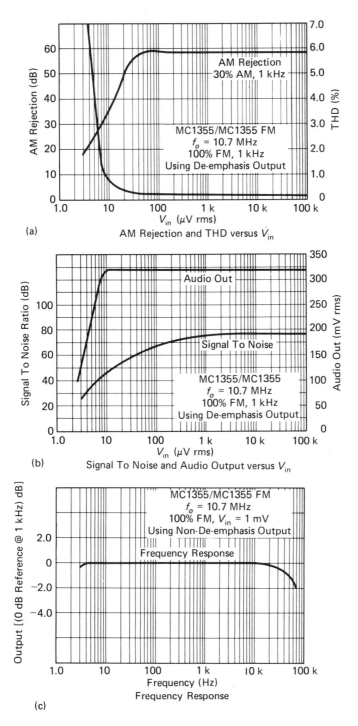

Fig. 7-49. High performance FM IF characteristics

Complete circuit.   The complete circuit for the dual MC1355 FM IF is shown in Fig. 7-48. The characteristics are as follows:

> Supply voltage: $+15$ V
>
> Supply current: 35 mA
>
> Sensitivity (using de-emphasis output)
>> 5.6 μV for 3%THD (Fig. 7-49a)
>> 5 μV for -3 dB limiting (Fig. 7-49b)
>
> Selectivity ($V_{in} = 4$ μV)
>> 70 dB at ± 400 kHz
>
> Total harmonic distortion (Fig. 7-49a)
>> 0.18%THD for 1000 μV input
>> <0.5%THD for 16 μV to$>$ 100 mV input
>
> Frequency response (Fig. 7-49c)
>> <5 Hz to 50 kHz for<-1 dB
>> (using non-de-emphasis output)
>
> Signal to noise (Fig. 7-49b)
>> 76 dB for 1000 μV input, $\Delta f = 75$ kHz
>> (using de-emphasis output)
>
> AM rejection (Fig. 7-49a)
>> 48 dB for $V_{in} = 18$ μV
>> ($V_{in} = 10$ dB above the required for 3% THD)
>> (using de-emphasis output)

7.5.2.   Quadrature detector FM IF

A design which features the MC3310P and MC1357 is ideal for FM auto radio applications because of excellent 8 V operation, and the low cost of the devices. Alignment is simplified by the ease of adjusting the MC1357 quadrature detector. A 4-pole ceramic filter provides most of the filtering in the IF amplifier.

Block diagram.   The block diagram for the overall system is shown in Fig. 7-50. The first filter block in this IF amplifier is simply a double-tuned circuit, while the second filter block consists of a 4-pole Vernitron (Clevite) ceramic filter. The terminating impedances required by this filter are 330 ohms. The MC3301P is a three-stage common emitter amplifier, ideal for

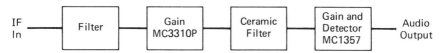

Fig. 7-50. Quadrature detector system block diagram

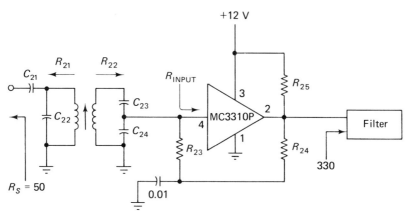

Given: R_{input} 1.5 kΩ; R_S = 50 Ω; $R_{21} = R_{22}$ = 25 kΩ (from Miller)

Transformed source impedance: $R_{21} = (Q_{C_{21}})^2 R_S$

Series Q of input: $Q_{C_{21}} = (X_{C_{21}})/R_S$

Substituting: $R_{21} = \dfrac{X^2 C_{21}}{R_S}$; $X^2 C_{21} = R_S R_{21}$; $X_{C_{21}} = \dfrac{1}{6.28 \, F C_{21}} = 1.12 \times 10^3$

and $C_{21} = \dfrac{1}{(1.12 \times 10^3)(6.28)(10.7 \times 10^6)} = 13.3 \times 10^{-12}$; Value used

C_{21} = 15 pF

From Miller: $C_{21} + C_{22}$ = 50 pF; $C_{22} = 50 - C_{21}$ = 35 pF; value used

C_{22} = 33 pF

If $R_{input}/X_{C_{24}} > 3$, then $R_{22} = \left[1 + \dfrac{C_{24}}{C_{23}}\right]^2 R_{input}$;

$\dfrac{R_{22}}{R_{input}} = \dfrac{25\ \text{k}\Omega}{1.5\ \text{k}\Omega} = \left[1 + \dfrac{C_{24}}{C_{23}}\right]^2$; $1 + \dfrac{C_{24}}{C_{23}} = 4.2$; $C_{24} = 3.2\,C_{23}$

From Miller: $(C_{23} C_{24})/(C_{23} + C_{24})$ = 50 pF

Substituting for C_{24}: $\dfrac{C_{23}(3.2\,C_{23})}{4.2\,C_{23}}$ = 50 pF; $C_{23} = \dfrac{50(4.2)}{3.2}$ pF $\approx$ 65 pF

C_{23} = 65 pF; C_{24} = 3.2 (65) $\approx$ 210 pF;

Values used: C_{23} = 56 pF; C_{24} = 180 pF

Fig. 7-51. Input filter and MC3310P circuit and calculations

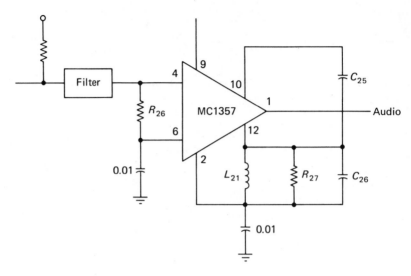

Fig. 7-52. Ceramic filter and MC1357

driving the low impedance ceramic filter. While the MC1357 shows good stability, care must be taken in laying out a board to ensure a good ground plane.

Design of blocks. Design of the first two is shown in Fig. 7-51. To keep the gain high and to terminate the filter, let $R_{25}=470$ ohms. Also, let R_{23} be approximately equal to 10 times R_{input}, where R_{input} is the input resistance of the MC3310P. So that the bias resistors will not load the MC3310P, let $R_{23}=20$ kΩ.R_{24} is adjusted to put the MC3301P in the center of its active region. Typically $R_{24}=390$kΩ

Design of the last two blocks is shown in Fig. 7-52. Resistor R_{26} must be 330 ohms to terminate the filter. However, the designer has some freedom with L_{21}, R_{27}, C_{26} and C_{25}. An IF bandwidth of about 240 kHz is necessary for good stereo reproduction, although the bandwidth of the detector must be greater than 240 kHz to allow linear operation. For a bandwidth of 500 kHz, the operating range of the detector will remain quite linear, resulting in low distortion. This requires that the Q of the quadrature coil be approximately 20. Either pin 9 or pin 10 may be used for coupling to the quadrature coil; however, care must be taken in choosing C_{25} and C_{26} to prevent overloading of the quadrature detector. Using pin 10 and the values shown in Fig. 7-53 results in very satisfactory operation.

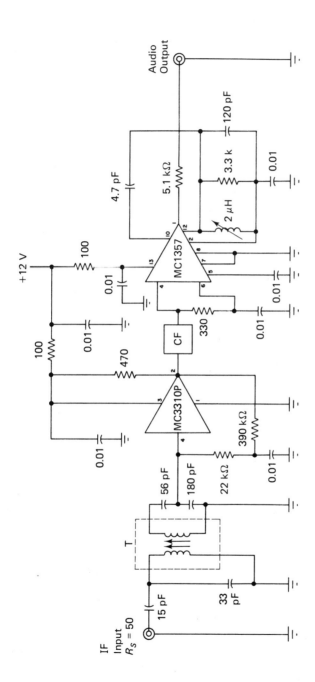

T = Miller # 8851A
or Equivalent
CF = Vernitron Ceramic Filt, Model FM4

Fig. 7-53. Quadrature detector FM IF schematic diagram

239

Complete circuit. The complete circuit for the quadrature detector FM is shown in Fig. 7-53. The characteristics are as follows:

Supply voltage: $+15$ V

Supply current: 32 mA

Sensitivity:
   18 μV for 3% THD (Fig. 7-54a)
   4.5 μV for -3 dB limiting (Fig. 7-54b)

Selectivity ($V_{in} = 4\ \mu$V)
   36 dB at $+400$ kHz
   42 dB at -400 kHz

Total harmonic distortion (Fig. 7-54a)
   0.6% THD for 1000 μV input
   $< 1.5\%$ THD for 35 μV to >100 mV input

Frequency response (Fig. 7-54c)
   <5 Hz to 40 kHz for < -3 dB

Signal to noise (Fig. 7-54b)
   68 dB for 1000 μV input $\Delta f = 75$ kHz

AM rejection (Fig. 7-54a)
   43 dB for V_{in} 57 μV
   ($V_{in} = 10$ dB above that required for 3% THD)

7.5.3. Composite AM/FM IF

The built-in AGC capability of the MC1350 makes a compostie AM/FM IF strip possible. The composite IF amplifier for this system incorporates the MC1350 and MC1355.

Block diagram. The block diagram of the overall system is shown in Fig. 7-55. The two input filters are standard, double-tuned Miller coils while the FM block filter is a TRW linear phase filter. The AM detector coil is hand wound, although a standard AM detector coil can be used.

The source impedance of the MC1350 is set by the filters used and the 10.7 MHz load impedance is set by the block filter used. For a Stern stability factor of 5 at 10.7 MHz, the required source and load impedances are 300 Ω and 13 kΩ, respectively. These load conditions are met by using the TRW filter and the input coils.

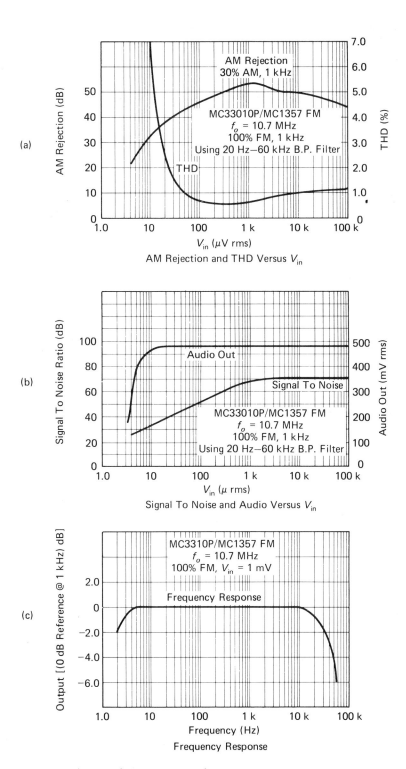

Fig. 7-54. Quadrature detector FM IF characteristics

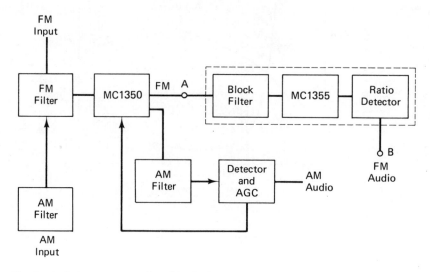

Fig. 7-55. Composite AM/FM block diagram

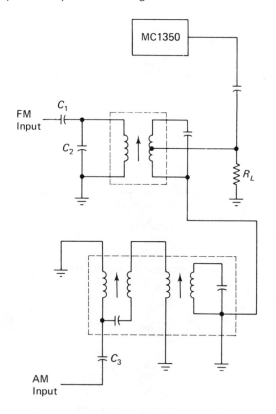

Fig. 7-56. Input filter circuits

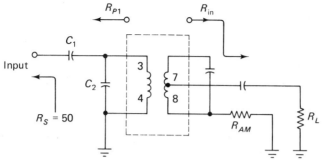

Neglect R_{AM} since $R_L \gg R_{AM}$
From Miller: $R_{P1} \approx 25 \text{ k}\Omega$; $R_{in} \approx 500 \ \Omega$; $C_1 + C_2 = 50 \text{ pF}$
Transformed source impedance: $R_{P1} = Q_S^2 R_S$

Input series Q: $Q_S = \dfrac{X_{C_1}}{R_S}$

Substituting: $(X_{C_1})^2 = R_{P1} R_S = \sqrt{(25 \times 10^3)(50)} = 1120 \ \Omega$

and $C_1 = \dfrac{1}{X_{C_1} \, 6.28F} = 13.3 \text{ pF}$; Value used 15 pF

From Miller: $C_2 = 50 - C_1 = 35 \text{ pF}$
$R_{in} = 300 \ \Omega$ per Miller specification, thus for high gain: $R_L = 510 \ \Omega$.

Fig. 7-57. FM input coil circuit and calculations

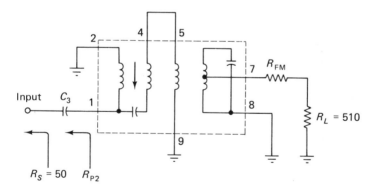

Neglect R_{FM} since $R_L \gg R_{AM}$
From Miller $R_{P2} \approx 25 \text{ k}\Omega$
Transformed source impedance: $R_{P2} = Q_S^2 R_S$

Input series Q: $Q_S = \dfrac{X_{C_3}}{R_S}$

Substituting: $(X_{C_3})^2 = R_{P2} R_S = \sqrt{(25 \times 10^3)(50)} = 1120 \ \Omega$

and $C_3 = \dfrac{1}{X_{C_3} \, 6.28F} = 317 \text{ pF}$; Value used 330 pF

For high gain: $R_L = 510 \ \Omega$.

Fig. 7-58. AM input coil circuit and calculations

The two double-tuned tank circuits used for the input to the MC1350 may be wired in series since the 10.7 MHz coil shows a low impedance at 455 kHz, and the 455 kHz coil appears as a low impedance at 10.7 MHz. A schematic of the two input filter circuits is shown in Fig. 7-56. Figures 7-57 and 7-58 show the equivalent circuits and calculations for component values of the FM and AM coils, respectively.

AM detector and AGC section. The AM detector is a standard diode detector circuit. To give linear operation at low signal levels, the RF voltage appearing at the diode should be as large as possible. This can be accomplished by using a high impedance load on the MC1350. When used with a 15 V supply, an output of approximately 14 V (peak-to-peak) can be obtained before the device begins to saturate.

One output of the MC1350 is used for the FM signal component and the other output for the AM component. An alternate method is to use one output with stacked filters. As shown in Fig. 7-59, the audio detector drives an emitter follower. This arrangement allows a 10 kΩ variable resistor to be used as the volume control and furnishes an AGC voltage for the MC1350. An external transistor is needed because the MC1350 requires 0.1 to 0.2 mA of AGC drive current, which is more than can be furnished by the detector diode. R_1, R_2, C_1 and C_2 are used for RF filtering, and represent the detector

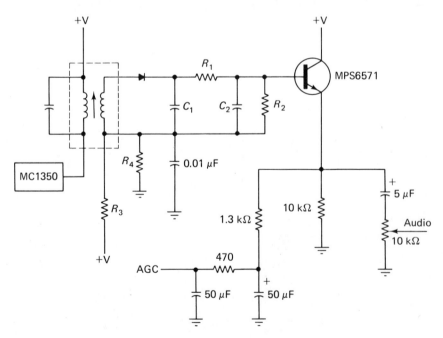

Fig. 7-59. AM detector and AGC circuit

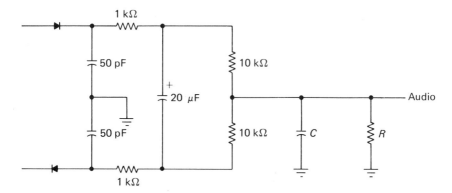

Fig. 7-60. Ratio detector circuit

load. The values for R_3 and R_4 are set at 6.2 kΩ and 5.6 kΩ, respectively, to bias the AGC system for AGC gain redution beginning at about 200 μV.

Design of the FM section from the block filter to the detected audio (shown as points A and B in Fig. 7-55) is the same as the FM IF system described in Sec. 7.5.1, and uses the same filter. However, the terminating loads for the filter in this system are raised to 1 kΩ (from 470 Ω) to increase the gain. (The 1 kΩ value of terminating resistance can be used without excessive filter pole shifting.) The FM ratio detector circuitry is shown in Fig. 7-60. The *RC* network on the output of the ratio detector is to filter out the high-frequency components. With *R* at 22 kΩ and *C* at 100 pF, rolloff occurs at about 70 kHz. The FM section in this IF system has no de-emphasis network. However, the de-emphasis network shown in Fig. 7-47 can be used if desired.

Complete circuit. The complete circuit for the composite AM/FM IF is shown in Fig. 7-61g. The characteristics are as follows:

> Supply voltage: +15 V
>
> Supply current: 32 mA
>
> AM DATA
> > Sensitivity (Fig. 7-61a)
> > > 15 μV input for 10% THD
> >
> > Selectivity (V_{in} = 50 μV)
> > > 6 dB at ±6 kHz
> > > 15 dB at ±10 kHz
> > > 25 dB at ±16 kHz
> >
> > Total harmonic distortion (Fig. 7-61a)
> > > 0.4% THD for 100 μV input
> > > <3% THD for 60 μV to 150 mV

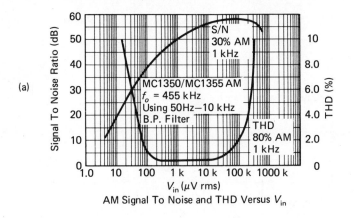

(a)

AM Signal To Noise and THD Versus V_{in}

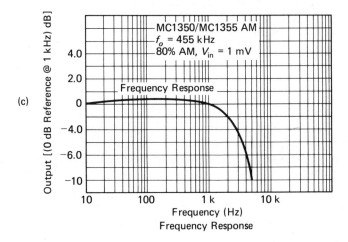

(c)

Frequency Response

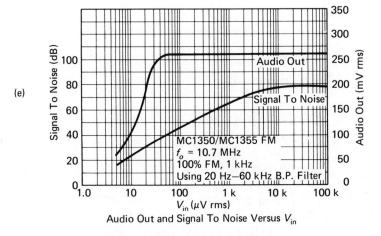

(e)

Audio Out and Signal To Noise Versus V_{in}

Fig. 7-61a thru f. Composite AM/FM characteristics

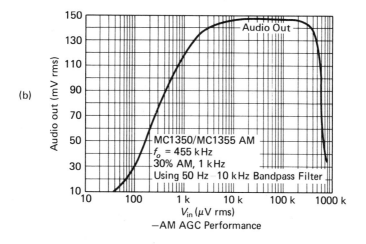

(b)

−AM AGC Performance

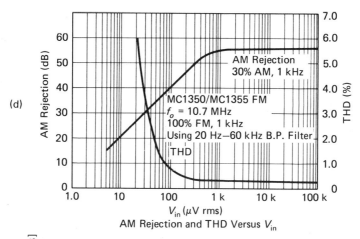

(d)

AM Rejection and THD Versus V_{in}

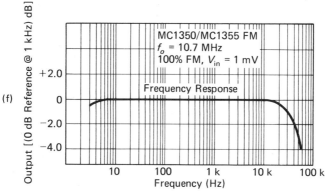

(f)

Fig. 7-61 (*continued*)

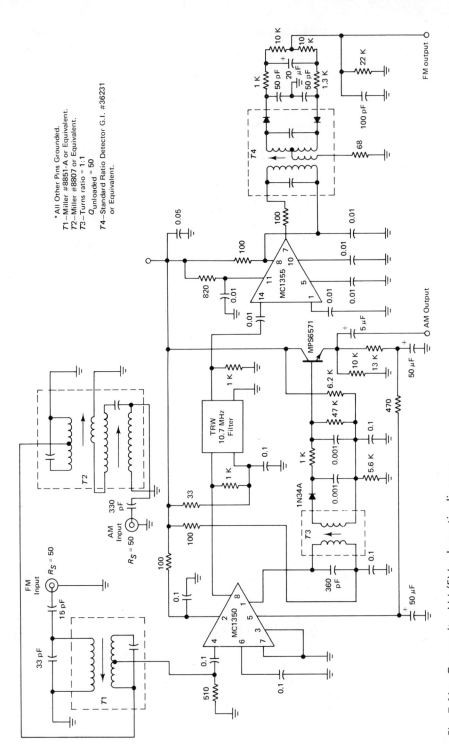

Fig. 7-61g. Composite AM/FM schematic diagram

248

AGC Figure of merit (Fig. 6-61b)
   60 dB (150 mV reference)
Audio out (Fig. 7-61b)
Frequency response (Fig. 7-61c)
   < 10 Hz to 2 kHz for < -2 dB
   -12 dB at 5 kHz

FM DATA
  Sensitivity
     35 μV for 3% THD (Fig. 7-61d)
     20 μV for -3 dB limiting (Fig. 7-61e)
  Selectivity ($V_{in} = 13$ μV)
     48 dB at $+400$ kHz
     57 dB at -400 kHz
  Total harmonic distortion (Fig. 7-61d)
     0.3% THD for 1000 μV input
     < 0.5% THD for 180 μV to > 100 mV input
  Frequency response (Fig. 7-61f)
     < 5 Hz to 40 kHz for < -2 dB
  Signal to noise (Fig. 7-61e)
     65 dB for 1000 μV input, $\Delta f = 75$ kHz
  AM rejection (Fig. 7-61d)
     42 dB for $V_{in} = 110$ μV
     ($V_{in} = 10$ dB above that required for 3% THD)

7.6.   IC LINEAR FOUR-QUADRANT MULTIPLIERS

The four-quadrant multiplier is an analog device used to produce an output voltage that is a linear product of two input voltages. Any control or instrumentation problem that requires the product, square, square root, or ratio of two analog quantities can be solved easily using a four-quadrant multiplier, provided that the quantities can be converted into voltages. For example, when used with transducers and op-amps, the multiplier can provide for such applications as monitoring power, brake horsepower, fluid flow, and the solving of complex nonlinear equations (using analog computer techniques).

    This section is concerned with the use of multiplier as a building block to perform a few of the basic arithmetic operations such as multiply, divide, square, and square root. Using these techniques, the circuit designer can then model and construct very complicated systems using combinations of these functions in conjunction with op-amps that may be used to add or subtract.

7.6.1. Typical multiplier circuit

Figure 7-62 is the schematic of a Motorola MC1595. Another multiplier, the MC1594, is available with internal level shift and voltage regulator circuits. Both multipliers operate on the *variable transconductance principle*. A multiplier of the variable transconductance type is based on the idea that the output of a transistor amplifier depends upon the input signal and the magnitude of the effective emitter resistance (assuming a common-emitter configuration) which can be controlled by the magnitude of the emitter current. Hence, the output at the collector is proportional to the input signal, times a function of the emitter current.

The multiplier of Fig. 7-62 provides a differential output voltage V_O which is proportional to the product of the input voltages, V_X and V_Y. This is shown in Fig. 7-63 which illustrates the basic multipler configuration. Note that

$$V_O = K V_X V_Y$$

where, $$K = \frac{2R_L}{I_3 R_X R_Y}$$

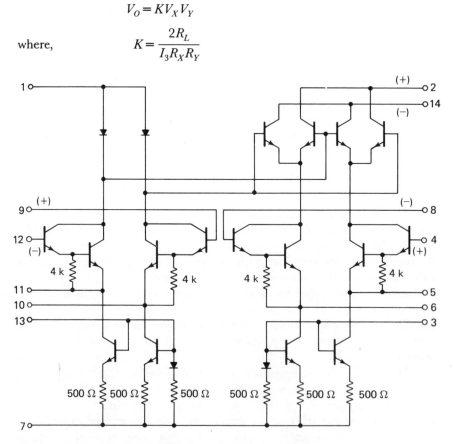

Fig. 7-62. Motorola MC1595L four quadrant multiplier

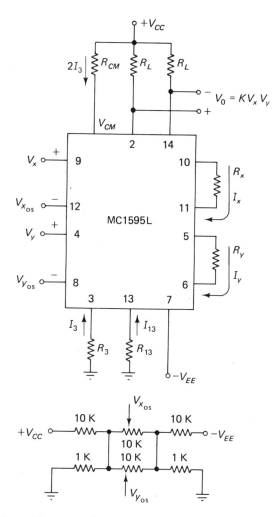

Fig. 7-63. Basic multiplier configuration

and

$$R_L = \text{the load resistor (at pins 2 and 14)}$$

$$I_3 = \text{the current flowing into pin 3}$$

$$R_X = \text{the resistor between pins 10 and 11}$$

$$R_Y = \text{the resistor between pins 5 and 6}$$

Usually, K is set to equal $\frac{1}{10}$ (by means of an external resistance). However, the constant K can easily be made to equal $\frac{1}{2}$, $\frac{1}{5}$, or any other fraction which is compatible with the other system constraints.

7.6.2. Basic multiplier design calculations

To illustrate basic multiplier design, consider the following theoretical example. Assume that it is desired to produce an output voltage that swings from -10 V to $+10$ V, with two input voltages that also swing from -10 V to $+10$ V. In effect, the multiplier must multipy one input voltage by the other, and then divide the product by 10. (If both inputs are $+10$ V, the product is $+100$ V, which must be reduced by a factor of 10 to keep within the desired output swing of $+10$ V.) The division by a factor of 10 is accomplished by setting the emitter currents (I_3 and I_{13} of Fig. 7-63). In a practical multiplier circuit, one emitter current is fixed (by a fixed resistance) with the other emitter current adjustable (by a potentiometer).

Figure 7-64 shows the basic multiplier circuit, together with typical circuit values. The following discussion illustrates how the external circuit values are calculated.

Current source values. The values of the current sources I_3 and I_{13} are determined by operating voltage and the external resistances. Select a value of current that will keep the chip power dissipation to an acceptable level, and still maintain operation in a good exponential portion of the diode curve. Current values between 0.5 and 2 mA are feasible for this particular multiplier. A value of 1 mA is selected to simplify design calculations. The negative supply voltage must be greater than -10 V (since that is the desired input capability), but must be less than 30 V (since that is the maximum specified on the datasheet for a voltage between any input pin and the negative supply pin). A negative supply of -15 V allows a -10 V input capability and ensures linear operation in the current source transistor.

With V_{EE} at -15 V, the current sources are each set to 1 mA by putting a resistor to ground from pins 3 and 13. The resistor values are calculated as follows: There is an assumed drop of 0.7 V across the diodes and emitter-base junctions (Fig. 7-62). 15 V -0.7 V $= 14.3$ V, and 14.3 V at 1 mA requires 14.3 kΩ resistances. However, note that there are 500 Ω resistances in each of the current source lines (Fig. 7-62). 14.3 kΩ $-$ 500 Ω $= 13.8$ kΩ. Thus, both R_3 and R_{13} should be 13.8 kΩ. In practice, R_{13} is a fixed resistance of 13.8 kΩ, and R_3 is a 20 kΩ potentiometer used to set the multiplier or K factor, as described in the following paragraphs.

Emitter resistor values. The values of emitter resistors R_X and R_Y are determined by the maximum input voltages, and the emitter currents, I_X and I_Y. In this example, the input voltages are -10 V and $+10$ V. The emitter currents are considered to be $\frac{2}{3}$ of currents I_3 and I_{13}, or 0.66 mA.

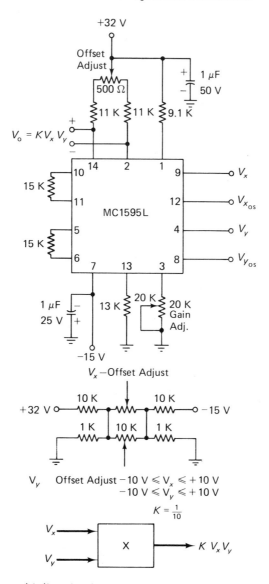

Fig. 7-64. Basic multiplier circuit

Thus, the values of R_Y and R_X can be calculated as follows:

$$R_X = \frac{V_X \,(\text{max})}{I_X \,(\text{max})} = \frac{+10 \text{ V}}{0.66 \text{ mA}} = 15 \text{ k}\Omega$$

$$R_Y = \frac{V_Y \,(\text{max})}{I_Y \,(\text{max})} = \frac{-10 \text{ V}}{0.66 \text{ mA}} = 15 \text{ k}\Omega$$

Note that R_X and R_Y are made equal in this example because the maximum voltages at both inputs are equal. Such is not always the case in all applications.

Load resistor values. The values of load resistors R_L are determined by the K factor, the emitter current I_3, and the values of emitter resistors R_X and R_Y. The values of both R_L resistors can be calculated as follows:

$$R_L = \frac{KI_3R_XR_Y}{2} = \frac{(10^{-1})(10^{-3})(15\times 10^3)(15\times 10^3)}{2} = 11.25 \text{ k}\Omega$$

If I_3 is varied slightly, a standard value of 11 kΩ can be used for R_L. As discussed in the following paragraphs, I_3 is adjusted to provide a given K factor.

Common-mode resistor and $+V_{CC}$ values. With a $+10$ V input at V_Y, the voltage at the collectors of the V_Y input differential amplifier should be about 13 V to ensure linear operation. Thus, the voltage at pin 1 should be 13.7 V (allowing 0.7 V drop across each of the diodes in series with the collectors). The $+13$ V collector potential appears at the bases of the cross-coupled differential pair where the minimum collector potential (again to ensure linear operation) should be about $+16$ V. With a minimum of 16 V and a 10 V swing, the quiescent collector potential is about 21 V (16 V$+\frac{1}{2}$ the 10 V swing). In the quiescent condition, the current source value (1 mA) is seen in each load resistor R_L, which are 11 kΩ. With 1 mA through 11 kΩ, the drop across each R_L is 11 V. The 11 V drop, plus the desired 21 V quiescent collector voltage, requires a $+V_{CC}$ of 32 V, which is obtained through the common-mode resistor R_{CM}. The value of R_{CM} is found by:

$$R_{CM} = \frac{(32-13.7)\text{ V}}{2 \text{ mA}} = 9.15 \text{ k}\Omega$$

Generally, the collector potential is not that critical, so a standard 9.1 kΩ resistor is acceptable for R_{CM}.

Input offset voltage. As in the case of op-amps, multipliers will have some input offset voltage and current (and a corresponding amount of output offset). The offset problem can be corrected using an input offset adjust circuit as shown in Figs. 7-63 and 7-64. The values of the offset network resistors are not critical, but should be selected to provide a minimum of drain on the power supply (something in the order of 1 or 2 mA).

Bypass capacitors. The values of the bypass capacitors shown in Fig. 7-64 are not critical. The 1 μF values shown are satisfactory for operating frequencies from dc up through the audio range.

7.6.3. Using the multiplier for analog multiplication

The circuit of Fig. 7-64 can be used directly to multiply voltages. However, there are two basic problems with the circuit. First, the output is differential,

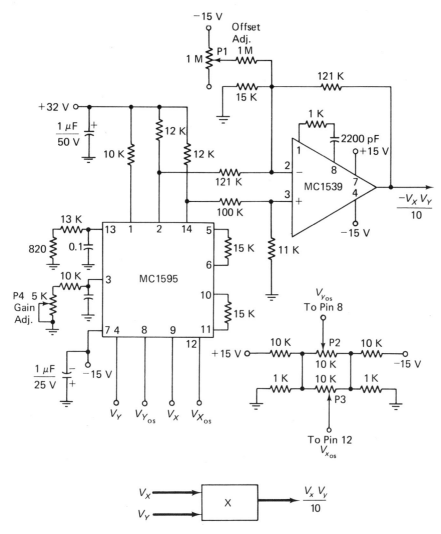

Fig. 7-65. Multiplier with op-amp level shifting circuit

whereas most analog systems require a single-ended output. Second, the output of the Fig. 7-64 circuit is a differential output that rides on a large common-mode voltage (about 21 or 22 V). For most applications, there should be no common-mode voltage. The multiplier manufacturer recommends several circuits to overcome these problems. Figure 7-65 shows the simplest of the multiplication circuits. This circuit is used where the input and output voltages to be multiplied are direct current or low frequency, and provides both level shifting (from 22 V to zero) and a single-ended output (from the op-amp). The circuit of Fig. 7-65 is relatively temperature insensitive, but is limited in frequency to about 50 kHz for large signal swings (± 10 V), because of the op-amp slew rate (refer to Chapter 3).

Setup of the multiplication circuit.   The circuit of Fig. 7-65 is set-up for use as follows:

1. Set $V_X = V_Y =$ zero volts. Adjust offset potentiometer P_1 until the output V_O is exactly zero volts.
2. Set V_X to 5.0000 V and V_Y to 0.0000 V. Adjust potentiometer P_2 until the output V_O is exactly zero volts.
3. Set V_Y to 5.0000 V and V_X to 0.0000 V. Adjust potentiometer P_3 until the output V_O is exactly zero volts.
4. Repeat step 1.
5. Set $V_X = V_Y = 5.000$ V. Adjust gain potentiometer P_4 until the output reads -2.5000 V (for a K of $\frac{1}{10}$). With V_X and V_Y at 5 V, the output will be 25 V, divided by a factor of 10 or $25/10 = 2.5$.
6. Set $V_X = V_Y = -5.000$ V and note the output V_O. The output should again be -2.5000. If the error is appreciable (greater than 1 or 2%) repeat steps 1 through 6.

7.6.4. Using the multiplier for squaring

If the V_X and V_Y inputs are connected together in the basic multiplier circuit, the resulting output is given by:

$$V_O = KV^2$$
where
$$V_X = V_Y = V$$

Thus, the output is proportional to the square of the input voltage. The proportionality constant is still given by

$$K = \frac{2R_L}{I_3 R_X R_Y}$$

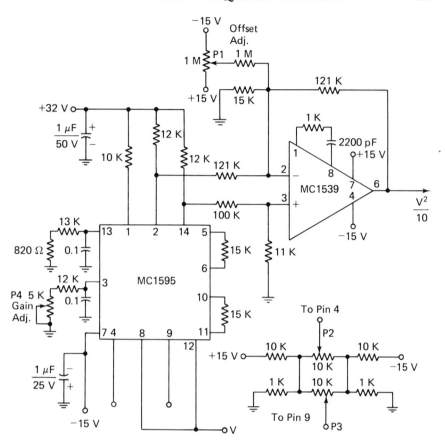

Squaring Circuit Results

V Volts	V_D Calculated	V_O Measured	% Error*
0	0.000	0.000	0
1	0.100	0.0994	0.6
2	0.400	0.3996	0.1
3	0.900	0.9002	0.022
4	1.600	1.601	0.063
5	2.500	2.503	0.12
6	3.600	3.602	0.055
7	4.900	4.905	0.1
8	6.400	6.408	0.125
9	8.100	8.105	0.062
10	10.000	10.051	0.51

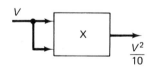

*% Error is Relative—Not Full Scale.

Fig. 7-66. Squaring circuit using multiplier

Figure 7-66 shows the multiplier used in a squaring circuit. Note that this circuit is similar to that of Fig. 7-65 (multiplication), except for the input voltage connections. In the Fig. 7-66 circuit, the multiplier inputs are tied together. The K factor still remains at $\frac{1}{10}$. Figure 7-66 also shows the squaring function in block form, and test results of the circuit (in chart form). Setup for the squaring circuit is identical to that for the basic multiplication circuit (Sec. 7.6.3).

7.6.5.   Using the multiplier for division

If the multiplier is placed in the feedback path of an op-amp as shown in Fig. 7-67a, the op-amp will try to maintain a "virtual ground" at the inverting $(-)$ input. Assuming that the bias current of the op-amp is negligible, then $I_1 = I_2$

and

$$\frac{KV_X V_Y}{R_1} = \frac{-V_Z}{R_2}$$

Solving for V_X,

$$V_X = \frac{-R_1}{R_2 K} \cdot \frac{V_Z}{V_Y}$$

If

$$R_1 = R_2$$

$$V_X = \frac{-V_Z}{K_{VY}}$$

If

$$R_1 = KR_2$$

$$V_X = \frac{-V_Z}{V_Y}$$

Thus, the output voltage is a ratio of V_Z to V_Y and provides a divide function. This analysis is, of course, the ideal condition. If the multiplier error is taken into account, the output voltage is found to be

$$V_X = -\left(\frac{R_1}{R_2 K}\right)\frac{V_Z}{V_Y} + \frac{\Delta E}{KV_Y}$$

where ΔE is the error voltage at the multiplier output.

Thus, divide accuracy is strongly dependent upon the accuracy at which the multiplier can be set, particularly at small values of V_Y. For example, assume that $R_1 = R_2$, and $K = \frac{1}{10}$. For these conditions, the output

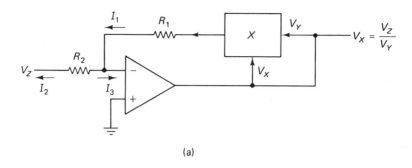

(a)

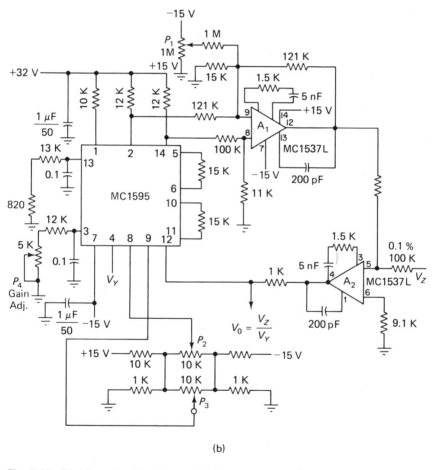

(b)

Fig. 7-67. Division circuit using multiplier

Divide Performance, V_Y = 8.00 Volts

V_Z Volts	V_O Calculated	V_O Measured	% Error*
1	0.125	0.141	13.6
2	0.250	0.266	6.4
4	0.500	0.517	3.4
8	1.000	1.019	1.9
10	1.250	1.270	1.6
12	1.500	1.521	1.4
16	2.000	2.023	1.15
20	2.500	2.526	1.04
24	3.000	3.031	1.04
32	4.000	4.042	1.05
40	5.000	5.060	1.20
64	8.000	8.090	1.12
80	10.000	10.120	1.2

*Note % Error is Relative Error—Not Referenced to Full Scale.

(c)

Fig. 7-67 (*continued*)

of the divide circuit is given by:

$$V_X = \frac{-10 V_Z}{V_Y} + \frac{10 \Delta E}{V_Y}$$

From this equation, it is seen that only when $V_Y = 10$ V is the error voltage of the divide circuit as low as the error of the multiply circuit. When V_Y is small, say 0.1 V, the error voltage of the divide circuit is 100 times the error voltage of the basic multiplier circuit.

In terms of percentage error,

$$\text{true percentage error} = \frac{\text{error}}{\text{actual}} \times 100\%$$

$$\text{calculated percentage error} = \frac{\dfrac{\Delta E}{KV_Y}}{\left(\dfrac{R_1}{R_2 K}\right) \dfrac{V_Z}{V_Y}} = \frac{R_2}{R_1} \cdot \frac{\Delta E}{V_Z}$$

From this equation, it can be seen that percentage error is inversely related to voltage V_X (for increasing values of V_Z, the percentage error decreases).

The actual circuit that performs the divide function is shown in Fig.

7-67b. For this circuit $R_1/R_2 = K$. As a result, the output voltage is given by

$$V_O = \frac{-V_Z}{V_Y}$$

Setup of the division circuit.   The circuit of Fig. 7-67b is set-up for use as follows:

1. Remove or disconnect amplifier A_2 from the circuit. Use the adjustment procedure given in Sec. 7.6.3 (multiplication).
2. Connect A_2 into the circuit. Set $V_Y = 10.000$ V and $V_Z = 0.0000$ V (connect to ground). Adjust potentiometer P_1 to read zero volts at the output (pin 12) of the multiplier.
3. Set $V_Y Y = V_Z = 10.000$ V and adjust gain potentiometer P_4 to give -1.000 V at the output.

Figure 7-67c shows the actual measured performance of the Fig. 7-67b circuit.

7.6.6.   Using the multiplier for square root

If both the V_X and V_Y inputs are tied together, the result is

$$\frac{V_Z}{R_2} = \frac{KV_X{}^2 + \Delta E}{R_1}$$

or

$$V_X = \sqrt{\frac{R_1 V_Z}{R_2 K} - \frac{\Delta E}{K}}$$

where ΔE is the multiplier error.
If $R_1 = R_2$, then

$$V_X = \sqrt{\frac{V_Z}{K} - \frac{\Delta E}{K}}$$

and if $R_1/R_2 = K$, then

$$V_X = \sqrt{V_Z - \frac{\Delta E}{K}}$$

Assume the ideal case for which $\Delta E = 0$, then

$$V_X = \sqrt{V_Z}$$

and the square root function is accomplished. Several points should be considered concerning this circuit. First, with the op-amp level shift (Fig. 7-65), the output of the multiplier is given by $V_O = (-V_X V_Y)/10$. For the squaring circuit and the square root circuit, the output of the level shift op-amp is $V_O = -V^2/10$, where $V_X = V_Y = V$.

Thus, the output of the level shift op-amp is negative, regardless of the polarity of the input signal V. Using this information, and referring to Fig. 7-68a, the obvious direction of current flow for I_1 and I_2 is actually reversed from that shown. This means that V_Z must always be positive. If it is required that V_Z be negative, pins 9 and 12 can be interchanged. However, operation at or around $V_Z = 0$ V is not permitted for either connection (pin 9 or pin 12).

Care must be taken to ensure that pins 8 and 4 are not interchanged. This would cause positive feedback and the op-amp circuit will latch to one of the supply voltages. Also, the setup procedure changes slightly because of the restriction that $V_Z = 0$.

Percentage of error. The percentage of error in the square root mode of operation is

$$\text{Percentage error} = \left(1 - \sqrt{1 - \frac{\Delta E}{KV_Z}}\,\right) \times 100$$

for $R_1/R_2 = K$.

For $K = \frac{1}{10}$, the equation becomes

$$\text{Percentage error} = \left(1 - \sqrt{1 - \frac{10\Delta E}{V_Z}}\,\right) \times 100.$$

If $10\Delta E/V_Z$ is much less than 1, then the percentage error is given approximately by

$$\text{Percentage error} \approx \frac{10\Delta E}{2V_Z} \times 100$$

or, in the general case, for any K

$$\text{Percentage error} \approx \frac{R_2 \Delta E}{2R_1 V_Z}$$

which is exactly half of the percentage error found for the divide circuit (Sec. 7.6.5). The measured performance of the square root circuit is found in Fig. 7-68c.

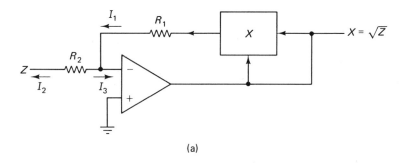

(a)

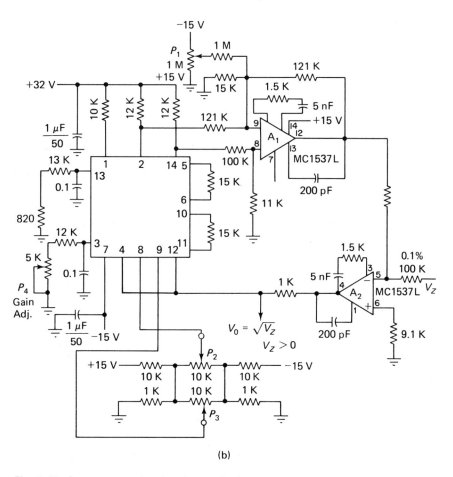

(b)

Fig. 7-68. Square root circuit using multiplier

Square Root Performance

V_Z Volts	V_0 Calculated	V_0 Measured	% Error*
1	1.000	0.924	7.6
2	1.414	1.371	3.1
4	2.000	1.966	1.7
8	2.828	2.800	1.0
16	4.000	3.982	0.45
25	5.000	4.993	0.14
36	6.000	6.008	0.13
49	7.000	7.028	0.40
64	8.000	8.058	0.73
81	9.000	9.098	1.1

*Note % Error is Relative Error—Not Referenced to Full Scale.

(c)

Fig. 7-68 (*continued*)

Setup of the square root circuit. The circuit of Fig. 7-68b is setup for use as follows:

1. Remove or disconnect amplifier A_2 from the circuit. Use the adjustment procedure given in Sec. 7.6.3 (multiplication).
2. Connect A_2 into the circuit. Set $V_Z = 25.000$ V and adjust gain potentiometer P_4 until the output reads -5.000 V.
3. Set $V_Z = 1.000$ V and adjust potentiometer P_1 until the output reads -1.000 V.
4. Repeat steps 2 and 3 until the desired accuracy is achieved.

7.6.7. Oscillation in multipliers

As in the case of op-amps, IC multipliers may tend to oscillate, particularly if long line lengths are used at the inputs. In a multiplier, the presence of oscillation is usually indicated by the inability to adjust the multiplier to better than 1% accuracy. Oscillations can be eliminated by placing Q-reducing networks at the inputs. An effective technique is to place a 510 ohm resistor in series with each input.

7.6.8. Using the multiplier as a true RMS sampling voltmeter

To illustrate one of the many applications for an IC multiplier, Fig. 7-69 shows a multiplier/op-amp circuit that detects true RMS. The circuit requires two multipliers and one op-amp, and functions as a sampling

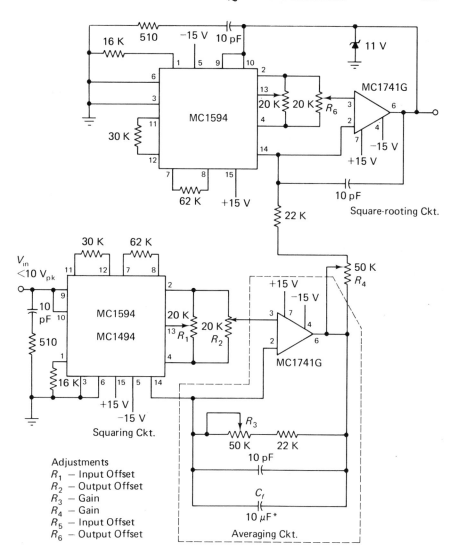

*C_f determined by lowest input frequency

Fig. 7-69. True RMS sampling voltmeter using multiplier

voltmeter for signals up to about 600 kHz with an accuracy over the input voltage range of about 1 or 2%. The unique feature is that a heating element (commonly used in true RMS voltmeters) is not needed.

Mathematically, the RMS value of a function is obtained by squaring the function, averaging it over a time period T, and then taking the square root:

$$V_{\mathrm{RMS}} = \sqrt{\frac{1}{T} \int_0^t V^2 \, dt}$$

In a practical sense, this same technique can be used to find the RMS value of a waveform. In the circuit of Fig. 7-69, the first multiplier is used to square the input waveform. Since the output of the multiplier is a current, the first op-amp is used to convert this output to a voltage. The same op-amp is also used to perform the averaging function by placing a capacitor in the feedback path (C_f in Fig. 7-69). The second op-amp is used with the multiplier as the feedback element to produce the square root configuration. Since no heating element is used, the multiplier/op-amp technique eliminates the thermal-response time that is prevalent in most RMS measuring circuits.

The input voltage range for the circuit is from 2 to 10 V (peak). Input scaling can be used for other voltage ranges. Since the input is dc-coupled, the output voltage includes the dc components of the input waveform. The overall calculation of the output voltage is

$$V_O = \sqrt{\frac{1}{RC} \int_0^t V_{\mathrm{in}}^2 \, dt}$$

where
$$R = R_3 + 22 \text{ k}\Omega$$
$$C = C_f + 10 \text{ pF}$$

7.7. IC VOLTAGE REGULATORS

As in the case of discrete component voltage regulators, IC voltage regulators provide a dc voltage that remains constant (within certain limits) in spite of changes in load, input voltage, power supply voltage, etc. In most cases, IC voltage regulators function as the control element in a solid-state voltage regulator circuit. However, there are many IC regulators that operate without external components.

There are three basic types of IC voltage regulators. The *fixed output voltage regulators* maintain the output of a power supply at a fixed voltage, and provide positive and/or negative regulation at various currents. The typical voltage range for the fixed output regulators is from 2 to 30 V, with currents from 100 mA to 3 A. The fixed output voltage regulators normally require no external add-on components. However, some manufacturers recommend that an input capacitor be used if the regulator is located an appreciable distance from the power-supply filter, and an output capacitor could improve transient response. Fixed output regulators are ideal for on-card regulation of subsystems, affording possible money-saving and performance-improvement advantages in applications where total system regulation is not required.

The *variable output voltage regulators* can be tailored to regulate at any specific output voltage (within a given range) through the use of external resistors. A specific output current is available directly from the IC regulator. Increased output current can usually be obtained through the use of external current-boosting circuits.

In *switching regulators*, the IC voltage regulator functions as the control element. Typically, the IC regulator includes the reference voltage, oscillator, pulse-width modulator, phase-splitter, and output sections found in a typical switching regulator power supply. IC switching regulators are used as the control circuit in pulse-width modulation (PWM), push-pull, bridge, and series type switch-mode power supplies.

In addition to the three basic types of IC regulators, there are a number of special purpose regulators. For example, there are *floating voltage and current regulators* designed for laboratory types of power supplies. Such floating regulators can deliver hundreds of volts, and are limited only by the breakdown voltage of associated, external, series-pass transistors. There are *dual-tracking regulators* that provide balanced positive and negative output voltages. Generally, the dual-tracking regulators are set to provide a specific output voltage (such as ± 15 V) but an external adjustment can change both outputs simultaneously (say from 8 to 20 V). Also, there are special *low-temperature drift*, low-voltage IC regulators where precision regulation is required.

In this section, we describe the basic, typical IC voltage regulator, and show how the various features affect power-supply design. Also included are discussions covering typical adjustable regulators and switching regulators.

7.7.1. The basic IC voltage regulator

Figure 7-70 shows a combined block diagram and simplified schematic of a basic IC voltage regulator. As shown, the basic regulator is divided into four parts. The control section provides a means of starting and stopping regulator action. Not all IC regulators are provided with a control section. This makes for a less expensive unit, but somewhat limits the regulator's usefulness in today's sophisticated equipment. The bias section provides a fixed, temperature-compensated reference voltage (V_{ref}). Typically, this reference

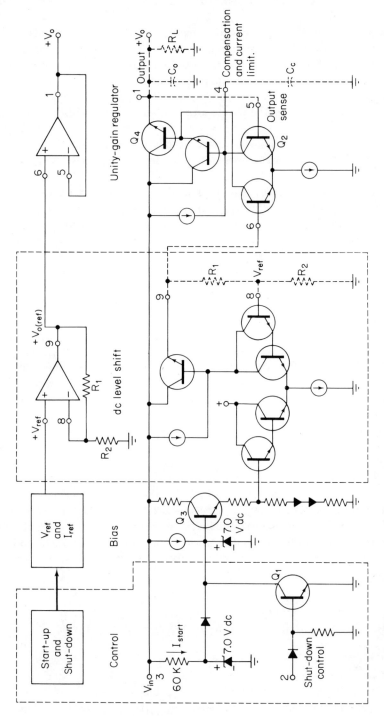

Fig. 7-70. Block diagram and corresponding simplified schematic of Motorola MC1569 and MC1469 voltage regulators

voltage is in the order of 3–4 V. Since the desired output voltage is rarely some exact value between 3–4 V, it is necessary to "multiply" the reference voltage to some higher exact value. This is done in the dc level shift section. The available output reference voltage $V_{O(ref)}$ is determined by the ratio of resistors R_1 and R_2, which set the feedback (and thus the gain) of the reference amplifier. The $V_{O(ref)}$ can be higher or lower than the fixed V_{ref} supplied by the bias section.

The output section is essentially a unity-gain, differential amplifier, followed by a pair of *npn* transistors. The differential amplifier inputs consist of $V_{O(ref)}$ and the output voltage V_O. These voltages are the same when the regulator is functioning normally, as it is at the selected voltage. Any change in output voltage unbalances the differential amplifier, and changes the bias of Q_1. The bias change increases or decreases the current through Q_1 as necessary to offset the initial change in output voltage. The maximum output voltage is determined by minimum input-output differential; that is, the smallest amount of voltage across Q_1 that allows normal operation.

Note that the differential amplifier has a frequency compensation capacitor connected at the collector. This external capacitor serves the same purpose as the frequency compensation capacitors in an op-amp (Chapter 3). The value of the compensating capacitor is specified on the datasheet, and is in the order of 0.001 to 0.1 μF. Also note that the basic regulator circuit is provided with an external capacitor at the output. This capacitor is a conventional filter capacitor, and is typically on the order of 10 μF.

7.7.2. Short-circuit and overload protection

Short-circuit current limiting is provided by either an external transistor or a diode string. This is shown in Fig. 7-71. In some IC regulators, the diode string is internal. Either way, the diode string (or transistor) is forward-biased when the load current creates a drop across the series resistor equal to the total diode voltage drops (or the base-emitter drop of the transistor). When the series resistor is of the proper value, the drop across the resistor will equal the total diode voltage drops only if the regulator's maximum current limit is reached (or is approached).

From the user's standpoint, external off-the-chip short-circuit diodes (or transistors) offer some advantages. If either the diodes or transistors are on the chip, the threshold voltage is altered by IC temperature rise. That is, if the temperature rises, the diodes or transistors will require a lower voltage drop before they pass current. As more load current is drawn, even normal current, IC temperature can rise quite rapidly (particularly if the case is not on a heat sink) and current limiting may set in long before the design value is reached.

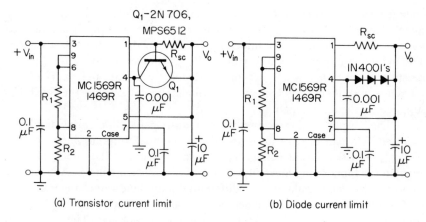

(a) Transistor current limit   (b) Diode current limit

Fig. 7-71. Short-circuit protection methods

No matter what system is used, when the threshold point is reached, the diodes (or transistor) draw base current away from the *npn* transistor pair, and thus limit maximum output current.

7.7.3. Current boosting

Current boosting is often used for higher load current or better regulation. Even though the basic IC may be capable of delivering the design current, boosting will reduce junction temperature rise, since the current-carrying transistor (that dissipates most of the heat) is separate from the IC. Thus, any temperature-related problems such as reference voltage drift are minimized.

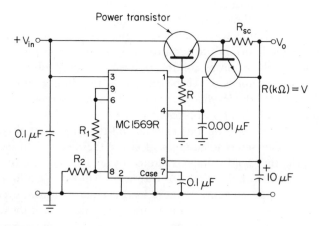

Fig. 7-72. *npn* current boosting

A typical *npn*-boosted current circuit is shown in Fig. 7-72. Here, load current is passed through a series-connected power transistor that, in turn, is controlled by the IC regulator output voltage. This type of circuit can also be used to increase efficiency at low output voltages since an external series-pass element requires only a volt or so to remain active.

7.7.4.   Current regulator

IC voltage regulators can also be used as current regulators. The basic circuit is shown in Fig. 7-73a. The output current passed through the load

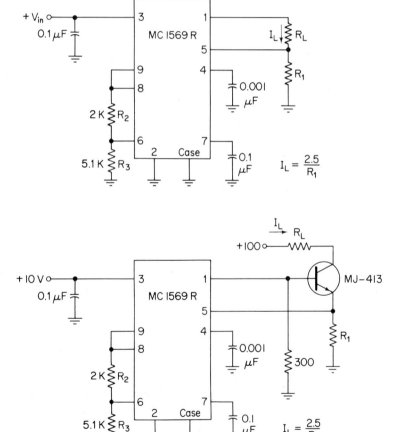

Fig. 7-73. Current regulators

R_L also passes through R_1, and produces a corresponding drop across R_1. The output current level sensed across R_1 is compared with the reference voltage (set by the ratio of R_2/R_3). Any deviations are amplified and corrected to maintain load current constant to within a fraction of one percent (typically 0.05%). Both load current and voltage may be extended by using an external power transistor as shown in Fig. 7-73b.

7.7.5. Shutdown circuits

As discussed in Sec. 7.7.1, and shown in Fig. 7-70, a well-designed IC regulator will have some means for shutdown and start-up. In the circuit of Fig. 7-70, control of the IC is accomplished by transistor Q_2 and the associated circuitry. In normal operation, when the IC regulator is functional, pin 2 (the shutdown control pin) is open, or connected to a voltage close to ground. In this condition, transistor Q_2 is nonconducting and its presence has no effect on the normal operation of the IC.

Transistor Q_1 shunts the internal reference zener Z_2. Should Q_1 be turned on, and saturated, the internal reference Z_2 is shorted to ground. Since all internal current sources are biased to the internal reference, they will not function, and the regulator will stop operation (Q_3 will shut off). Shutdown of the I is thus accomplished by applying a voltage to pin 2 of sufficient value to turn on and saturate Q_1. The regulator returns to operation once this potential is removed.

Logic control.   Figure 7-74 illustrates how the shutdown of the IC regulator can be controlled by a logic gate. Here, it is assumed that the regulator is operating in its normal mode—as a positive regulator reference to ground, and that the logic gate is of the saturating type, operating from a $+ V$ supply to ground. The gate can be of the TTL type (or even the older DTL or RTL type), where the output stage uses an active pull-up resistor. (Refer to Chapter 8.)

One advantage of this technique is for remote systems where power is to be conserved until a particular subsystem is needed to function. At such a time, the regulator is turned on via a computer control, it performs its task, data is collected and stored, and the subsystem is shut down.

Junction temperature control.   An IC regulator has one problem not found in the discrete component regulators. Most of an IC regulator chip is occupied by the series-pass transistor (since this transistor handles most of the current). The temperature of the entire IC is thus set by the series-pass transistor. Variations in load cause the transistor temperature to change, thus changing the temperature of remaining components on the IC. This "thermal feedback" or "temperature feedback" can cause thermal runaway.

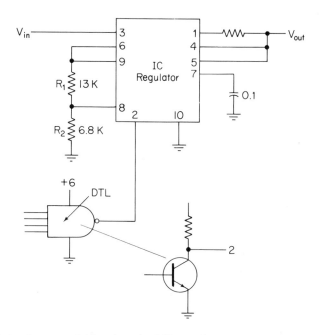

Fig. 7-74. Logic control (shutdown) of IC regulator

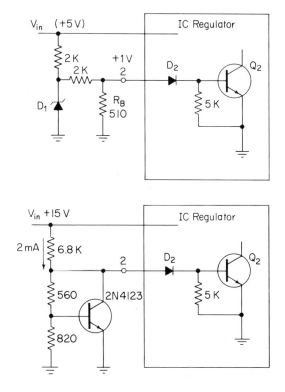

Fig. 7-75. Junction temperature control (shutdown) of IC regulator

It is possible to use the thermal feedback to control the IC, and to initiate shutdown if an unsafe temperature is reached. Several methods are illustrated in Fig. 7-75.

Figure 7-75a shows a method of shutdown which is controlled by junction temperature (T_J). For practical purposes, the temperature of the D_2 and Q_2 junctions can be considered the same as the output series transistor. As temperature increases, the voltage required for turn-on of D_2 and Q_2 (regular turn-off) drops. Thus, if a fixed voltage is applied, D_2 and Q_2 turn on when a predetermined temperature is reached. The Zener diode provides the fixed dc voltage that can be varied by varying the value of R_B.

Figure 7-75 shows a shutdown method which is controlled by ambient temperature (T_A). Here, the fixed reference control is set by the drop across the junction of an external transistor. Since this drop is controlled by ambient temperature, the control voltage (and turn-on point) is set by the ambient temperature.

In the circuit of Fig. 7-75a, the Zener diode is usually mounted away from the IC so that heat from the IC will not affect the Zener. On the other hand, the transistor of Fig. 7-75b is usually mounted near the IC, so that a large increase in IC heat will change the ambient temperature around the transistor.

These temperature control systems are often used in conjunction with the short circuit and overload circuits described in Sec. 7.7.2. When a short circuit or overload occurs, increased current is drawn through the IC component, and the IC can heat, possibly to a dangerous level. By using both overload and temperature control, the IC will turn off when the temperature reaches a predetermined point during overload. When the IC is cool enough, the regulator returns to the on condition. If the short or overload still exists, the heating process begins again, and the IC continues in a cycling mode as long as the output is shorted. This technique provides control of both the maximum short-circuit current, and the maximum junction temperature. This complete protection is often considered superior to the "current foldback" system described in Sec. 7.7.6.

Current boost temperature control. The thermal shutdown techniques can be used when the IC is operating in a current boost circuit as described in Sec. 7.7.3. The external series-pass transistor must be mounted on a common heat sink with a monitor transistor, as shown in Fig. 7-76. In this configuration, Q_2 heats when the output is shorted or there is an overload, while Q_1 monitors the heat sink temperature. Q_1 conducts for a predetermined temperature increase above the normal ambient, as set by the values of R_2 and R_3. Resistor R_4 is used to limit the drive into pin 2.

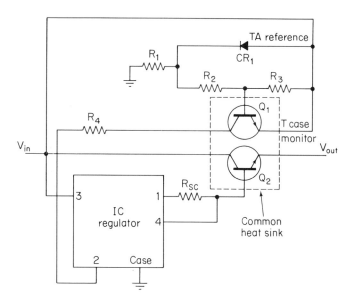

Fig. 7-76. Thermal shutdown using external phase transistor

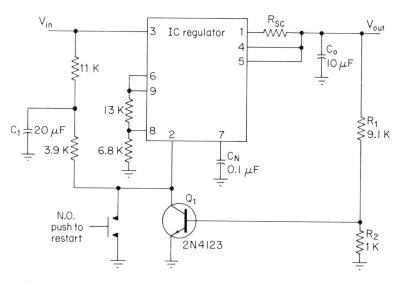

Fig. 7-77. Output short-circuit shutdown

Output short circuit shutdown.   Not all shutdown systems are based on temperature. A standard circuit that does not use T_J or T_A control for shutdown is shown in Fig. 7-77. In this circuit, Q_1 is normally saturated because of the base drive supplied through the resistor divider R_1 and R_2. With Q_1 in saturation, the voltage on pin 2 is below the threshold of the IC control circuit. If the output of the IC is short circuited, Q_1 turns off, the voltage at pin 2 rises, and shutdown of the regulator occurs. R_1 and R_2 can be chosen to saturate Q_1 as well as to sink the minimum regulator load current (typically about 1 mA). Capacitor C_1 is necessary to provide an RC time constant that prevents shutdown when V_{in} is initially applied (before Q_1 can saturate). When the short circuit is removed, the regulator must be manually reset before it will return to full regulation.

7.7.6.   Current foldback technique

It is not absolutely necessary for an IC regulator to have a shutdown or control section. There are circuits that will, in effect, shut the regulator off should an overload or short circuit occur. These circuits reduce the regulator output voltage and current to zero (or near zero) when the load current

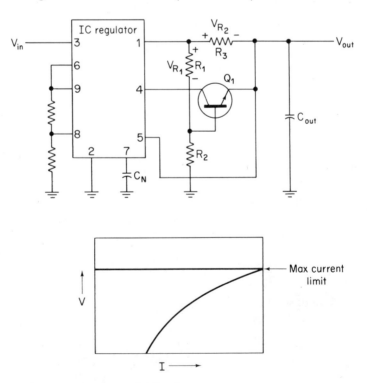

Fig. 7-78. Short-circuit current foldback

reaches a certain limit. Such a circuit, and corresponding current-voltage graph, are shown in Fig. 7-78. Note that when output current reaches a certain limit, both the voltage and current drop back (or "foldback") to some low (safe) level.

In the circuit of Fig. 7-78, the voltage drop across R_2 is set to about twice the base-emitter drop of Q_1 (or about 1.4 V) when the output voltage is normal. For Q_1 to conduct, the voltage across R_3 (that senses the output load) must be about equal to the drop across R_1, plus the base-emitter voltage. By proper selection of resistance values, this will occur only if the current limit has been reached. When Q_1 conducts, the collector connected to pin 4 diverts current drive from the series pass elements (Fig. 7-70), producing current foldback.

7.7.7. Specifying an IC voltage regulator

Unfortunately, the specifications of IC regulators from all manufacturers are not consistent. The following is a list of the terms used in Motorola datasheets, with a brief explanation for each. If the designers fully understand these terms, they should have no difficulty in specifying an IC regulator, or understanding the datasheets of other manufacturers.

Output voltage range. This is the range of output voltages over which the specifications apply. A particular output voltage is established by a user-selected external resistive divider network (R_1 and R_2 of Fig. 7-70).

Output current. All IC regulators are capable of supplying a certain amount of current to a load without the use of external transistors. The magnitude of current is specified on the datasheet as a maximum and a minimum for each device. For example, a particular IC regulator might be capable of supplying 500 mA for any output voltage from 2.5 to 37 V. However, the same regulator must supply at least 1 mA to maintain its regulator characteristics. For this reason, the IC datasheets often specify that a resistance be placed across the output to draw a minimum of 1 mA at the selected voltage.

Input-Output voltage differential. The input-output voltage differential ($V_{in} - V_O$) is the voltage required to bias the circuitry that supplies the drive current to the series-pass transistor. As the input voltage drops in magnitude and approaches the output voltage dc level, the regulator drops out of regulation at the minimum value of this specification.

Input voltage range. The input voltage range is specified on the datasheet as a minimum and a maximum. This indicates that at least the minimum voltage must be present to properly bias the Zener diode (or other reference) on the IC. This input voltage must also exceed the output voltage by at least the specified input-output voltage differential. For example, if the $V_{in} - V_O$ is 2 V minimum and 3 V maximum, and the desired output is 10 V, the input voltage must be at least 12 V, and possibly 13 V.

Load regulation and output impedance. Load regulation is the percentage change in output voltage for a dc steady-state change in load current from the minimum to maximum value specified. This is expressed as:

$$\text{load regulation} = \frac{V_{NL} - V_{FL}}{V_{NL}} \times 100 = (\% V_O)$$

where V_{NL} is the output voltage with minimum load current
V_{FL} is the output voltage when the full load current is being drawn, and the units are a percentage of $V_O (\% V_O)$.

Output impedance. Output impedance Z_O is a small-signal ac parameter. Z_O indicates the ability of the regulator to prevent common power-supply voltage changes created by fluctuating currents drawn at signal frequencies by circuit load. High values of Z_O can create undesirable coupling between circuits powered by a common voltage regulator, and thereby cause system oscillation. Typical output impedances for an IC regulator are less than 1 Ω and often as low as 0.025 Ω.

Input regulation. Input regulation is the percentage change in output voltage per volt change in the input voltage and is expressed as:

$$\text{input regulation} = \frac{V_O}{V_O(\Delta V_{in})} = 100 \ (\% / V_{in})$$

where ΔV_O is the change in the output voltage V_O for the input change ΔV_{in} and $\% / V_{in}$ are the units.
Completely packaged, line-operated voltage regulators usually use the term "line regulation" to show the dependence of the output voltage on the ac power-line variations. Since an IC regulator does not operate directly from the line, the term "input regulation" has been used to be more accurate.

Temperature coefficient. This is the stability of the output voltage over a change in operating temperature.

7.7.8. Adjustable shunt regulator

Figure 7-79 shows the schematic and characteristics of an adjustable shunt regulator, the Texas Instruments TL430. The IC is a three-terminal "programmable" shunt regulator, and may be thought of as a programmable

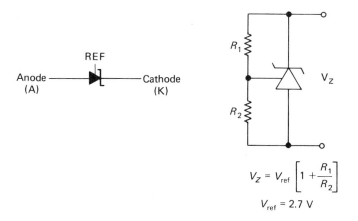

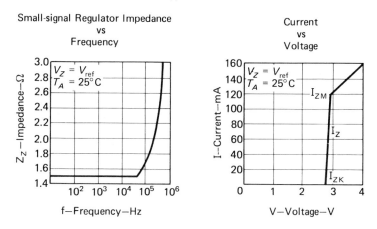

Fig. 7-79. Texas Instruments TL430 adjustable shunt regulator

Zener diode. The IC has a thermal stability of 100 ppm/°C which makes it suitable as a replacement for high-cost, temperature-compensated Zeners. The IC also offers improved performance over low-voltage Zeners by virtue of the sharp breakdown characteristics, even at low temperatures, as shown in Fig. 7-79. The IC is programmable over a voltage range of about 3 V to 30 V, and is capable of shunting up to 100 mA.

The IC is programmed by an external resistor ladder network connected across (anode-to-cathode), with the reference terminal connected to the node of R_1–R_2, as shown in Fig. 7-79. The resistor values are selected such that the voltage developed across R_2 is 2.7 V for the desired programmed Zener voltage.

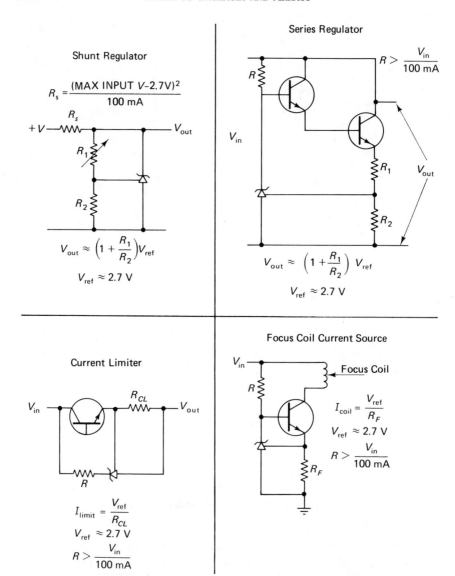

Fig. 7-80. Texas Instruments TL430 used as shunt regulator, series regulator, current limiter, and focus coil current source

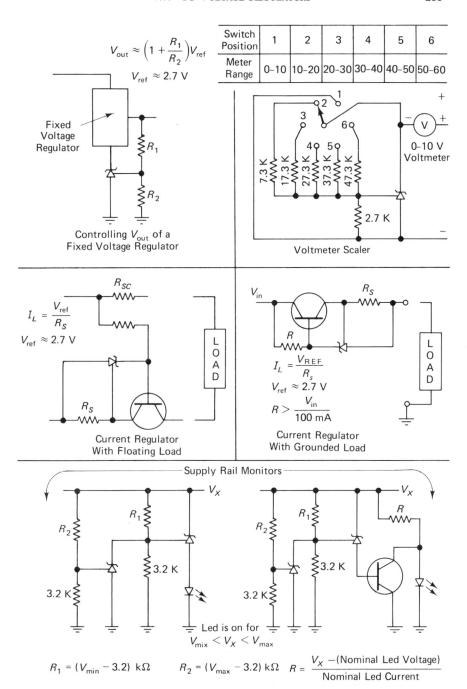

$$V_{out} \approx \left(1 + \frac{R_1}{R_2}\right)V_{ref}$$

$$V_{ref} \approx 2.7 \text{ V}$$

Switch Position	1	2	3	4	5	6
Meter Range	0–10	10–20	20–30	30–40	40–50	50–60

Fixed Voltage Regulator

R_1

R_2

Controlling V_{out} of a Fixed Voltage Regulator

0–10 V Voltmeter

7.3 K   17.3 K   27.3 K   37.3 K   47.3 K

2.7 K

Voltmeter Scaler

$$I_L = \frac{V_{ref}}{R_S}$$

$$V_{ref} \approx 2.7 \text{ V}$$

R_{SC}

R_S

LOAD

Current Regulator With Floating Load

V_{in}

R_S

R

$$I_L = \frac{V_{REF}}{R_s}$$

$$V_{ref} \approx 2.7 \text{ V}$$

$$R > \frac{V_{in}}{100 \text{ mA}}$$

LOAD

Current Regulator With Grounded Load

Supply Rail Monitors

V_X

R_1

R_2

3.2 K

3.2 K

V_X

R_1

R_2

R

3.2 K

3.2 K

Led is on for
$$V_{mix} < V_X < V_{max}$$

$$R_1 = (V_{min} - 3.2) \text{ k}\Omega \qquad R_2 = (V_{max} - 3.2) \text{ k}\Omega \qquad R = \frac{V_X - (\text{Nominal Led Voltage})}{\text{Nominal Led Current}}$$

Fig. 7-81. Additional applications for Texas Instruments TL430

The IC is not limited to only shunt regulator functions. This is illustrated in Fig. 7-80 and 7-81 which show the IC in a variety of regulator functions. Figures 7-80 and 7-81 also show the calculations necessary to determine the values of external components in each regulator system.

7.7.9. IC switching regulators

As in the case of a conventional voltage regulator, a switching regulator functions to convert an unregulated dc input to a regulated dc output. However, the method used to accomplish the function differs from the conventional regulator. In a switching regulator, the power transistor is used in a switching mode rather than a linear mode. Switching regulator efficiencies are in the order of 60 or 70% or better, which is about double that of the conventional voltage regulator. High-frequency switching regulators offer considerable weight and size reductions, and better efficiency at high power, over conventional 60 Hz transformer-coupled series-regulator power supplies.

Design of switching-regulator power supplies, with or without ICs, is a specialized subject, and will not be covered in detail here. (Switching-regulator power-supply design is discussed fully in the author's *Handbook of Simplified Solid-State Circuit Design*, second edition (Prentice-Hall, Inc., Englewood Cliffs, New Jersey, 07632, 1978). Instead, the following paragraphs concentrate on design of a specific power supply, using a specific IC. This is followed by a discussion of a specific IC switching regulator, and the calculations necessary to select the external components for typical applications.

7.7.10.   Controlling a dc-dc converter switching regulator with an IC multivibrator

This section describes design of a dc-dc converter, using the switching regulator principle, with an IC multivibrator. The IC selected is a Motorola MC3380, which is described in the catalog as a special-purpose astable multivibrator.

Device operation.   Figure 7-82 is a schematic of the MC3380 in an oscillator circuit. Q_1 and Q_2 form an astable multivibrator whose pulse width and repetition are determined by R_{EXT}, C_{EXT} and the magnitude of I_4, which is proportional to the current into pin 6, I_{FB}. To understand operation of the circuit, begin by assuming that a voltage exists on C_{EXT} sufficient to turn Q_1 off. Under this condition, Q_2 is on and the oscillator output V_O is high. The charge in C_{EXT} flows to ground at a rate determined

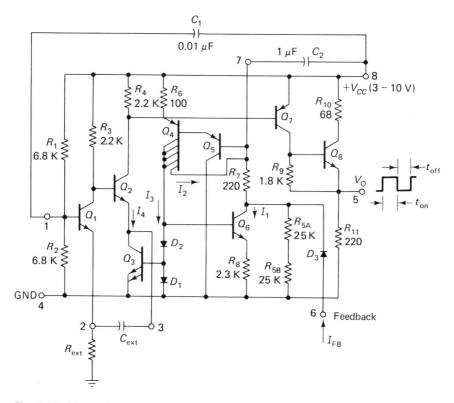

Fig. 7-82. Motorola MC3380 schematic

by R_{EXT} until the voltage across C_{EXT} is low enough to allow Q_1 to turn on. This period is t_{ON}. When Q_1 turns on, Q_2 turns off, the output goes low, and C_{EXT} is now charged at a rate determined by C_{EXT} and I_4 until the voltage across C_{EXT} is sufficient to turn off Q_1 again. This period is t_{OFF}.

Transistor Q_6, in conjunction with D_1, D_2, and R_8, forms a constant-current source whose output is I_1 (nominally 400 μA). The value of I_2 is approximately $I_1 - I_{FB}$. Q_4 sources a current I_3 whose value is $4I_2$ or $4(I_1 - I_{FB})$. Since the emitter area of Q_3 is twice that of D_1, the combination of Q_3, D_1, and D_2 forms a current mirror that sinks a current I_4 equal to $2I_3$ or $8(I_1 - I_{FB})$. By increasing I_{FB}, the magnitude of I_4 is decreased, the charging time of C_{EXT} is increased and t_{OFF} becomes longer. As I_{FB} approaches I_1, I_4 approaches zero and oscillation ceases. The value of I_{FB} at which this occurs is specified to be between 250 μA and 600 μA.

Two additional external capacitors are required for proper operation of the oscillator circuit. C_1 assures initial startup, whereas C_2 serves as a noise filter for the current-summing node of Q_6.

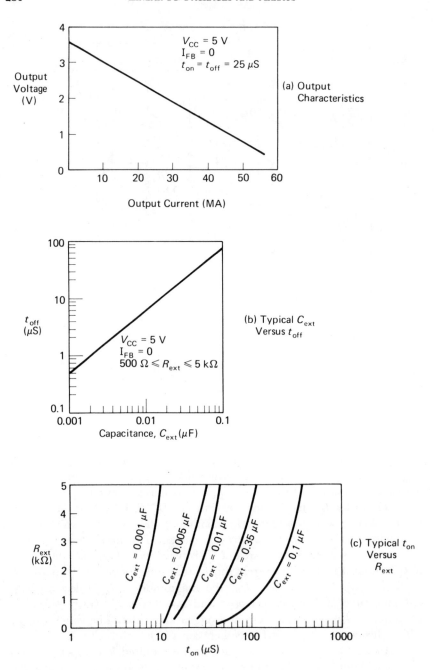

Fig. 7-83. MC3380 characteristics

Device characteristics.   The output characteristics of the MC3380 are shown in Fig. 7-83a. Figure 7-83b shows a plot of t_{OFF} versus C_{EXT}, illustrating the minimum t_{OFF} that can be obtained with a given value of C_{EXT} that occurs when I_{FB} is equal to zero. A plot of t_{OFF} versus I_{FB} is not shown since this characteristic varies widely between individual devices and is not normally necessary for proper circuit design if the MC3380 is used in a closed-loop control system. The following example shows that knowledge of t_{OFF} versus C_{EXT} is necessary only for an I_{FB} equal to zero. A graph of t_{ON} versus R_{EXT} and C_{EXT} is shown in Fig. 7-83c. Note that t_{ON} is not dependent upon the value of I_{FB}, but only on the value of R_{EXT} and C_{EXT}.

Free-running oscillator.   If a fixed-frequency, free-running oscillator is required, the circuit of Fig. 7-84 can be used. In this circuit, the current sink Q_3 has been disabled by connecting pin 7 to pin 8. Q_4 and Q_5 are off, and $I_4 = 0$. R_2 provides a charging path for C during t_{OFF}. The frequency range of this oscillator is from approximately 1 kHz to greater than 300 kHz. Plots of frequency and duty cycle versus R_1, R_2, and C are shown in Fig. 7-85.

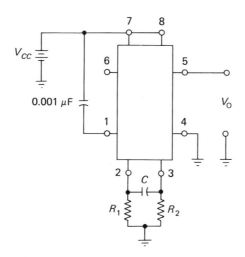

Fig. 7-84. MC3380 used as fixed-frequency, free-running oscillator or multivibrator

200 V switching regulator circuit design.   The circuit shown in Fig. 7-86 converts 5 V into 200 V and provides up to 15 mA of output current. The circuit was designed as a supply for gas discharge displays, and provides sufficient current to drive the usual 10 to 15 digits of most displays.

Operation of the circuit is as follows. During t_{ON}, Q_1 and Q_2 are saturated and the primary current of T_1 increases. This reverse-biases D_1,

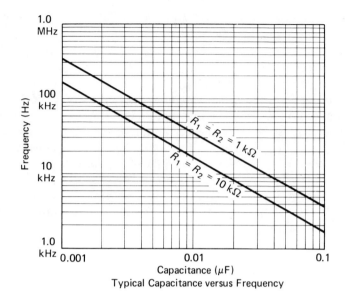

Typical Capacitance versus Frequency

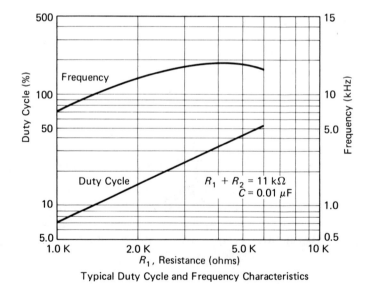

Typical Duty Cycle and Frequency Characteristics

Fig. 7-85. Characteristics of MC3380 used as fixed-frequency, free-running oscillator

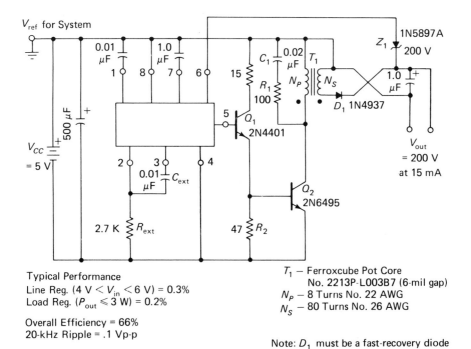

Typical Performance
Line Reg. (4 V < V_{in} < 6 V) = 0.3%
Load Reg. (P_{out} ≤ 3 W) = 0.2%

Overall Efficiency = 66%
20-kHz Ripple = .1 Vp-p

T_1 — Ferroxcube Pot Core
    No. 2213P-L003B7 (6-mil gap)
N_P — 8 Turns No. 22 AWG
N_S — 80 Turns No. 26 AWG

Note: D_1 must be a fast-recovery diode
to minimize circuit losses

Fig. 7-86. MC3380 used in 200 V switching regulator circuit

preventing secondary current flow during this period. Q_1 and Q_2 are then switched off, and the energy stored in the primary of $T_1(1/2LI^2)$ is transferred to the secondary during $_{OFF}$ Feedback to the MC3380 is accomplished by a Zener diode Z_1. If the output voltage tends to increase, the Zener current I_{FB} increases and lengthens t_{OFF}. The rate of energy transfer is decreased, and the output voltage tends to decrease.

Design procedure. The general design procedure for the circuit of Fig. 7-86 is as follows:
    1. Design specifications.

        Input voltage V_{in} 5 V

        Output voltage V_O 200 V ± 10%

        Output current I_O 15 mA

        Conversion efficiency η > 70%

        Full load operating frequency > 20 kHz

2. Determine input power P_{in}

$$P_{in} = \frac{P_O}{\eta} = \frac{V_O I_O}{\eta} = \frac{2 \times 10^2 (15 \times 10^{-3})}{0.7} = 4.3 \text{ W}$$

3. Arbitrarily choose $t_{ON} = 30 \mu s$, $t_{OFF} = 10 \mu s$, for a full-load frequency of 25 kHz.

4. Determine R_{EXT} and C_{EXT}

Choose C_{EXT} such that $t_{OFF} < 10 \mu s$ when $I_{FB} = 0$. This guarantees that $I_{FB} > 0$ at full load, assuring proper regulation. From Fig. 7-83b, $C_{EXT} < 0.02 \mu F$; $0.01 \mu F$ is a convenient value.

R_{EXT} is found from Fig. 7-83c. If $C_{EXT} = 0.01 \mu F$, then $R_{EXT} = 2.5 \text{ k}\Omega$ for $t_{ON} = 30 \mu s$. $2.7 \text{ k}\Omega$ is the nearest standard value, making $t_{on} = 31 \mu s$.

5. Determining I_{PK} through T_1 primary

$$P_{in} = \frac{V_{in} I_{PK} (t_{ON})}{2(t_{ON} + t_{OFF})}$$

$$I_{PK} = \frac{2 P_{in} (t_{ON} + t_{OFF})}{V_{in} t_{ON}}$$

$$= \frac{2(4.3) 41 \times 10^{-6}}{5(31 \times 10^{-6})} = 2.3 \text{ A}$$

Note that a 500 μF capacitor is used to decouple the power supply from the 2.3 A peak current.

6. Determine primary inductance, neglecting the transistor saturation voltage.

$$L_P = \frac{V_{in} t_{ON}}{I_{PK}} = \frac{5(31 \times 10^{-6})}{2.3} = 65 \mu H$$

7. Using manufacturer's design equations, determine inductor specifications. A Ferroxcube pot core (No. 2213-L00387) with the following winding specifications was used:

$N_P = 8$ turns of No. 22 AWG magnet wire
Gap $= 0.006$ inch

8. Determine maximum number of secondary turns allowable for complete decay of secondary current during the minimum t_{OFF} at rated output.

$$L_{S(max)} = \frac{V_O t_{OFF}}{I_{S(PK)}}$$

where

$$I_{S(PK)} = \frac{2I_O\left(t_{OFF} + t_{ON}\right)}{t_{OFF}}$$

$$I_{S(PK)} = \frac{2(15 \times 10^{-3})41 \times 10^{-6}}{10 \times 10^{-6}} = 123 \text{ mA}$$

therefore

$$L_{S(\max)} = \frac{200(10 \times 10^6)}{123 \times 10^{-3}} = 16.26 \text{ mH}$$

$$N_{S(\max)} = \frac{N_P}{\sqrt{\dfrac{L_P}{L_{S(\max)}}}}$$

$$= \frac{8}{\sqrt{\dfrac{65 \times 10^{-6}}{16.26 \times 10^{-3}}}} = 125 \text{ turns}$$

9. Determine turns ratio of T_1. From the standpoint of secondary current decay, the number of secondary turns could be from 1 to 125. However, winding space limits the number of turns, while the breakdown rating necessary for Q_2 increases as the turns ratio decreases. Also, the peak diode current increases with decreasing secondary turns. A good compromise between space limitations and Q_2 breakdown rating is 80 turns. The necessary BV_{CER} is:

$$BV_{CER} > V_R + V_{in} \quad \text{where} \quad V_R = V_O \frac{N_P}{N_S}$$

$$= \frac{200 \times 8}{80} = 20 \text{ V}$$

Therefore, a minimum BV_{CER} of 25 V is required.

10. Power switch. Q_2 must have the following characteristics: Low $V_{CE(sat)}$ at $I_C = 2.3$ A and $I_C/I_B = 10$, high switching speed, and a $BV_{CER} > 25$ V. Q_1 is to be chosen on a similar basis. R_2 is chosen to be as low as possible to aid turn-off of Q_2. R_1 and C_1 form a snubber network to dampen any high voltage spikes generated by the leakage inductance of T_1.

11. Feedback. The type of feedback scheme depends upon the polarity of the input and output voltages with respect to system ground. A positive 200 V output is desired, and a negative input voltage is available. Therefore, the system ground reference must be pin 8 and a Zener diode can be used as a feedback element. The output voltage is given by:

$$V_O \approx V_Z = 200 \pm 10\% \text{ V (for a 10\% tolerance Zener)}$$

Multiple-output regulator.   Multiple-output voltages can be obtained by using additional secondary windings. Figure 7-87 illustrates the use of this technique. The regulation provided by these additional outputs depends upon the degree of their coupling to the regulated 5 V winding.

The system ground reference for this circuit is pin 4. Feedback is accomplished by amplifying the output error with op-amp A_1, and applying this voltage to pin 6 of the MC3380. Referring back to Fig. 7-82, it can be seen that if the voltage pin 6 increases to within one diode drop of the voltage at pin 8, Q_4 and Q_5 will begin to turn off, decreasing I_4 and increasing t_{OFF}.

Design of the transformer T_1, and the power circuit consisting of Q_1 and Q_2 is accomplished using the same methods discussed in the previous section.

7.7.11.   Designing with an IC switching regulator

This section describes an IC switching regulator used with a few external components to form specific voltage conversion circuits (step-up, step-down, inversion). The IC selected is a Texas Instruments TL497 switching regulator.

A block diagram of the TL497, together with typical external components to form a step-down regulator circuit, is shown in Fig. 7-88. The IC is a fixed on-time, variable frequency, switching voltage regulator control circuit. The on-time is programmed by a single external capacitor. Capacitor C_T is charged by an internal constant-current generator to a predetermined threshold. The charging current and the threshold vary proportionally with V_{CC}. Thus, the on-time remains constant over the specified range of input voltage (5 to 15 V).

The output voltage is programmed by an external resistor ladder network R_1-R_2 that attenuates the desired output voltage to 1.22 V. This feedback voltage is compared to the 1.22 reference by a high-gain comparator. When the output voltage decays below the programmed voltage, the comparator enables the oscillator circuit which charges and discharges C_T. The internal-pass transistor is driven on during the charging portion of C_T. The transistor may be used directly for switching currents up to 500 mA. The internal transistor collector and emitter are uncommitted, and the transistor is current-driven to allow operation from the positive rail or ground.

An internal diode matched to the current characteristics of the internal transistor is also available for blocking or commutating purposes. The IC also contains on-chip current limit circuitry that senses the peak currents in the switching regulator and protects the inductor against saturation and the pass transistor against overstress. The current limit is adjustable, and is

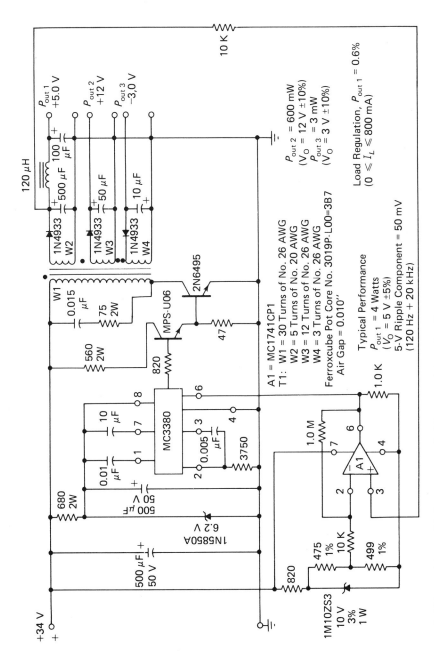

Fig. 7-87. MC3380 used in multiple-output switching regulator circuit

291

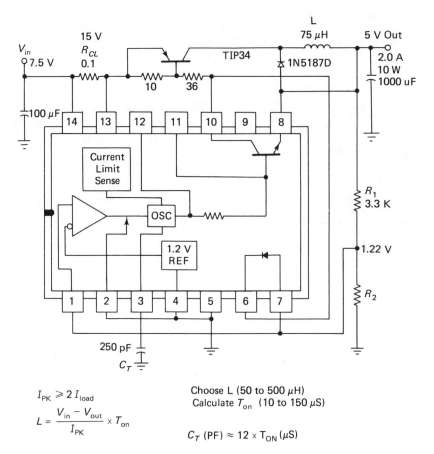

$$I_{PK} \geqslant 2\,I_{load}$$

$$L = \frac{V_{in} - V_{out}}{I_{PK}} \times T_{on}$$

Choose L (50 to 500 μH)
Calculate T_{on} (10 to 150 μS)

C_T (PF) $\approx$ 12 $\times$ T_{ON} (μS)

Fig. 7-88. Texas Instruments TL497 switching regulator block diagram with pinout showing external connections for typical step-down regulator

programmed by a single sense resistor between pin 14 and pin 13. The current-limit circuitry is activated when 0.7 V is developed across the program resistor R_{CL}.

The IC regulator datasheet shows the necessary interconnections to form other specific voltage conversion circuits (dual supply, microprocessor supply, current limiting, etc.). Such information will not be repeated here.

7.8. IC BALANCED MODULATORS FOR COMMUNICATIONS CIRCUITS

The main application for an IC balanced modulator is a "building block" for high-frequency communications equipment. The IC functions as a broadband, double-sideband, suppressed-carrier balanced modulator without transformers or tuned circuits. The IC can also be used as an SSB product detector, AM modulator/detector, FM detector, mixer, frequency doubler, and phase detector.

7.8.1. Basic balanced modulator

Figure 7-89 is the schematic of the basic IC balanced modulator. The circuit consists of differential amplifier Q_5-Q_6 driving a dual differential amplifier composed of transistors Q_1 through Q_4. Transistors Q_7 and Q_8 form constant-current sources for the lower differential amplifiers Q_5-Q_6.

In operation, a high-level input signal is applied to the carrier input, and a low-level input is applied to the signal input. This results in saturated switching operation of carrier dual differential amplifiers Q_1-Q_4, and linear operation of the modulating differential amplifier Q_5-Q_6. The resulting output signal contains only the sum and difference frequency components and amplitude information of the modulating signal. This is the desired condition for most balanced modulator applications.

Saturated operation of the carrier-input dual differential amplifiers also generates harmonics. Reducing the carrier input amplitude to its linear

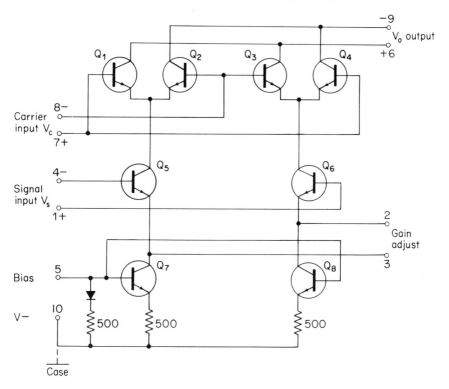

Fig. 7-89. Motorola MC1596 balanced modulator

range greatly reduces these harmonics in the output signal. However, this reduces gain, causing the output signal to contain carrier signal amplitude variations.

The carrier input differential amplifiers have no emitter feedback. Therefore, the carrier input levels for linear and saturated operation are readily calculated. The crossover point is typically in the range of 15–20 mV, with linear operation below this level, and saturated operation above it.

The modulating-signal differential amplifier has its emitters brought out to pins 2 and 3. This permits the designer to select his own value of emitter feedback resistance, and thus tailor the linear dynamic range of the modulating signal input to a particular requirement. The resistor also determines device gain.

7.8.2. Balanced modulator

Figure 7-90 shows a typical balanced modulator circuit. Typical input signal levels are 60 mV for the carrier, and 300 mV for the modulating signal. The modulation input must be kept at a level to ensure linear operation of Q_5–Q_6. If the signal level is too high, harmonics of the modulating signal are generated, and appear in the output as undesired sidebands of the suppressed carrier.

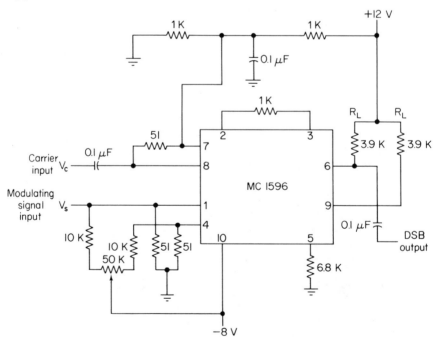

Fig. 7-90. Typical balanced modulator circuit for double sideband output

Operating with a high-level carrier input has the advantages of maximum gain and ensuring that any amplitude variations present on the carrier do not appear on the output sidebands. However, there is the disadvantage of increasing some of the undesired signals.

The decision to operate with a low- or high-level carrier input will, of course, depend on the application. For a typical filter-type SSB generator, the filter will remove all undesired outputs except some undesired sidebands of the carrier. For this reason, operation with a high-level carrier will probably be selected for maximum gain and to ensure that the desired sideband does not contain any undesired amplitude variations present on the carrier input signal.

On the other hand, in a low-frequency broadband balanced modulator, undesired outputs at any frequency may be unacceptable. Thus, low-level carrier operation may be the best choice.

7.8.3. Amplitude modulator

Figure 7-91 shows the balanced modulator used as an amplitude modulator. Modulation for any percentage from zero to over 100 is possible. The circuit operates by unbalancing the carrier null to insert the proper amount of

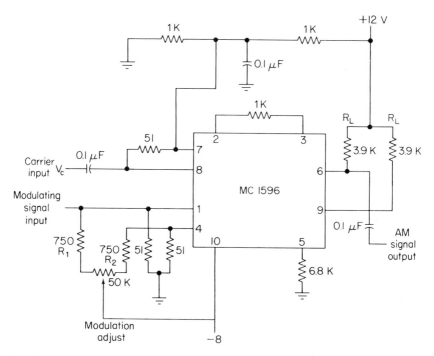

Fig. 7-91. Amplitude modulator using balanced modulator

carrier into the output signal. Note that the circuit for amplitude modulation is essentially the same as for balanced modulation, except for the values of resistors R_1 and R_2. These resistors are of lower value for the amplitude modulator, thus permitting a wider range for the modulation adjust potentiometer. This increase in range permits the circuit to be unbalanced to a point where some carrier appears at the output. In use, the potentiometer is adjusted until the desired percentage of modulation is measured at the output.

7.8.4. Product detector

Figure 7-92 shows the IC in an SSB product detector configuration. For this application, all frequencies except the desired demodulated audio are in the RF spectrum, and can be easily filtered at the output. As a result, the usual carrier null adjustment need not be included.

Upper differential amplifiers Q_1-Q_4 are driven with a high-level signal. Since carrier output level is not important in this application (carrier is filtered at the output) carrier input level is not critical. A high-level carrier input is desirable for maximum gain, and to remove any carrier amplitude variations from the output. Typical carrier inputs are from 100 to 500 mV.

The modulated signal (single-sideband, suppressed-carrier) input level to differential amplifier pair Q_5-Q_6 is maintained within the limits of linear operation. Typically, the modulated input is less than 100 mV. No trans-

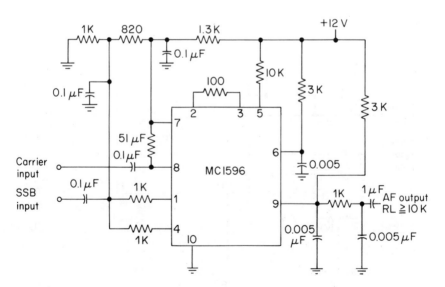

Fig. 7-92. Product detector using balanced modulator

formers or tuned circuits are required for excellent product detector performance from very low frequencies up to 100 MHz.

Note that dual outputs are available from the product detector, one from pin 6 and another from pin 9. One output can drive the receiver audio amplifiers while a separate output is available for the AGC system.

7.8.5.   AM detector

The product detector circuit of Fig. 7-92 can also be used as an AM detector. The modulated signal is applied to the upper differential amplifier, while the carrier signal is applied to the lower differential amplifier. Ideally, a constant-amplitude carrier signal would be obtained by passing the modulated signal through a limiter ahead of the carrier input terminals. However, if the upper input signal is at a high enough level (typically greater than 50 mV), the upper signal's amplitude variations do not appear in the output signal.

For this reason, it is possible to use the product detector circuit shown in Fig. 7-92 as an AM detector simply by applying the modulated signal *to both inputs* at a level of about 600 mV on modulation peaks, without using a limiter ahead of the carrier input. A small amount of distortion is generated as the signal falls below 50 mV during modulation valleys, but the distortion is not significant in most applications. Advantages of the IC as an AM detector include linear operation and the ability to have a detector stage with gain.

7.8.6.   Mixer

Since the IC generates an output signal consisting of the sum and difference frequencies of the two input signals only, the IC can be used as a double-balanced mixer. Figure 7-93 shows the IC used as a high frequency mixer with a broadband input, and a tuned output at 9 MHz. The 3 dB bandwidth of the 9 MHz output tank is 450 kHz.

The local oscillator signal is injected at the upper input with a level of 100 mV. The modulated signal is injected at the lower input with a maximum level of about 15 mV. Since the input is broadband, the mixer can be operated at HF and VHF input frequencies. The same circuit can be used with a 200 MHz input signal, and a 209 MHz local oscillator signal, as an example. At the higher frequency, the circuit shows an approximate 9 dB conversion gain and a 14 μV sensitivity.

Greater conversion gains can be obtained by using tuned circuit with impedance matching on the signal input. Of course, the bandwidth is narrower when tuned circuits are used. The nulling circuit permits the local oscillator signal to be nulled from the output. The local oscillator signal can be eliminated with a tuned output in many applications. Likewise, the tuned

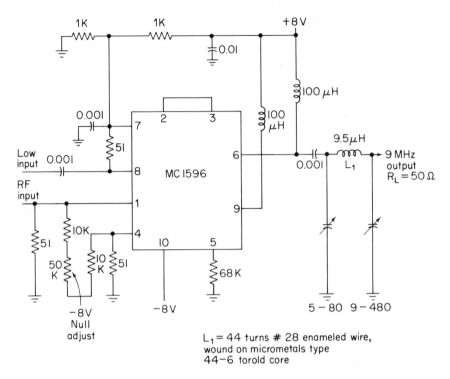

Fig. 7-93. Mixer using balanced modulator

output tank can be replaced with a resistive load to form a broadband input and output doubly balanced mixer. The magnitude of the output load resistance becomes a simple matter of trade-off between conversion gain and output signal bandwidth.

7.8.7. Doubler

The IC balanced modulator can function as a frequency doubler when the same signal is injected in both inputs. For operation as a broadband low-frequency doubler, the balanced modulator circuit of Fig. 7-90 need be modified only by adding ac coupling between the two inputs, and reducing the lower differential amplifier emitter resistance between pins 2 and 3 to zero (tieing pin 2 to pin 3). This latter modification increases the circuit sensitivity and doubler gain.

A low-frequency doubler with this modification is shown in Fig. 7-94. This circuit will double in the range below 1 MHz. For best results, both upper and lower differential amplifiers should be operated within their linear ranges. Typically, this limits input signals to about 15 mV.

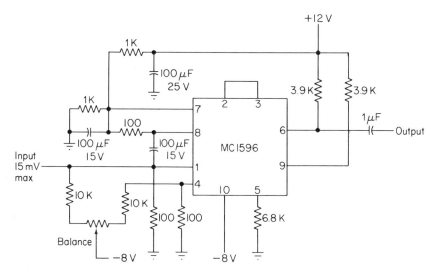

Fig. 7-94. Doubler using balanced modulator

7.8.8.   FM detector and phase detector

A balanced modulator produces a dc output which is a function of the phase difference between two input signals of the same frequency. Thus, when one signal is fixed frequency, and the other signal is frequency-modulated, the dc output corresponds to the *phase difference* between the two signals, or to the modulation. The balanced modulator can therefore be used as a phase detector (or FM detector).

7.9.   IC ARRAYS

An IC array provides several diodes, transistors, and other active devices on a single semiconductor chip. Because all of the active devices are fabricated simultaneously on the same chip, they have (nearly) identical characteristics. Particularly important in many applications, their parameters track each other with temperature variations as a result of their close proximity, and the good thermal conductivity of silicon. Consequently, IC arrays are particularly useful in circuits that require balance (such as a balanced diode bridge, a balanced two-channel amplifier, etc.). IC arrays are also particularly helpful where a number of active devices must be interconnected with external parts not possible to fabricate in IC form (tuned circuit, large-value resistors, variable resistors, large-value capacitors, etc.).

Figure 7-95 shows some typical IC arrays. They are described by the manufacturer (RCA) as linear arrays. Figure 7-95a is a 5-transistor array

(a)

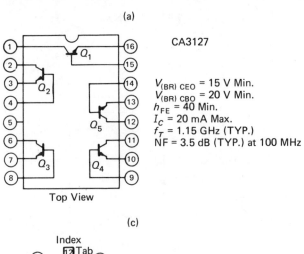

CA3127

$V_{(BR)\,CEO}$ = 15 V Min.
$V_{(BR)\,CBO}$ = 20 V Min.
h_{FE} = 40 Min.
I_C = 20 mA Max.
f_T = 1.15 GHz (TYP.)
NF = 3.5 dB (TYP.) at 100 MHz

Top View

(c)

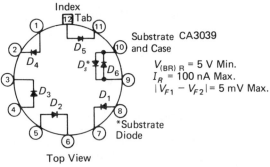

Substrate CA3039
and Case

$V_{(BR)\,R}$ = 5 V Min.
I_R = 100 nA Max.
$|V_{F1} - V_{F2}|$ = 5 mV Max.

*Substrate
Diode

Top View

(e)

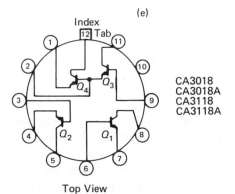

CA3018
CA3018A
CA3118
CA3118A

Top View

	$V_{(BR)\,CEO}$	$V_{(BR)\,CBO}$	h_{FE}	I_C
CA3018	15 V Max.	20 V Max.	30 Min.	50 mA Max.
CA3018A	15 V Max.	30 V Max.	60 Min.	50 mA Max.
CA3118	30 V Max.	40 V Max.	30 Min.	50 mA Max.
CA3118A	40 V Max.	50 V Max.	30 Min.	50 mA Max.

Fig. 7-95. Typical RCA IC arrays

(b)

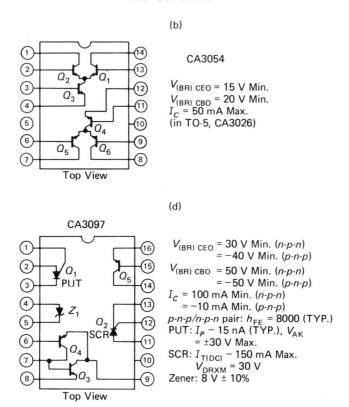

CA3054

$V_{(BR) CEO}$ = 15 V Min.
$V_{(BR) CBO}$ = 20 V Min.
I_C = 50 mA Max.
(in TO-5, CA3026)

(d)

CA3097

Top View

$V_{(BR) CEO}$ = 30 V Min. (n-p-n)
         = −40 V Min. (p-n-p)
$V_{(BR) CBO}$ = 50 V Min. (n-p-n)
         = −50 V Min. (p-n-p)
I_C = 100 mA Min. (n-p-n)
    = −10 mA Min. (p-n-p)
p-n-p/n-p-n pair: h_{FE} = 8000 (TYP.)
PUT: I_P − 15 nA (TYP.), V_{AK}
      = ±30 V Max.
SCR: $I_{T(DC)}$ − 150 mA Max.
     V_{DRXM} = 30 V
Zener: 8 V ± 10%

Fig. 7-95 (*continued*)

tor array designed for high-frequency applications, as indicated by the
specifications (f_T = 1.15 GHz typical). The IC of Fig. 7-95b includes two
differential amplifier pairs (Q_1-Q_2 and Q_5-Q_6), each with a constant
current source transistor (Q_3-Q_4). The diode array of Fig. 7-95c includes 6
identical diodes, plus a substrate diode. The mixed array of Fig. 7-95d
includes transistor Q_5, programmable unijunction transistor Q_1, Zener diode
Z_1, control rectifier Q_2, and a *pnp*/*npn* pair Q_3-Q_4. The IC of Fig. 7-95e
includes two isolated transistors, plus two transistors with an emitter-base
common connection.

7.9.1. Applications for IC arrays

There are almost unlimited applications for IC arrays. The following
paragraphs describe two typical applications.

Application for diode arrays. The diode array of Fig. 7-95c can be
used in any circuit where from one to six diodes are required, provided that

the voltage and current are within the ICs capabilities. However, the IC is most useful when the circuit requires diodes of matched characteristics. The *high-speed gate* of Fig. 7-96 is a classic example. Here all six diodes are used, and the most is made of their matching characteristics.

In high-speed gates, the gating signal often appears at the output and causes the output signal to rise on a "pedestal." A diode-quad bridge circuit can be used to balance out the undesired gating signal at the output and reduce the pedestal to the extent that the bridge is balanced.

The diode-quad gate functions as a variable impedance between a source and a load, and can be connected either in series or in shunt with the load. The circuit configuration used depends upon the input and output impedances of the circuits to be gated. A series gate is used if the source and load impedances are low compared to the diode back resistance. A shunt gate is used if the source and load impedances are high compared to the diode forward resistance.

The circuit of Fig. 7-96 uses the six diodes as a series gate in which the diode bridge, in series with the load resistance, balances out the gating signal to provide a pedestal-free output. With a proper gating voltage (1 to 3 V, 1 to 500 kHz) diodes D_5 and D_6 conduct during one half of each gating cycle,

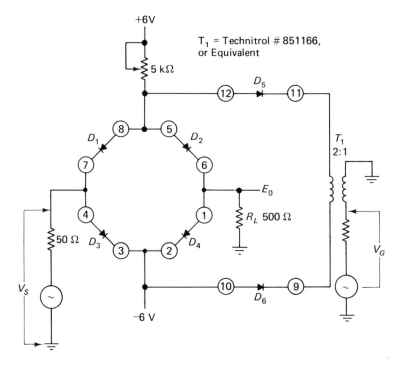

Fig. 7-96. Series gate using matched diode array

and do not conduct during the other half of the cycle. When diodes D_5 and D_6 are conducting, the diode bridge (D_1-D_4) is not conducting, and the high diode back resistance prevents the input signal V_S from appearing across the load resistance R_L. When diodes D_5 and D_6 are not conducting, the diode bridge conducts and the low diode forward resistance allows the input signal to appear across the load resistance. Resistor R_1 may be adjusted to minimize the gating voltage present at the output.

Application for transistor arrays. The transistor array of Fig. 7-95e can be used in any circuit where from one to four transistors are required, provided that the voltage, current and gain are within the ICs capabilities. However, there are certain design considerations. These considerations are best illustrated by reference to Fig. 7-97 which shows the substrate connections. Note that diodes are formed between the substrate pin (10) and the collectors of the four transistors. Also, two transistor terminals are connected to a common lead. The particular configuration is useful in emitter-follower and Darlington circuit connections. The four transistors can be used almost independently if terminal 2 is grounded (or ac grounded) so that Q_3 can be used as a common-emitter amplifier, and Q_4 as a common-base amplifier.

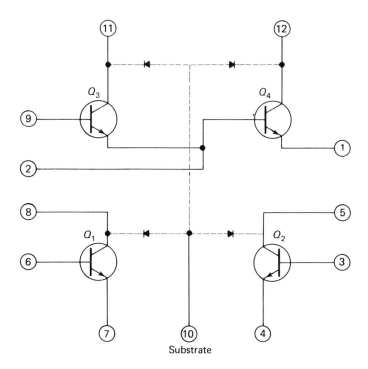

Fig. 7-97. RCA CA3018 transistor array

In pulse video amplifiers and line-drivers, Q_4 can be used as a forward-biased diode in series wth the emitter of Q_3. Likewise, transistor Q_3 can be used as a diode connected to the base of Q_4 or, in a reverse-biased connection, Q_3 can serve as a protective diode in RF circuits connected to antennas. The presence of Q_3 does not inhibit the use of Q_4 in a large number of circuits.

In transistors Q_1, Q_2 and Q_4, the emitter lead is interposed between the base and collector leads to minimize package and lead capacitances. In Q_3, the substrate lead serves as the shield between base and collector. This lead arrangement reduces feedback capacitance in common-emitter amplifiers, and thus extends video bandwidth and increases tuned-circuit amplifier gain stability. Q_3

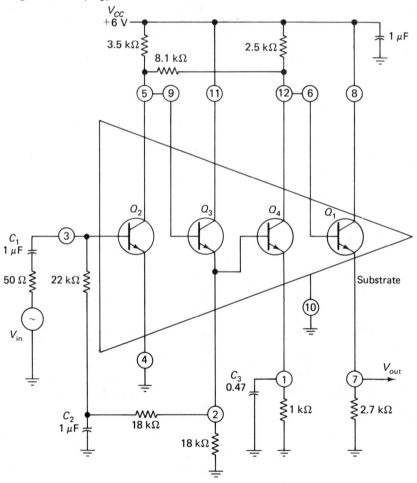

Fig. 7-98. Broadband video amplifier using RCA CA3018 IC transistor array

Broadband video amplifier. Figure 7-98 shows a broadband video amplifier design using the four-transistor array. This amplifier can be considered as two dc-coupled stages, each consisting of a common-emitter, common-collector configuration. The common-collector transistor provides a low-impedance source to the input of the common-emitter transistor, and a high-impedance, low capacitance load at the common-emitter output.

Two feedback loops provide dc stability of the broadband video amplifier and exchange gain for bandwidth. The feedback loop from the emitter of Q_3 to the base of Q_2 provides dc and low-frequency feedback. The loop from the collector of Q_4 to the collector of Q_2 provides both dc feedback and ac feedback at all frequencies.

7.10. IC PHASE-LOCKED LOOPS

Phase-locked loops (PLLs) are used in both linear and digital applications. FM demodulation, FSK demodulation, tone decoding, frequency multiplication, signal conditioning, clock synchronization, and frequency synthesis are some of the many applications of a PLL. This section describes some typical PLLs available in IC form, and shows how the PLLs can be used in a variety of linear and digital applications.

7.10.1. Review of PLL fundamentals

The basic PLL system is shown in Fig. 7-99. As shown, a basic PLL consists of three parts: phase comparator, low-pass filter, and voltage-controlled oscillator (VCO), all connected to form a closed-loop, frequency-feedback system.

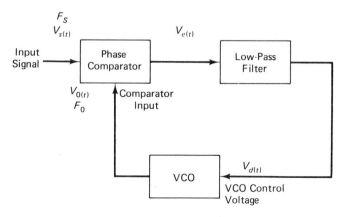

Fig. 7-99. Basic phase-locked loop system

With no signal input applied to the PLL system, the error voltage at the output of the phase comparator is zero. The voltage $V_{d(t)}$ from the low-pass filter is also zero, which causes the VCO to operate at a set frequency, f_o, called the center frequency. When an input signal is applied to the PLL, the phase comparator compares the phase and frequency of the signal input with the VCO frequency and generates an error voltage proportional to the phase and frequency difference of the input signal and the VCO.

The error voltage $V_{e(t)}$ is filtered and applied to the control input of the VCO. $V_{d(t)}$ varies in a direction that *reduces* the frequency difference between the VCO and signal-input frequency. When the input frequency is sufficiently close to the VCO frequency, the closed-loop nature of the PLL forces the VCO to *lock* in frequency with the signal input. When the PLL is in lock, the VCO frequency is identical to the signal input, except for a finite phase difference. The range of frequency over which the PLL can maintain this locked condition is defined as the *lock range* of the system. The lock range is always larger than the band of frequencies over which the PLL can acquire a locked condition with the signal input. This latter band of frequencies is defined as the *capture range* of the PLL system.

7.10.2. Technical description of typical IC PLLs

Figure 7-100 shows a block diagram of an RCA PLL (the CD4046A). The PLL structure consists of a low-power, linear VCO and two different phase comparators having a common signal input amplifier and a common comparator input. A 5.4 V Zener is provided for supply regulation, if necessary. The VCO can be connected either directly or through frequency dividers to the comparator input of the phase comparators. The low-pass filter is implemented through external parts because of the radical configuration changes from application to application, and because some components are noninterchangeable.

Phase comparators.   Both phase comparators are driven by a common-input amplifier configuration composed of a bias stage and four inverting-amplifier stages. The phase-comparator signal input (terminal 14) can be direct-coupled provided the signal swing is within COS/MOS logic levels. (COS/MOS is discussed in Chapter 8.) For smaller input signal swings, the signal must be capacitively coupled to the self-biasing amplifier at the signal input to ensure an overdriven digital signal into the phase comparators.

Phase comparator I is essentially an overdriven balanced mixer. When using phase comparator I, the signal and comparator input frequencies must

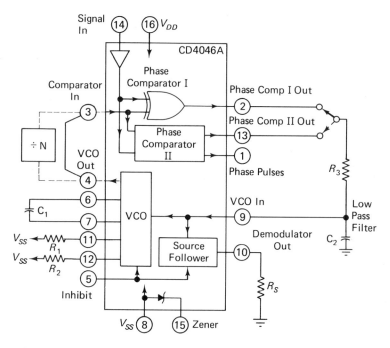

Fig. 7-100. RCA CD4046A PLL block diagram

have 50% duty cycles. With no signal or noise input phase comparator I has an average output voltage equal to $V_{DD}/2$. The low-pass filter connected to the output of phase comparator I supplies the average voltage to the VCO input, and causes the VCO to oscillate at the center frequency f_o. The frequency range over which the PLL remains locked to the input frequency (lock range) is close to the theoretical limit of $\pm f_c$. The range of frequencies over which the PLL can acquire lock (capture range) is dependent upon the low-pass filter characteristics, and can be made as large as the lock range. Phase comparator I enables a PLL system to remain in lock in spite of high amounts of noise in the input signal.

One characteristic of this type of phase comparator is that it may lock onto input frequencies that are close to harmonics of the VCO center-frequency. A second characteristic is that the phase angle between the signal and the comparator input varies between 0° and 180°, and is 90° at the center frequency. Figure 7-101a shows the typical triangular, phase-to-output, response characteristic of phase-comparator I. Typical waveforms for a PLL using phase-comparator I in locked condition of f_o is shown in Fig. 7-101b.

Phase-comparator II acts only on the positive edges of the signal and comparator-input signals. The duty cycles of the signal and comparator

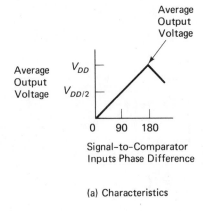

(a) Characteristics

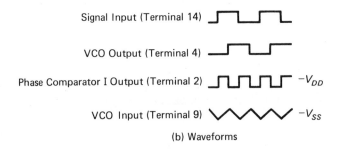

(b) Waveforms

Fig. 7-101. Phase comparator I characteristics and waveforms

inputs are not important since positive transitions control the PLL system. If the signal input frequency is higher than the comparator input frequency, a p-type output driver is maintained ON continuously. If the signal input frequency is lower than the comparator input frequency, an n-type output driver is maintained ON continuously. If the signal and comparator input frequencies are the same, but the signal input lags the comparator input in phase, the n-type output driver is maintained ON for a time equal to the phase difference. If the signal and comparator output frequencies are the same, but the signal input leads the comparator input, the p-type output driver is maintained ON for a time equal to the phase difference.

With this configuration, the capacitor C_2 voltage of the low-pass filter is continuously adjusted until the signal and comparator input are equal in both phase and frequency. At this stable operating point, both p- and n-type output drivers remain OFF, and the signal at the phase-pulses output (terminal 1) is at a logic 1, indicating a locked condition. For phase comparator II, no phase difference exists between signal and comparator

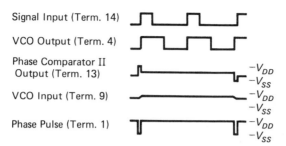

Fig. 7-102. Phase comparator II waveforms

input over the full VCO frequency range. It should be noted that the PLL lock range for this type of phase comparator is equal to the full VCO range. With no signal present at the signal input, the VCO is adjusted to its lowest frequency for phase comparator II. Figure 7-102 shows typical waveforms for the PLL using phase comparator II.

Voltage-controlled oscillator.   To assure low system-power dissipation, it is desirable that the low-pass filter consume little power. For example, in an *RC* filter, this requirement dictates that a high-value *R* and low-value *C* be used. The VCO input must not, however, load down or modify the characteristics of the low-pass filter. To solve this problem, the VCO uses a configuration with almost infinite input resistance, allowing a great degree of freedom in selection of the low-pass filter components. Further, in order not to load the low-pass filter, a source-follower output of the VCO input voltage is provided (demodulated output). If this output is used, a load resistor R_S of 10 kilohms or more should be connected from terminal 10 to ground. If unused, terminal 10 should be left open. A logic 0 on the inhibit input (terminal 5) enables the VCO and the source follower, whereas a logic 1 turns off both to minimize standby power consumption.

Other IC PLLs.   Figure 7-103 shows four Motorola IC PLLs. The PLL of Fig. 7-103a is for general-purpose digital applications at frequencies up to 1.4 MHz. Note that the PLL is similar to the PLL of Fig. 7-100.

The PLL of Fig. 7-103b is for use with an external VCO. This PLL has both fixed and programmable counters to set the input signals (which are to be compared) at some precise frequency.

The PLL of Fig. 7-103c is for use in CB and FM transceivers. This PLL includes a reference oscillator with dividers, plus a programmable divider, and operates with the transceiver's crystal-controlled oscillators as a frequency synthesizer (to produce signals of appropriate frequency for all 40 CB channels, etc.). A further discussion of IC PLLs used as frequency synthesizers is found in the author's *Handbook of Practical CB Service* (Prentice-Hall, Inc., Englewood Cliffs, New Jersey 07632, 1978).

Phase Comparators/VCO

MC14046B — for general-purpose, digital
applications at frequencies to 1.4 MHz.

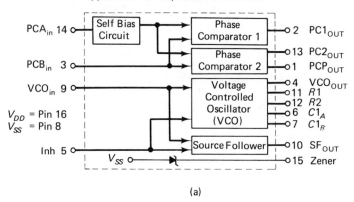

(a)

Frequency Synthesizers

MC145104
MC145106   MC145109 } —for CB and FM
MC145107   MC145112 }   transceivers

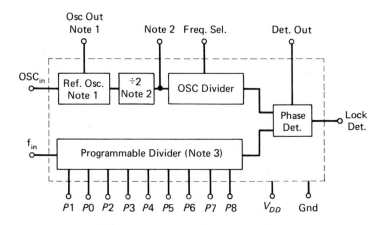

Note 1 — MC145107, MC145109 have amplifier, no output.
          Other types have Reference Oscillator, with output.
Note 2 — MC145106, MC145107 have ÷2 stage with output.
          MC145112 has ÷2 stage, no output.
          Other types have ÷2 omitted.
Note 3 — MC145104, MC145107 have 2^8-1 Divider.
          Other types have 2^9-1 Divider.

(c)

Fig. 7-103. Motorola MC14046B, MC14568B, MC14504-7, and MLM565C PLL
devices

310

Phase Comparator/Programmable Counters

MC14568B — for use with external
VCO for PLL applications.

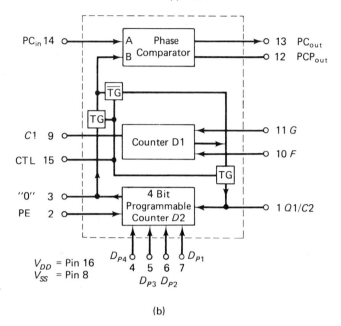

(b)

Linear PLL
MLM565C — for general-purpose
analog applications at frequencies
to 500 kHz.

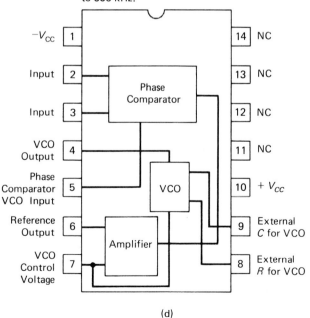

(d)

Fig. 7-103 (*continued*)

The PLL of Fig. 7-103d is for general-purpose linear analog applications at frequencies up to 500 kHz. This PLL includes a VCO, phase comparator, and an amplifier.

7.10.3.   PLL design examples

The following paragraphs describe some typical applications for the PLL shown in Fig. 7-100. These applications are based on the design equations of Fig. 7-104. When using Fig. 7-104, the frequency values are in kHz, resistance values in $k\Omega$, and capacitance values in μF. The VCO conversion

(a)

Characteristics	Using Phase Comparator 1		Using Phase Comparator II	
	VCO without Offset $R_2 = \infty$	VCO with Offset	VCO without Offset $R_2 = \infty$	VCO with Offset
VCO Frequency				
Resistor R_1 and R_2 Approximations	$R_1 = \dfrac{K_1}{f_O C_1}$	$R_1 = \dfrac{K_1}{f_L C_1}$ $R_2 = \dfrac{2K_1}{(f_O - f_L)C_1}$	$R_1 = \dfrac{2K_1}{f_{max} C_1}$	$R_1 = \dfrac{2K_1}{(f_{max} - f_{min})C_1}$ $R_2 = \dfrac{2K_1}{f_{min} C_1}$
Center Frequency, f_O	$f_O \approx \dfrac{K_1}{R_1 C_1}$	$f_O \approx \dfrac{K_1}{R_1 C_1} + \dfrac{2K_1}{R_2 C_1}$	Not applicable—for no signal input VCO is adjusted to lowest possible comparator frequency	
Frequency Lock Range, f_L	$f_L \leqslant \pm f_O$	$f_L \leqslant \pm \dfrac{1}{1 + 2\dfrac{R_1}{R_2}} \cdot f_O$	f_L Full VCO Frequency Range	
Frequency Capture Range, f_C	$f_c \leqslant f_L$ $f_c \approx \pm \dfrac{1}{2\pi}\sqrt{\dfrac{2\pi f_L}{R_3 C_2}}$		$f_c - f_L$	
Phase Angle between Signal and Comparator	90 at Center Frequency (f_O), approximating 0 and 180 at ends of lock range (f_L)		Always 0 in-lock	
Locks on Harmonics of Center Frequency	Yes		No	
Signal Input Noise Rejection	High		Low	

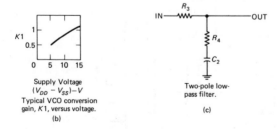

Typical VCO conversion gain, $K1$, versus voltage.
(b)

Two-pole low-pass filter.
(c)

Fig. 7-104. Basic PLL design equations

gain factor K_1 shown in Fig. 7-104a is based on supply voltage, and is found using the graph of Fig. 7-104b. Figure 7-100 shows where the external components outlined in Fig. 7-104a are to be connected in the PLL system. Figure 7-104c describes the low-pass filter used with the system. As a guideline, R_1, R_2 and R_S should be 10 $k\Omega$ or larger, and C_1 should be 50 pF or larger.

7.10.4.   FM demodulation with a PLL

Figure 7-105 shows connections for the PLL as an FM demodulator. When a PLL is locked on an FM signal, the VCO tracks the instantaneous frequency of that signal. The VCO input voltage, that is the filtered error voltage from the phase detector, corresponds to the demodulated output.

For this example, an FM signal consisting of a 10 kHz carrier frequency is modulated by a 400 Hz audio signal. The total FM signal amplitude is 500 mV. Thus, the signal is ac coupled to the signal input, terminal 14. Phase comparator I is used for this application because a PLL system with a center frequency equal to the FM carrier frequency is needed.

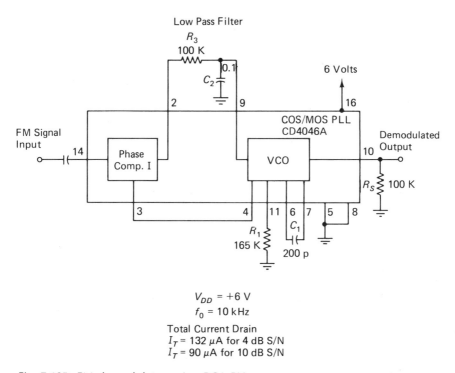

Fig. 7-105. FM demodulator using RCA PLL

Phase comparator I lends itself to this application also because of its high signal-input-noise-rejection characteristics.

The equations shown in Fig. 7-104a for phase comparator I (with R_2 = infinity) are used in the following considerations. The center frequency of the VCO is set to be equal to the carrier frequency of 10 kHz. The VCO center frequency is adjusted experimentally as follows:

1. From Fig. 7-104b, find the typical VCO conversion gain factor K_1 at the PLL system supply voltage. In this case, the supply is 6 V, and the K_1 factor is about 0.6. Keep in mind that the VCO will be operated at about one-half of V_{DD} in this application.
2. Arbitrarily choose the timing capacitor C_1 to determine the approximate value of R_1 (which sets the input voltage to the VCO). To simplify calculations, use 200 pF for C_1.
3. Connect an adjustable resistor close to the value of R_1, calculated from the equation: $R_1 = K_1/(f_o C_1)$, between terminal 11 and ground.
4. Adjust R_1 until the VCO input voltage is one-half of V_{DD} (or at about 3 V).
5. Then adjust the value of R_1 until the exact center frequency of 10 kHz is obtained from the VCO output at terminal 4. Use a precision frequency counter at terminal 4 for best results. Disconnect R_1 and measure its value. As shown in Fig. 7-105, a value of 165 kΩ for R_1, and a value of 200 pF for C_1 produce an output of 10 kHz.

The capture range f_c of the system is set by the values of the low-pass filter R_3-C_2, as shown by the equations of Fig. 7-104a. Using the values shown, the capture range is about ±0.4 kHz. With an FM signal amplitude of about 150 mV the demodulated output is about 30 mV.

7-10.5   Frequency synthesizer using a PLL

Figure 7-106 shows connections for the PLL as a frequency synthesizer. The PLL system can function as a frequency-selective frequency multiplier when a *frequency divider* or *counter* is inserted into the feedback loop between the VCO output and the comparator input. (Note that PLLs designed specifically for frequency synthesizer applications include a programmable frequency divider, such as shown in Fig. 7-103c.) In the circuit of Fig. 7-106, an *external* frequency divider consisting of three decades is used. N, the external frequency divider modulus, can vary from 3 to 999 in steps of 1. When the PLL system is in lock, the signal and comparator inputs are at the same frequency, and the frequency range of the synthesizer circuit is 3 to 999 kHz, in 1 kHz increments (which is programmable by the switch position of the Divide-by-N counter).

Phase comparator II is used for this application because it will not lock

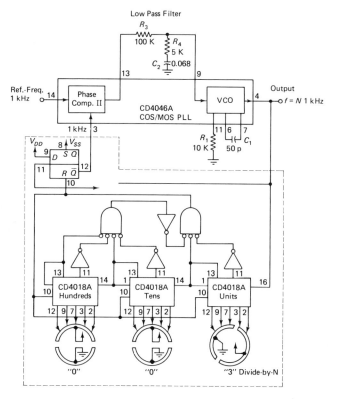

Fig. 7-106. Frequency synthesizer using RCA PLL

on harmonics of the signal input reference frequency (as discussed, phase comparator I does lock on harmonics). Since the duty cycle of the output of the Divide-by-N frequency divider is not 50 percent, phase comparator II lends itself directly to this application.

Using the equations for phase comparator II shown in Fig. 7-104a, the VCO is set up to cover a range of 0 to 1.1 MHz. In this case, the steps for choosing the VCO external components are the same as those outlined in Sec. 7.10.4, FM demodulation, with the following exceptions: adjust the VCO input voltage to V_{DD}; the initial setting for $R_1 = 2K_1/(f_{max} C_1)$. The low-pass filter for this application is a two-pole, lag-lead filter that enables faster locking for step changes in frequency.

7.10.6.   Split-phase data synchronization and decoding with a PLL

Figure 7-107 shows a digital application of the PLL, split-phase data synchronization and decoding. A split-phase data signal consists of a series of binary digits that occur at a periodic rate, as shown in waveform A of Fig. 7-107. The weight of each bit, 0 or 1, is random, but the duration of each bit, and thus the periodic bit-rate, is essentially constant. To detect and process the incoming signal, it is necessary to have a clock that is synchronous with the data-bit rate. This clock signal must be derived from the incoming data signal. Phase-lock techniques can be used to recover the clock and the data.

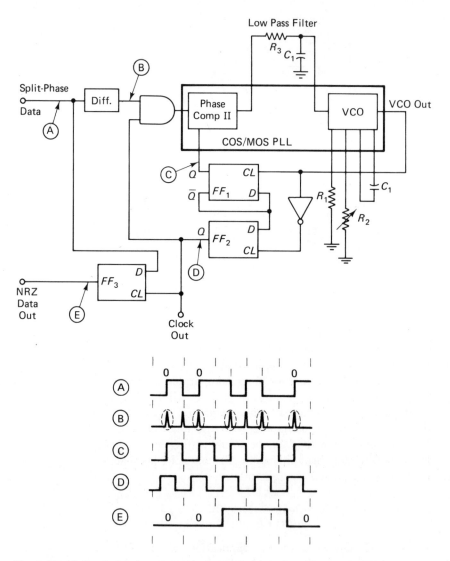

Fig. 7-107. Split-phase data synchronization and decoding using RCA PLL

Timing information is contained in the data transitions, which can be positive or negative in direction, but both polarities have the same meaning for timing recovery. The phase of the signal determines the binary bit weight. A binary 0 or 1 is a positive or negative transition, respectively, during a bit interval in split-phase data signals.

As shown in Fig. 7-107, the split-phase data input (A) is first differentiated to mark the locations of the data transitions. The differentiated signal (B), which is twice the bit rate, is gated into the PLL. Phase-comparator II in the PLL is used because of its insensitivity to duty cycle on both the signal and comparator inputs. The VCO output is fed into the clock input of FF_1, that divides the VCO frequency by two.

During the ON intervals, the PLL tracks the differentiated signal (B); during the OFF intervals the PLL remembers the last frequency present and still provides a clock output. The VCO output is inverted and fed into the clock input of FF_2 the data input of which is the inverted output of FF_1. FF_2 provides the necessary phase shift in signal (C) to obtain signal (D), which is the recovered clock signal from the split-phase data transmission. The output of FF_3 (E) is the recovered binary information from the phase information contained in the split-phase data. Initial synchronization of the PLL system is accomplished by a string of alternating 0 and 1 bits that precede the data transmission.

7.10.7.   PLL lock-detection systems

In some applications that use a PLL, it is sometimes necessary to have an output indication of when the PLL is in lock. One of the simplest forms of lock-condition indicator is a binary signal. For example, a 1 or a 0 output from a lock-detection circuit would correspond to a locked or unlocked condition, respectively. This signal could, in turn, activate circuitry using a locked PLL signal. The condition could also be used in frequency-shift-keyed (FSK) data transmissions in which digital information is transmitted by switching the input frequency between either of two discrete input frequencies, one corresponding to a digital 1 and the other to a digital 0.

Figure 7-108 shows a lock-detection scheme for a PLL. The signal input is switched between two discrete frequencies of 20 kHz and 10 kHz. The PLL system uses phase-comparator II; the VCO bandwidth is set up for an f_{min} of 9.5 kHz and an f_{max} of 10.5 kHz. Therefore, the PLL locks and unlocks on the 10 kHz and 20 kHz signals, respectively. When the PLL is in lock, the output of phase comparator I is low except for some very short pulses that result from the inherent phase difference between the signal and comparator inputs; the phase-pulses output (terminal 1) is high except for some very small pulses resulting from the same phase difference. This low condition of phase comparator I is detected by the lock-detection circuit shown in Fig. 7-108.

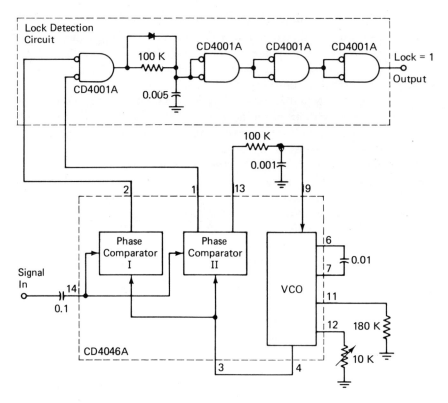

Fig. 7-108. PLL lock-detection system

8

DIGITAL
IC BASICS

Unlike linear ICs, which can be used in a great variety of applications, digital ICs are used primarily for logic circuits. To make full use of digital ICs, the user must be familiar with digital logic, including simplification and manipulation of logic equations, working with logic maps, and the implementation of basic logic circuits (decoders, encoders, function generators, parity networks, data distributors and selectors, multiplexers, flip-flops, counters, registers, A/D and D/A converters, arithmetic units, memory units, line drivers and receivers, and numeric display systems). All of these subjects, and more, are discussed in the author's *Logic Designer's Manual* (Reston Publishing Company, Reston, Virginia, 22090, 1977).

In this chapter, we concentrate on what types of digital ICs are available, logic forms, their relative merits, selecting logic ICs, interpreting logic IC datasheets, basic interfacing problems common to all logic ICs, and troubleshooting devices for logic ICs. Chapter 9 describes how digital and linear ICs can be combined to solve a variety of design problems.

8.1. LOGIC FORMS

The following paragraphs describe the various logic forms in use today. Some of these forms appeared as discrete component circuits. However, most of the forms are the result of packaging logic elements as integrated circuits.

The following discussions not only describe what is available to the logic designer, but how (in brief) the circuits operate. It is essential that logic designers understand these operating principles to use and interconnect the IC elements in logic systems properly. The IC logic elements discussed here represent a cross section of the entire logic field, and do not necessarily represent the products of any particular manufacturer.

The author makes no attempt to promote one form over another, but simply summarizes the facts (capabilities and limitations) about each form. Thus, designers can make informed and intelligent comparisons of the logic forms, and select those that are best suited to their needs.

8.1.1. Resistor-transistor logic (RTL)

RTL was derived from Direct-Coupled Transistor Logic, or DCTL, and was the first IC logic form introduced around 1960. The basic circuit is a direct translation from the discrete design into integrated form. This circuit was the most familiar to logic designers, easy to implement, and therefore the first introduced by IC manufacturers.

The basic RTL gate circuit shown in Fig. 8-1 is presented here to illustrate the basic building block type of logic. The most complex elements are constructed simply by the proper interconnection of this basic circuit. There are many electronic devices using RTL still in operation. Thus, RTL is of interest to the student and service technician. However, because of the problems discussed in Sec. 8.2, RTL is not generally used in the design of new logic circuits.

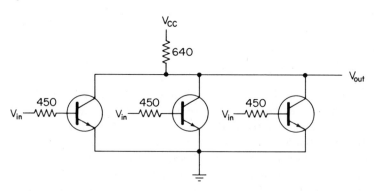

Fig. 8-1. Basic RTL gate

8.1.2. Diode-transistor logic (DTL)

DTL is another logic form that was translated from discrete design into IC elements. DTL was very familiar to the discrete component logic designer in that the form used diodes and transistors as the main components (plus a

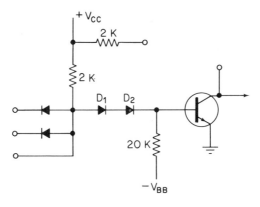

Fig. 8-2. Basic DTL gate

minimum number of resistors). The diodes provided higher thresholds than could be obtained with RTL.

Figure 8-2 shows the basic DTL gate. Note that it requires two power supplies (to improve turn-off time of the transistor inverter). As in the case of RTL, the DTL logic form is primarily of interest to students and technicians, but is not generally used in the design of new devices.

8.1.3. High-threshold logic (HTL)

HTL is designed specifically for logic systems where electrical noise is a problem, but where operating speed is of little importance (since HTL is the slowest of all IC logic families). Because of this slow speed, and because other logic forms (such as the MOS logic described in Sec. 8.1.6) offer similar noise immunity, HTL is generally not being used for design of present-day logic systems (although HTL is found in many existing systems, particularly where logic must be connected to industrial and other heavy duty equipment).

Figure 8-3 shows a typical HTL gate along with the transfer characteristics. Note that the gate is identified as MHTL, or Motorola HTL. Also note that the MHTL gate is compared with an MDTL (or Motorola DTL) gate. This comparison is made since the HTL is essentially the same as the DTL, except that a Zener diode is used for D_1 in the HTL. The use of a Zener for D_1 (with a conduction point of about 6 or 7 V), and the higher supply voltage (a V_{CC} of about 15 V), produces the wide noise margins shown in the transfer characteristics of Fig. 8-3. That is, the HTL gate will not operate unless the input voltage swing is large. Noise voltages below about 5 V will have no effect on the gate. The problems of noise in logic systems are discussed further in Sec. 8.2.2.

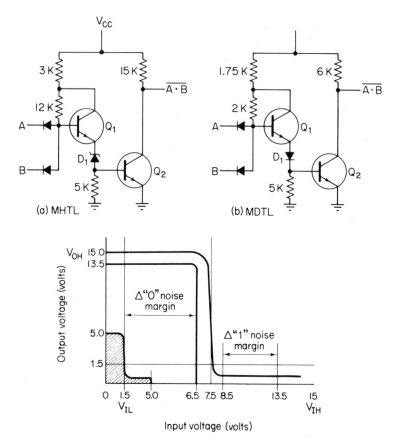

Fig. 8-3. Motorola MHTL gate and transfer characteristics

8.1.4.   Transistor-transistor logic (TTL or T²L)

TTL (or T^2L) has become one of the most popular logic families available in IC form. Most IC manufacturers produce at least one line of TTL, and often several lines. This fact gives TTL the widest range of logic functions.

Figures 8-4 and 8-5 show the schematics of two typical logic gates for comparison. Figure 8-4 shows a conventional high-speed TTL gate, whereas Fig. 8-5 shows an MTTL, or Motorola TTL gate. One difference between the MTTL and TTL is the replacement of resistor R_4 (Fig. 8-4) with resistors R_{4A} and R_{4B}, and transistor Q_6, as shown in Fig. 8-5.

With the gate of Fig. 8-5, when there is a "low" on either A or B, Q_1 is forward-baised and no base drive is available for Q_2. This keeps Q_2, as well as Q_5 and Q_6 in the off condition. The collector of Q_2 is approximately at V_{CC}, and base current is supplied to Q_3 and Q_4, keeping Q_3 and Q_4 on.

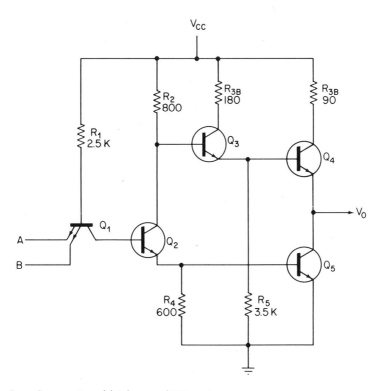

Fig. 8-4. Conventional high-speed TTL gate

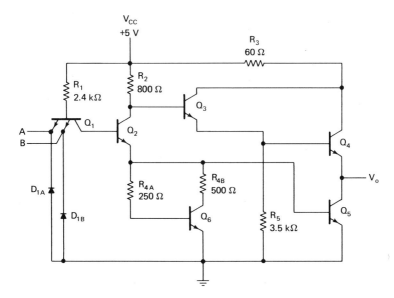

Fig. 8-5. Improved Motorola MTTL gate

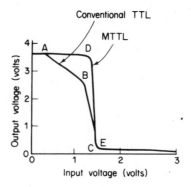

Fig. 8-6. Comparison of MTTL and conventional TTL characteristics

Now assume that input A is "high" and input B gradually goes from a low to a high. The base of Q_2 tracks the voltage at the input (the same as for conventional TTL) by the difference of $V_{BE} - V_{BC}$ of Q_1. At the point where Q_2 turns on in conventional TTL, Q_2 does not turn on in MTTL, since the equivalent of an open circuit exists at the Q_2 emitter.

With no current flow, the collector of Q_2 remains near V_{CC}, Q_5 saturates, and point E is reached. The resistors in the bypass network are chosen so that the network conducts the same current as resistor R_4 in Fig. 8-4 when transistor Q_5 is saturated. Figure 8-6 shows a comparison of MTTL and conventional TTL transfer characteristics.

Function of input diodes.   Because of high speeds of operation, TTL generates large values of current and voltage rates-of-change. A 1 V in approximately 1.3 ns rise time, and a 1 V in approximately 1 ns fall time produces dV/dt rates on the order of 10^9 volts per second. With these rates of change, undershoot exceeding 2 V can develop in the system. Such undershoot can cause two serious problems.

First, false triggering of the following stage is possible since a possible overshoot follows the large undershoot. This positive overshoot may act as a high signal, and turn on the following stage for a short period of time.

Second, if the unused inputs of a gate are returned to the supply voltage (as is usually recommended for TTL logic design), and a negative undershoot in excess of 2 V occurs, the reverse-biased emitters of the inputs may break down, and draw excessive current, generating noise in the system.

Typical TTL gates.   The basic gate in most TTL systems is the NAND gate. However, a full TTL logic line will include AND, OR, AOI, NOR, as well as EXCLUSIVE OR and NOR.

TTL logic elements. In addition to basic gates, there is an almost unlimited supply of TTL logic elements, such as flip-flops, counters, registers, decoders, multiplexers, line drivers and receivers, etc.

TTL power gates. In some logic systems, there are fan-out requirements that exceed the capability of a standard gate. Power gates are designed to meet these requirements with a minimum of additional circuitry. A typical power gate (an MTTL AND gate) is shown in Fig. 8-7. With this power gate, the output circuitry is designed to provide twice the fan-out of conventional gates (in this case, 20 standard gate loads, instead of 10). (Fan-out is discussed in Sec. 8.2.)

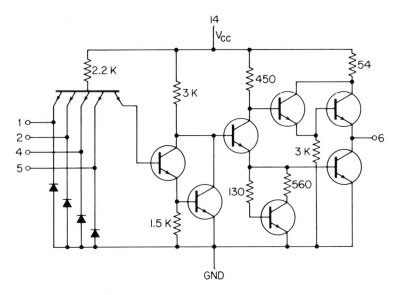

Fig. 8-7. Motorola MTTL power gate

TTL line drivers. IC line drivers are generally used as amplifiers to increase the fan-out capability of gates, without the use of power gates. Figure 8-8 shows the circuit of a NAND line driver; Fig. 8-9 shows a typical application of the circuit. Note that the gate output has 75 Ω resistors in series with the standard output (at pins 4 and 5), in addition to the direct output (at pin 6). These resistors provide for terminating the line.

Using an unterminated line driver, a line appears essentially as an open circuit at each end. Any pulse traveling down the line will see a reflection almost equal in magnitude to the original pulse. By terminating the loaded end, reflections and switching transients are minimized. For driving 93 Ω coaxial cable, or 120 Ω twisted pair, a good match can be made

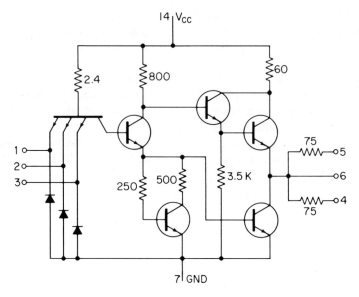

Fig. 8-8. Motorola MTTL terminated line driver

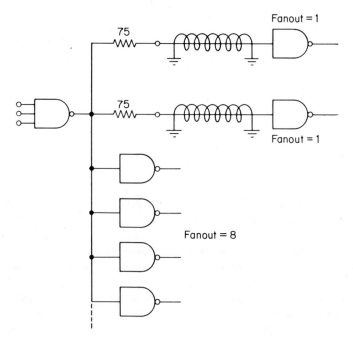

Fig. 8-9. Typical application of line driver

at the output of each resistor. For loads of 50 to 93 Ω, the two resistive outputs are shorted together for better impedance matching. The nonresistive output can be used to drive gates in a normal manner.

Open collector gates.   Most TTL gates have an active pull-up resistor at the output. This does not permit using a wired-OR operation. Special gates, such as the MTTL gate shown in Fig. 8-10, is included in many TTL product lines to overcome this limitation. The output of the Fig. 8-10 circuit can be used for wired-OR, or to drive discrete components.

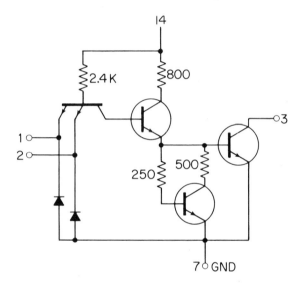

Fig. 8-10. Open collector Motorola MTTL gate for implementing the wired-OR function

Open impedance.   Some form of Darlington output is used for most TTL lines. The Darlington output configuration provides extremely low output impedance in the high state. The low impedance results in excellent noise immunity, and allows high-speed operation while driving large capacitive loads. Typically, the high state output impedance varies from about 10 Ω at outputs of 3.5 V, to about 64 Ω at lower output voltages.

The low state output impedance is typically 6 Ω at an output of 0.2 V, and goes up to about 500 Ω if the output increases to 0.5 V.

Totem-pole output.   One disadvantage of conventional TTL is the so-called "totem-pole" output. As shown in Fig. 8-4, both output transistors are on during a portion of the switching time. Since the turn-off time of a transistor is normally greater than the turn-on time, the following occurs:

In going from a high state to a low state on the output, transistor Q_4 is initially on, and is in the process of turning off. Transistor Q_5, at the same instant in time, is off and is attempting to turn on. Transistor Q_5 turns on before transistor Q_4 can turn off. The result is a current spike through both transistors and the load resistor. The same effect takes place when the conditions are reversed, and transistor Q_4 turns on before transistor Q_5 can turn off. The active bypass network in the MTTL line (transistor Q_6), shown in Fig. 8-5, helps to limit this problem.

8.1.5. Emitter-coupled logic (ECL)

ECL, shown in Fig. 8-11, operates at very high speeds (compared to TTL). Another advantage of ECL is that *both a true and complementary output is produced*. Thus, both OR and NOR functions are available at the output.

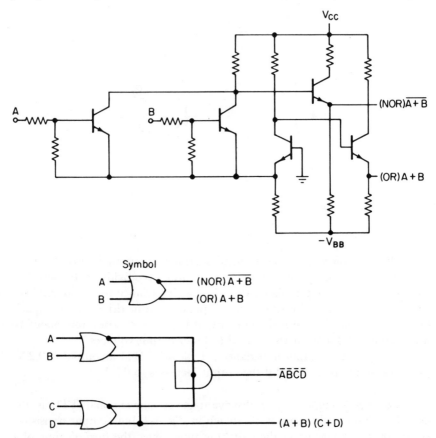

Fig. 8-11. Emitter-coupled logic (ECL)

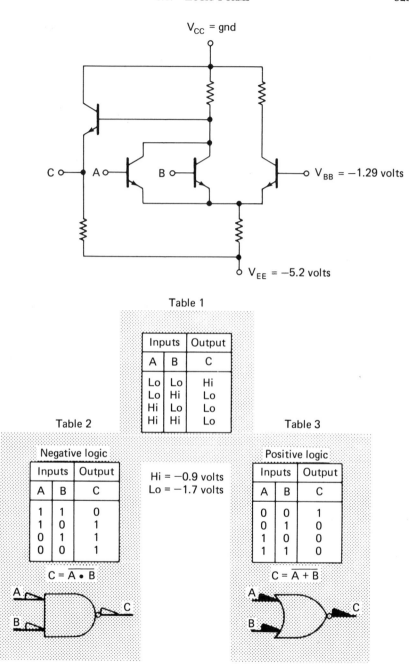

Fig. 8-12. Basic MECL gate circuit and logic function in positive and negative nomenclature

Note that when the NOR function of two ECL gates is connected in parallel, the outputs are ANDed, thus extending the number of inputs. For example, as shown in Fig. 8-11, when two two-input NOR gates are ANDed, the results are the same as a four-input NOR gate (or a four-input NAND gate in negative logic). When the OR functions of two ECL gates are connected in parallel, the outputs are ANDed, resulting in an OR/AND function.

The high operating speed is obtained since ECL uses transistors in the nonsaturating mode. That is, the transistors do not switch full-on or full-off, but swing above and below a given bias voltage. Delay times range from about 2 tc 10 ns. ECL generates a minimum of noise, and has considerable noise immunity. However, as a trade-off for the nonsaturating mode (which produces high speed and low noise), ECL is the least efficient. That is, ECL dissipates the most power for the least output voltage.

A typical ECL gate is shown in Fig. 8-12. The tables of Fig. 8-12 illustrate the logic equivalences of the ECL family. It is possible for some logic elements to have two equivalent outputs or functions, depending upon

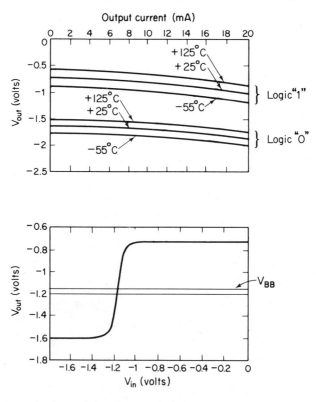

Fig. 8-13. Typical ECL transfer characteristics

logic definition (positive or negative logic). The ECL gate shown in Fig. 8-12 can be considered as a NAND in negative logic, or a NOR in positive logic.

Saturated logic families such as TTL have traditionally been designed with the NAND function as the basic logic function. However, in positive logic the basic ECL function is NOR. Thus, the designer may either design ECL systems with positive logic using the NOR, or design with negative logic using the NAND, whichever is more convenient. On one hand, TTL designers are familiar with positive logic levels and definitions. On the other hand, they are familiar with implementing systems using NAND functions.

For positive logic, a logic 1 for the circuit of Fig. 8-12 is about -0.9 V, which corresponds to one base-emitter voltage (V_{BE}) drop below ground. Logic 0 is -1.7 V, which yields a nominal voltage swing or switch of about 0.8 V ($1.7 - 0.9 = 0.8$). However, ECL lines are available with logic swings up to about 2 V. Some typical ECL transfer characteristics are shown in Fig. 8-13.

Bias problems. Unlike TTL (and other saturated logic), ECL requires a bias voltage V_{BB}. In the case of the Fig. 8-12 gate, the V_{BB} bias is -1.29 V, when the supply voltage V_{EE} is -5.2 V (with V_{CC} at ground or 0 V). With such a gate, if the power supply voltage is increased (V_{EE} increased by poor supply regulation, and so on), the 0 level will move more negative, while the 1 level remains essentially constant.

It is essential that the bias voltage V_{BB} track any variations in the power supply voltage. For this reason, some ECL manufacturers provide a *bias driver* with their ECL lines. An example of this is the Motorola bias driver shown in Fig. 8-14. The bias driver provides a temperature and

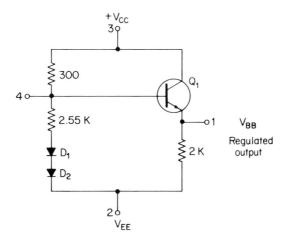

Fig. 8-14. Motorola MECL bias driver

voltage compensated reference for MECL logic. Any of the three MECL voltages may be grounded, but the common voltage of the bias driver must correspond to that of the logic system. If V_{BB} is obtained from the bias driver connected to the *same power supply* as the ECL logic element, the bias or reference voltage will track the supply voltage changes or temperature variations, thus keeping V_{BB} in the center of the logic levels.

Using ECL logic.   A fairly complete line of ECL devices is available in IC form. Here are the highlights or general rules for using ECL.

1. The maximum recommended ac fan-out (Sec. 8.2) for typical ECL is about 15 input loads. Direct-current fan-out is about 25 loads. The ac fan-out is lower than the dc fan-out because of the increase in rise time and fall time with high fan-out. Also, if high fan-outs and long leads are used, overshoot caused by lead inductances becomes a problem.

2. A circuit such as the bias driver (Fig. 8-14) will fan-out to about 25 loads. Note that a dual gate or half adder is equivalent to two gate input loads for a circuit such as the bias driver.

3. Each J or K input to a flip-flop is equivalent to one and one-half loads. For example, a J and K input tied together as a flip-flop clock input would be a load of three, allowing a gate (with an ac fan-out of 15) to drive five flip-flops. All other inputs are a load of unity (or one).

4. The output of two ECL gates may be tied together to perform the wired-OR function, in which case a maximum fan-out of 5 is allowed. If only one *pull-down* resistor (an emitter resistor in the output, rather than a *collector pull-up* resistor) is used, each additional common output is equivalent to one gate load. For example, if 6 gates are wired together with only one pull-down resistor connected, the fan-out is (15–5), or a fan-out of 10 remaining.

5. All unused inputs should be tied to V_{EE} for reliable operation (assuming that the power connections are as shown in Fig. 8-12). As seen from the gate input characteristics, the input impedance of a gate is very high when at a low level voltage. Any leakage to the input and/or wiring of the gate will gradually build up a voltage on the input. This may affect noise immunity of the gate or hinder switching characteristics at low repetition rates. Returning the unused inputs to V_{EE} ensures no buildup of voltage on the input, and a noise immunity dependent only upon the inputs used.

6. A recommended maximum of three input expanders should be used (assuming that each input expander provides 5 inputs). Thus, the recommended maximum input to any ECL gate is 15. If this is exceeded, the NOR output rise and fall times suffer noticeably be-

cause of the increased capacitance at the collector of the input transistors. For low frequencies, higher fan-ins may be used, if rise and fall times are of no significance.

7. Each gate in the IC package must have external bias supplies (except for certain ECL gates which have an internal bias scheme). ECL flip-flops do not normally require an external bias.

8.1.6. MOS logic ICs

The following paragraphs describe three present-day lines of MOS logic. These include: the McMOS, which is the trademark of Motorola Semiconductor Products, Inc.; complementary MOS devices; the COS/MOS, which is the RCA Solid State Division complementary MOS line; and the standard MOS/LSI line of Texas Instruments, Inc.

Keep in mind that these descriptions are for state-of-the-art equipment and are subject to change. That is, the details such as power requirements, propagation times, logic levels, and so on may change in the future. However, the basic principles for MOS logic devices, both standard lines and custom products, remain the same. A careful study of this section will enable you to interpret the future data of the manufacturers, as well as the data of other manufacturers.

Complementary MOS logic. Before going into the logic lines, let us consider how the complementary MOS principle (described in Sec. 1.2.2) is used to form a basic IC gate. Figure 8-15 shows how a NAND gate is formed in a MOS logic IC.

The p-channel devices are connected in parallel, and the n-channel complements are connected in series. The truth table for the three-input NAND gate is also given on Fig. 8-15. Note that the output swings from 0 V to $+V$ (supply voltage). If a supply voltage is 10 V or 15 V, the difference between a logic 0 and a logic 1 is approximately 10 V or 15 V. Thus the MOS logic device has a much greater noise immunity than the HTL described in Sec. 8.1.3. For this reason, and because MOS uses much less power and operates at higher speeds, HTL is generally being replaced by MOS.

For the NAND function, the output is always high unless all three inputs are high. If any one or any pair of inputs is high, one or more of the p-channel devices will be held ON by the remaining low inputs and the common output, but will be at $+V$. When all three inputs are high, all three series n-channels will be ON, and the output is low.

Note that the zero output level is developed across three series elements. However, the leakage current from all three of the p-channel devices is in the pA range, resulting in nV output levels (even for very large gates). For example, assume a p-channel leakage of 20 pA, and an n-channel

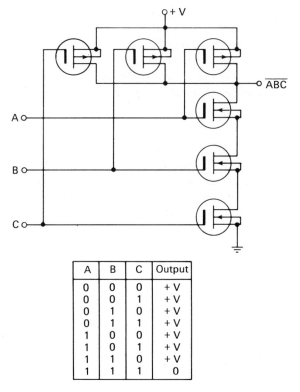

Fig. 8-15. Three-input NAND gate using MOS devices

resistance of 130 Ω. For a 50-input gate, the total leakage current is 1 nA (50 × 20 pA). Assuming a series output resistance of 6.5 kΩ (which is a very high output resistance), the resultant output voltage is 6.5 μV (which is an extremely low output voltage, particularly if the normal logic swing is from 0 to 10 or 15 V).

As with any solid-state logic device, the limitation on width of the NAND gate (or how long the gate may be held in the 1 or 0 condition) is set by decreasing switching speeds and increasing power dissipation (as the width increases).

8.1.7. Motorola Complementary MOS logic ICs

Figure 8-16 shows how the *p*-channel and *n*-channel devices are connected to form the basic element of McMOS. Figure 8-17 gives the V_{in} versus V_O transfer curve for the basic inverter at a power supply voltage of 10 V. Note that the logic pulse output is approximately equal to the supply voltage.

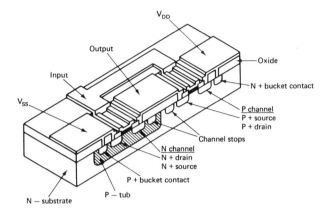

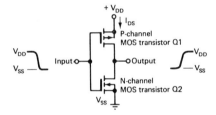

Fig. 8-16. Basic McMOS p-channel and n-channel devices connected to form an inverter

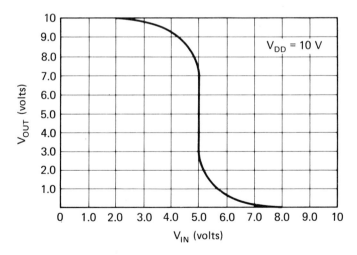

Fig. 8-17. McMOS inverter transfer curve

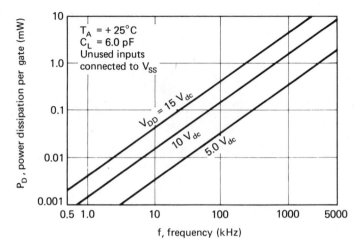

Fig. 8-18. Power dissipation of CMOS inverter

Dynamic characteristics. Because the dynamic power dissipation of a complementary MOS logic device results from capacitive loading, power dissipation is also a function of the frequency at which the capacitance is charged and discharged. Figure 8-18 illustrates this relationship for a basic inverter. It can be seen that power dissipation is linear with frequency. Figure 8-19 shows the effects of capacitive loading and power supply voltage on propagation delays. Higher operating speeds are possible at higher supply voltages, at the expense of power dissipation.

The threshold of the inverter is about 45 percent of the supply voltage. That is, the inverter will switch over when the input logic signal (or pulse) is about 45 percent of the supply voltage. The output voltage then switches from zero to almost 100 percent of the supply voltage. Thus, the MOS logic can be operated over a wide range of supply voltages (5, 10 and 15 V are typical). Also, since the output is equally isolated from both V_{DD} and V_{SS} terminals, McMOS can operate with negative as well as positive supplies. The only requirement is that V_{DD} be more positive than V_{SS}.

McMOS transmission gate. A second important building block for complementary McMOS circuits is the transmission gate shown in Fig. 8-20. When the transmission gate is ON, a low resistance exists between the input and the output, allowing current flow in either direction.

The voltage on the input line must always be positive with respect to the substrate V_{SS} of the n-channel device, and negative with respect to the substrate V_{DD} of the p-channel device. The gate is ON when the gate G_1 of the p-channel is at V_{SS}, and the gate G_2 of the n-channel is at V_{DD}. When G_2

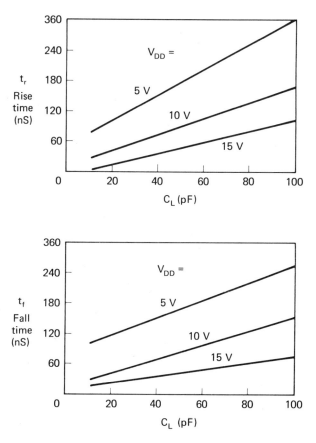

Fig. 8-19. Typical delay characteristics of two-input McMOS NOR gate

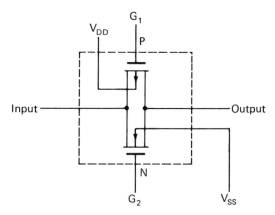

Fig. 8-20. Basic McMOS transmission gate

is at V_{SS} and G_1 is at V_{DD}, the transmission gate is OFF, and a resistance of greater than 10^9 ohms exists between input and output.

The resistance between the input and output of a basic transmission gate in the ON condition is dependent upon the voltage applied at the input, the potential difference between the two substrates ($V_{DD} - V_{SS}$), and the load on the output. Typically, R_{ON} is defined as the input-to-output resistance with a 10 kΩ load resistor from the output to ground. With voltages between the two extremes, both devices are partially ON and the value of R_{ON} is caused by the parallel resistance of the p- and n-channel devices.

Transmission gate applications.   An illustration of use for the basic transmission gate is given in the McMOS MC14013 flip-flop, shown in Fig. 8-21. The flip-flop works on the master-slave principle and consists of four

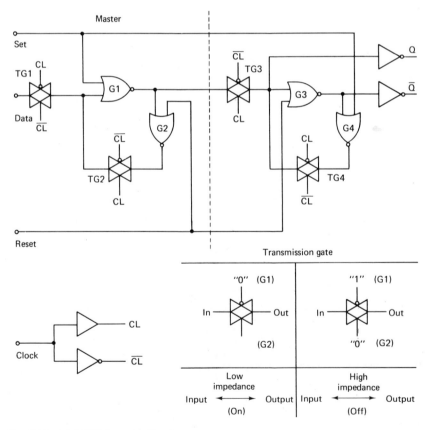

Fig. 8-21. McMOS type-D flip-flop

transmission gates, as well as four NOR gates, two inverters, and a clock buffer/driver.

When the clock is at a logic 0, transmission gates TG_2 and TG_3 are OFF; TG_1 and TG_4 are ON. In this case, the master is logically disconnected from the slave. With TG_4 ON, gates G_3 and G_4 are cross-coupled and latched in a stable state.

Assuming that the SET and RESET inputs are low, the logic states of G_1 and G_2 are determined by the logic changes to a logic 1. Under these conditions, TG_2 and TG_3 turn ON, and TG_1 and TG_4 turn OFF. Gates G_1 and G_2 are cross-coupled through TG_2, and the gates latch into the state in which they existed at the time the clock changed from a 0 to a 1. With TG_3 ON, the logic state of the master section (output of gate G_1) is fed through an inverter to the Q output, and through G_3 and another inverter to the $\bar{Q}$ output.

When the clock returns to a logic 0, TG_3 turns OFF, and TG_4 turns ON. This disconnects the slave from the master and latches the slave into the state that existed in the master when the clock changed from a 1 to a 0. Thus, information is entered into the master on the positive edge of the clock. When the clock is high, the output of the master is transmitted directly through the slave to Q and $\bar{Q}$. When the clock changes back to a low state, the state of the master is stored by the slave which then provides the output.

8.1.8. RCA COS/MOS complementary MOS logic devices

COS/MOS fundamentals can best be understood by reference to Fig. 8-22. The typical characteristics are shown in Fig. 8-23. The basic logic inverter (or logic gate) formed by use of p- and n-type devices in series is illustrated in Fig. 8-24.

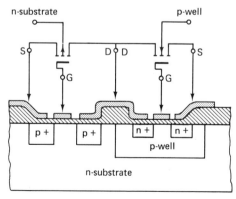

Fig. 8-22. Cross section of COS/MOS device

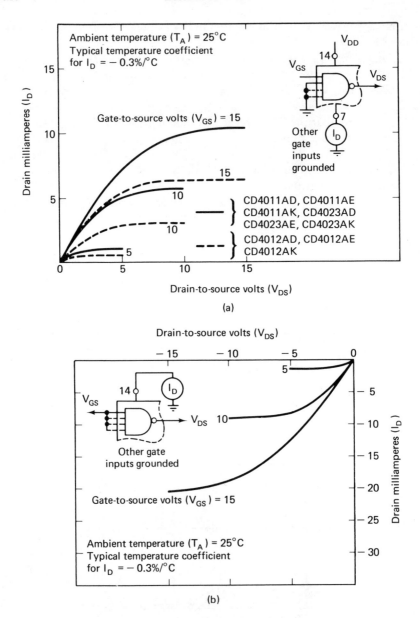

Fig. 8-23. Typical *n*-channel and *p*-channel characteristics of COS/MOS devices

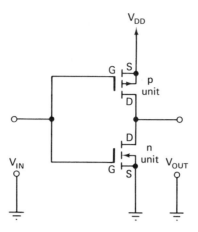

Fig. 8-24. Basic COS/MOS inverter

Quiescent device dissipation. When the input load is grounded, or otherwise connected to V (logic 0), the *n*-device is cut off, and the *p*-device is biased on. As a result, there is a low-impedance path from the output to V_{DD}, and an open circuit to ground. The resultant output becomes essentially zero volts (logic 0).

Note that one of the devices is always cut off at either logic extreme, and that no current flows into the insulating gates, resulting in negligible inverter quiescent power dissipation (equal to the product of V_{DD} times the leakage current).

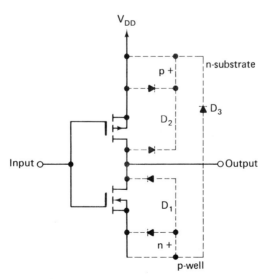

Fig. 8-25. Basic inverter showing parasitic diodes

A cross section of the COS/MOS inverter (as it is formed in an IC) on an n-type substrate is illustrated in Fig. 8-22. Compare this with Fig. 8-25. Note that the source-drain diffusions and the p-well diffusion form *parasitic diodes* (in addition to the desired transistors). (Keep in mind that the parasitic diodes are not to be confused with the protective diodes.) The parasitic diode elements are back-biased (across the power supply) and contribute, in part, to the device leakage current, and thus to the quiescent power dissipation.

Power dissipation of product line.   The RCA COS/MOS product line consists of circuits of varying complexity (from the dual 4-input logic gates that contain 16 MOS devices to the more complex 64-bit static shift registers that contain over 1000 devices). Some logic gates are specified to operate with a typical power dissipation of 5 nW ($V_{DD} = 10$ V). The more complex devices, such as a 7-stage counter or register, are specified to operate with a typical power dissipation of 5μW ($V_{DD} = 10$V). Published data include both typical device quiescent-current levels and maximum levels ($V_{DD} = 5$ V and $V_{DD} = 10$ V).

Switching characteristics.   The signal extremes at the input and output are approximately zero volts (logic 0) and V_{DD} (logic 1). The switching point is typically 45 to 55% of the magnitude of the power supply voltage (regardless of the magnitude) over the entire range (typically from 5 to 15 V).

Ac dissipation characteristics.   During the transition from a logic 0 to a logic 1, both devices are momentarily ON. This condition results in a pulse of instantaneous current being drawn from the power supply, the magnitude and duration of which depends upon the following factors:

1. The impedance of the particular devices being used in the inverter circuit;
2. The magnitude of the power supply voltage;
3. The magnitude of the individual device threshold voltages;
4. The input driver rise and fall times.

An additional component of current must also be drawn from the power supply to charge and discharge the internal parasitic capacitances and the load capacitances seen at the output.

The device power dissipation resulting from these current components is a *frequency-dependent parameter*. The more often the circuit switches, the greater the resultant power dissipation; the heavier the capacitive loading, the greater the resultant power dissipation. The power dissipation is not duty-cycle dependent. For practical purposes, power dissipation can be considered frequency (repetition rate) dependent.

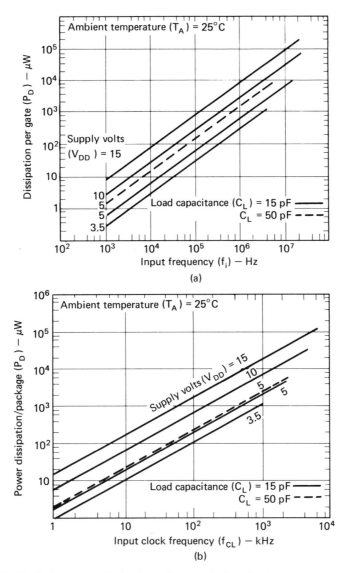

Fig. 8-26. Typical power dissipation characteristics: (a) basic gate power dissipation characteristics; (b) MSI device power dissipation characteristics

Because the COS/MOS product line ranges widely in circuit complexity from device to device, the ac device dissipations vary widely. The effect of capacitive loading on the individual devices also varies. Figure 8-26 shows a family of curves for a typical gate device, and a typical MSI (medium-scale integration) device. These curves illustrate how device power dissipation varies as a function of frequency, supply voltage, and capacitive loading. Note how the MOS ICs require more power at higher frequencies.

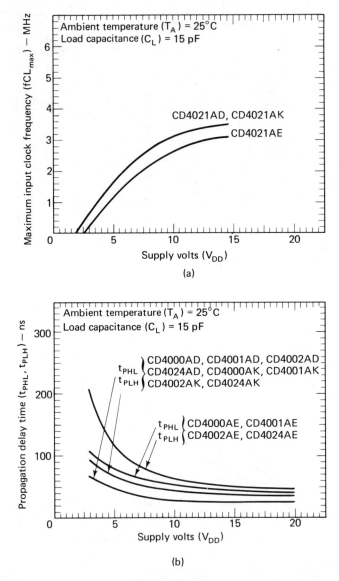

Fig. 8-27. Operating frequency and propagation delay as a function of power supply voltage: (a) maximum guaranteed operating frequency as a function of power-supply voltage; (b) propagation delay as a function of power-supply voltage for the basic gate

Ac performance characteristics. During switching, the capacitances within a given device and the load capacitances external to the circuit are charged and discharged through *p*- and *n*-type device conducting channels. As V_{DD} increases, the impedance of the conducting channel decreases accordingly. This lower impedance results in a shorter *RC* time constant (this nonlinear property of MOS ICs can be seen in the curves of Fig. 8-23). The result is that the maximum switching frequency of a COS/MOS device increases with increasing supply voltage (as seen in Fig. 8-27).

Figure 8-27b shows curves of propagation delay as a function of supply voltage for a gate IC. However, the trade-off of low supply voltage (lower output current to drive a load) is lower speed of operation.

Calculating system power. The following guidelines have been developed to assist the logic designer in estimating system power for the COS/ MOS line. The same general guidelines can be applied to similar MOS lines.

The total system power is equal to the sum of quiescent power and dynamic power. Therefore, system power can be calculated with the following approach:

1. Add all typical IC package power dissipations, using published data. Because quiescent power dissipation is equal to the product of quiescent device current multiplied by supply voltage, quiescent power may also be obtained by adding all typical quiescent IC currents, and multiplying the sum by the supply voltage V_{DD}. Quiescent device current is shown in the published COS/MOS data for supply voltages of 5 V and 10 V only.

2. Add all dynamic power dissipations using typical curves of dissipation per package (such as Fig. 8-26) as a function of frequency. In a fast-switching system, most of the power dissipation is dynamic, therefore quiescent power dissipation may be neglected. That is, since the inverters are in a transition state most of the time, the dynamic dissipations govern the total power dissipation.

8.1.9. Texas Instruments MOS/LSI products

The Texas Instruments line of MOS devices includes shift registers, read-only memories (ROM), programmable logic arrays (PLA), and random access memories (RAM), as well as special purpose ICs such as buffers, switches, and custom MOS/LSI ICs. The MOS/LSI shift registers are used as an example.

MOS/LSI shift registers.   Although MOS shift registers perform the same functions as TTL and other registers, the MOS registers operate on different principles. Flip-flops as such are not used in MOS registers. The following descriptions apply to the Texas Instruments MOS/LSI line. However, similar principles are used by other manufacturers of MOS registers.

Basic configuration.   MOS shift registers can be supplied in the following configurations: serial-in/serial-out, parallel-in/serial-out, and serial-in/parallel-out. The serial-in/serial-out configuration is by far the most popular. A MOS register is able to store *n* bits, each on a *basic cell* consisting of two MOS inverters and timing devices, as shown in Fig. 8-28.

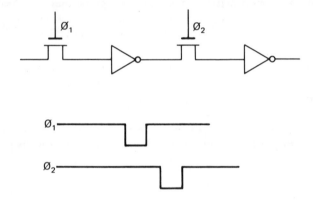

Fig. 8-28. Basic MOS/LSI shift register cell

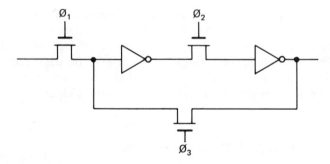

Fig. 8-29. Basic MOS/LSI static shift register

Static versus dynamic.   Dynamic shift registers use two independent inverters (not cross-coupled). The information is stored temporarily on a capacitor inherent to the MOS device (the gate capacitance). The device cannot be operated below a certain clock frequency or the data storage will be lost.

A *static shift register* (Fig. 8-29) operates in the same way as a dynamic shift register, as long as the frequency is high. The two inverters used in a static shift register are the static type (unclocked load). When the frequency falls below a certain level, a third phase is generated internally, and this signal is used to close a feedback loop between the output of the second inverter and the input of the first inverter. Comparing the two, the dynamic shift registers are faster and use less power than static shift registers. However, dynamic shift registers are not as flexible to use in a system.

Static shift registers. As shown in Fig. 8-30, a static shift register uses two static MOS inverters. Three phases (or clocks) are necessary for operation. The third-phase clock is always *generated internally*, and is used to time the feedback loop. The second-clock phase is often generated internally.

In the basic cell of Fig. 8-30, A and B are storage elements. The device operates dynamically except when phase 3 is on. Phase 3 is present only when phase 1 is at logic 0 and phase 2 is at logic 1 for more than 10 μs. This condition must be maintained for long-time data storage. Static shift registers typically operate in the 0 to 2.5 MHz clock range, are extremely flexible, and can hold data indefinitely (as long as power is supplied).

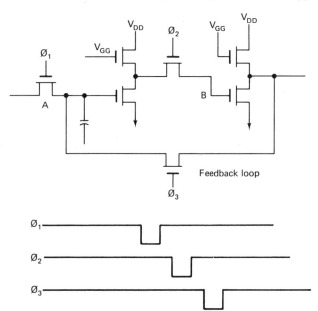

Fig. 8-30. Basic cell of MOS/LSI static shift register

Dynamic shift registers.   Dynamic shift registers use either two or four phases (or clocks). These phases can be generated on the chip or supplied externally. Two-phase shift registers can be classified as *ratio* or *ratioless* circuits.

The two-phase ratio-type shift register (Fig. 8-31) consists of two simple dynamic inverters and timing devices. When phase 1 is at a logic 1 (low), the capacitance C_1 charges at the inverse of the data input. Information is transferred out when phase 2 goes to 1.

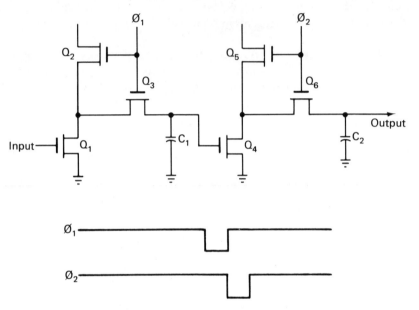

Fig. 8-31. Basic cell for MOS/LSI dynamic shift register

In a ratio-type circuit, current flows through the inverter when the clock and data input are at a logic 1 simultaneously. There must be a certain minimum ratio between the size of the two transistors in the inverters (typically 5 to 1), requiring more chip area than in a ratioless shift register in which the MOS devices are usually identical in size.

The *two-phase ratioless dynamic shift register* (Fig. 8-32) has been designed to decrease the power dissipation and chip area. This ratioless register uses identical transistors throughout, and can thus work at higher clock rates because the precharging paths are of lower impedance than those in the circuit. When phase 1 goes to 1, C_2 charges to 1 via Q_3, and C_1 charges to the data input level via Q_1. When phase 1 returns to zero, Q_2 turns on if the input level was a 1. This discharges C_2.

For a 0 input, Q_2 stays off and C_2 is not discharged. Under these conditions, phase 2 goes to a 1 and turns on Q_4 so that C_2 shares any charge

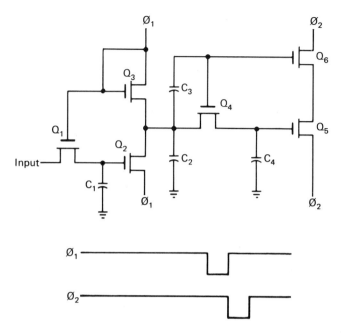

Fig. 8-32. MOS/LSI two-phase ratioless dynamic shift register

it has with C_4. Capacitor C_3 is used to compensate for the loss of potential across C_2 by introducing a small extra charge on the negative edge of phase 2. However, the small charge does not introduce enough energy to destroy a logic 0 on C_2. When phase 2 returns to a zero, the charge on C_4 transfers the data-input level to the output.

Register design characteristics. When selecting an IC register for some particular design purpose, several factors must be considered. Obviously, the power supply and interface characteristics must be compatible with the system in which the IC register is to be used. The following points should also be considered.

Keep in mind that interface (Secs. 8.4 through 8.8) not only includes logic levels, but pulse timing and spacing. For example, Fig. 8-33 shows the timing diagrams and recommended operating conditions for a Texas Instruments MOS/LSI 512-bit dynamic shift register. A quick study of this information shows that data is transferred into the register when the phase 1 clock is low (nominally −12 V). The data must set up at least 100 ns before the clock phase 1 goes high (+5 V) and held steady at least 20 ns after phase 1 reaches this state. Also, output delay time is defined as the time required for the output to reach the TTL changeover threshold after the phase 2 clock reaches 90 percent of its low voltage. This time is shorter than 100 ns.

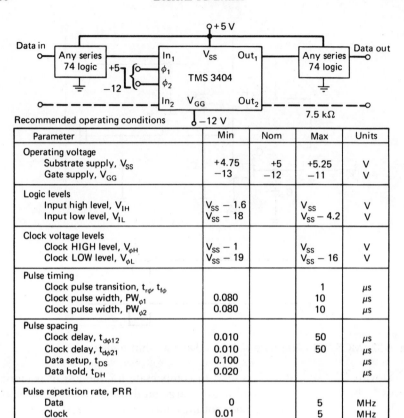

Recommended operating conditions

Parameter	Min	Nom	Max	Units
Operating voltage				
Substrate supply, V_{SS}	+4.75	+5	+5.25	V
Gate supply, V_{GG}	−13	−12	−11	V
Logic levels				
Input high level, V_{IH}	$V_{SS} - 1.6$		V_{SS}	V
Input low level, V_{IL}	$V_{SS} - 18$		$V_{SS} - 4.2$	V
Clock voltage levels				
Clock HIGH level, $V_{\phi H}$	$V_{SS} - 1$		V_{SS}	V
Clock LOW level, $V_{\phi L}$	$V_{SS} - 19$		$V_{SS} - 16$	V
Pulse timing				
Clock pulse transition, $t_{r\phi}$, $t_{f\phi}$			1	μs
Clock pulse width, $PW_{\phi 1}$	0.080		10	μs
Clock pulse width, $PW_{\phi 2}$	0.080		10	μs
Pulse spacing				
Clock delay, $t_{d\phi 12}$	0.010		50	μs
Clock delay, $t_{d\phi 21}$	0.010		50	μs
Data setup, t_{DS}	0.100			μs
Data hold, t_{DH}	0.020			μs
Pulse repetition rate, PRR				
Data	0		5	MHz
Clock	0.01		5	MHz

Timing diagram and voltage waveforms

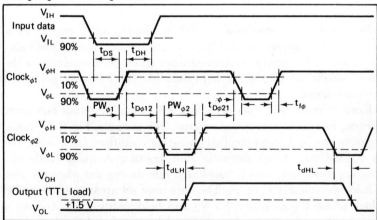

Fig. 8-33. Timing diagrams and operating conditions for a typical MOS/LSI 512-bit dynamic shift register

Often, characteristics such as those shown in Fig. 8-33 are critical to operation of registers (and counters) when used in a system. For this reason, the datasheets must always be consulted.

8.2. SELECTING LOGIC INTEGRATED CIRCUITS

One of the main problems for the logic IC user is to choose the right logic family for a given application. In some cases, price is the all-important factor. In other circumstances, a particular logic function (such as a complete electronic calculator board) may be available in only one logic family. In still other cases, there is a specific design problem (high-speed operation, noise immunity, and so on) that requires a certain logic line.

Before selecting any logic family, the designer should study the characteristics of all available types. At present, there are six basic logic families, three of which (TTL, ECL, and MOS) dominate the field for design of new equipment. Each of these is discussed in Sec. 8.1. It is assumed that the designer will read this material and the discussions showing applications of the various logic families in Chapter 9. Designers should also study all available datasheets for logic ICs that might suit their requirements. Then, and only then, will the designer be in a good position to choose the right logic IC. However, the designer should keep the following points in mind when reading all of the data.

8.2.1. Availability and compatibility

TTL is the most available of all logic families. That is, at present it is possible to obtain the greatest variety of off-the-shelf logic ICs in the TTL family. Some designers consider TTL as the "universal" IC logic family, since they can obtain an infinite number of gates (with a variety of input/output combinations), buffers, inverters, counters, registers, arithmetic units, and so on, from more than one manufacturer.

RTL and DTL were, at one time, the next most available logic ICs. Today, both RTL and DTL have been replaced by TTL. RTL offers no advantage over TTL. DTL has a slightly higher logic swing (voltage differential from a logic 0 to a logic 1) than TTL (typically 4.5 V for DTL compared to about 3.5 V for RTL). Also, DTL requires slightly less power than TTL, but the operating speeds of DTL are so much slower than TTL (typically TTL is three times faster) that TTL is the preferred family for any present-day application.

HTL is used only where a high logic swing (about 13 V) and high noise immunity is required. Except for this one advantage, HTL is generally of little value to the modern designer. HTL is the slowest and requires the most power of all logic families. In most applications, HTL can be replaced by MOS, since MOS can provide the same logic swing with far less power and at higher speeds. For example, if MOS is operated at a supply voltage of 15 V, the logic swing can also be almost 15 V.

ECL is used primarily where high speed is essential. ECL is still the fastest of all logic families. The disadvantages of ECL are high power consumption (the highest next to HTL), and a low logic swing (usually less than 1 V, but some ECL will provide nearly 2 V).

MOS is the newest of the commonly used logic families. The advantages are low power consumption (the lowest of all families), a logic swing equal to any family, and small size. Stated another way, you can get more MOS devices or functions on a given area than any other family. If large-scale integration (LSI) is required, MOS is the best choice. TTL and ECL are limited to MSI (medium-scale integration).

TTL and DTL are directly compatible with each other. Thus, some designers use only these two logic families. Or, they design with TTL only when they must adapt new logic circuits to existing DTL systems.

RTL is the most compatible with linear and analog systems, or any discrete transistor application. This is because RTL is essentially an IC version of conventional solid-state circuits.

Because of their special nature, ECL and HTL are the least compatible with other logic families and with external devices. ECL requires a large supply voltage for a comparatively small logic swing, while HTL produces a very large logic swing that is generally too high for other families. MOS can be made compatible with other families, even though the MOS operating principles are quite different.

Keep in mind, that barring some unusual circumstance any logic family can be adapted for use with other logic families or external equipment by means of *interface circuits*. These interface circuits are discussed in Secs. 8.4 through 8.8.

8.2.2. Noise considerations

HTL has the highest noise immunity of any logic family, with the possible exception of MOS. (Or, HTL has the least noise sensitivity, whichever term you prefer.) Typically, signal noise up to about 5 V will not affect HTL. Often, HTL can be used without shielding in noise environments where other families require extensive shielding. MOS devices trigger at about 45 to 50 percent of the supply voltage. If MOS logic is operated at 15 V, the devices will trigger at about 7 V, and will not be affected by lower voltages. Thus, MOS can have higher noise immunity than HTL, if the supply voltage can be kept at 15 V.

RTL has the lowest noise immunity (or is the most noise sensitive). Typically, signal noise in the order of 0.5 V can affect RTL. Considering that RTL operates at logic levels of 1 V, an RTL system is almost operating near the noise threshold. As a result RTL is not recommended for noisy environments.

DTL and TTL are about the same in regard to noise immunity or sensitivity. ECL can be operated so that the noise immunity is about equal to that of DTL and TTL. The input of ECL is essentially a differential amplifier. If one base is connected to a fixed-bias voltage about half way between the logic 1 and 0 levels, this sets the noise immunity at the level of the bias voltage.

Power supply and ground line noise. In addition to signal line noise, logic ICs are affected by noise on the power supply and ground lines. Most of this noise can be cured by adequate bypassing as described in Sec. 2.4.1. However, if there are heavy ground currents because of large power dissipation by the ICs, it may be necessary to use separate ground lines for power supply and logic circuitry.

Noise generation. In addition to noise immunity or sensitivity, the generation of noise by logic circuits must be considered. Whenever a transistor or diode switches from saturation to cutoff, and vice versa, large current spikes are generated. These spikes appear as noise on the signal lines, as well as the ground and power lines. Since ECL does not operate in the saturation mode, it produces the least amount of noise. Thus, ECL is recommended for use where external circuits are sensitive to noise. On the other hand, TTL produces considerable noise, and is not recommended in similar situations.

8.2.3. Propagation delay and speed

The speed of an IC logic system is inversely proportional to the propagation delay of the IC elements. That is, ICs with the shortest propagation delay can operate at the highest speed. Since ECL does not saturate, the delay is at a minimum (typically 2–4 ns), and speed is maximum. TTL is the next-to-fastest IC logic family, with delays of about 10 ns, and can be used in any application except where the extreme high speed of ECL is involved. MOS is the slowest of the three currently popular families (TTL, ECL, and MOS). However, MOS is considerably faster than HTL.

8.2.4. Power source and dissipation

MOS requires the least power consumption of all IC logic families, and can be operated over a wide range of power supply voltages (typically 5, 10 or 15 V). CMOS is well suited for battery-operated systems, since little standby power is required. CMOS logic uses power when switching from one state to another, but not during standby (except for some power consumed by leakage).

TTL and ECL generally operate with a 5 V supply, and consume about 15 and 25 mW (per gate), respectively. Power consumption of a MOS gate is generally figured in the μW range, rather than mW.

8.2.5. Fan-out

Fan-out, or the number of load circuits that can be driven by an output, is always of concern to logic designers. Some IC datasheets list fan-out as simple number. For example, a fan-out of three means that the IC output will drive three outputs or loads. While this system is simple, it may not be accurate. Usually, the term fan-out implies that the output will be applied to inputs of the same logic family, and the same manufacturer. Other datasheets describe fan-out (or load and drive) in terms of input and output current limits. (This system is discussed in Sec. 8.3.)

Aside from these factors, the following typical fan-outs are available from the logic families: RTL 4–5, DTL 5–8, TTL 5–15, HTL 10, ECL 25, MOS 10.

8-3. INTERPRETING LOGIC IC DATASHEETS

Logic (or digital) IC datasheets are presented in various formats. However, there is a general pattern used by most manufacturers. For example, most logic IC datasheets are divided into four parts. The first part usually provides a logic diagram and/or circuit schematic, plus truth tables, logic equations, general characteristics, and a brief description of the IC. The second part is devoted to test, and shows diagrams of circuits for testing the ICs. These two parts are fairly straightforward and are usually easy to understand.

This is not necessarily true of the remaining two parts (or one large part in some cases), which have such titles as "maximum limits" or "maximum ratings" and "basic characteristics" or "electrical characteristics." These parts are generally in the form of charts, tables, or graphs (or combination of all three), and often contain the real data needed to design with logic ICs. The terms used by manufacturers are not consistent. Likewise, a manufacturer may use the same term to describe two slightly different characteristics, or use two different terms to describe the same characteristic when different lines or different families are being discussed.

It is impractical to discuss all characteristics found on logic IC datasheets. However, the following notes should help the designer interpret the most critical values.

8.3.1. Electrical characteristics

It is safe to assume that any electrical characteristic listed on the datasheet has been tested by the manufacturer. If the datasheet also specifies the test conditions under which the values are found, the characteristic can be of immediate value to the designer. For example, if leakage current is measured under worst-case conditions (maximum supply voltage and maximum logic input signal), the leakage current shown on the datasheet can be used for design. However, if the same leakage current is measured with no signal input (inputs grounded or open), the leakage current is of little value. The same is true of such factors as output breakdown voltage and maximum power supply current.

To sum up, if an electrical characteristic is represented as being tested under typical (or preferably worst-case) operating conditions, it is safe to take that characteristic as a design value. If the characteristic is measured under no-signal conditions, it is probably included on the datasheet to show the relative merits of the IC. When in doubt, test the IC as described in the datasheet, and use the test results (not the datasheet values) for design.

In the case of quiescent power values for a complementary MOS device, always use the maximum value, rather than the typical value, if both are given. Also, consider clock times, or logic operating speed, when calculating power. In a complementary MOS system, most of the power is dissipated *during transition*. Thus, if transition is slow because of long clock pulses, average power consumption is increased.

8.3.2. Maximum ratings

Maximum ratings are values that must never be exceeded in any circumstance. They are not typical operating levels. For example, a maximum rating of 15 V for V_{CC} means that if the regulator of the supply system fails, and the V_{CC} source moves up from the normal 8–10 V to 15 V, the IC will probably not be burned out. But never design the system for a normal V_{CC} of 15 V. Allow at least a 10 to 20 percent margin below the maximum, and preferably a 50 percent margin for power supply (voltage and current) limits. Of course, if typical operating levels are given, these can be used even though they are near the maximum ratings.

8.3.3. Drive and load characteristics

Generally, the most important characteristics of logic ICs (from the designer's standpoint) are those that apply to the output drive capability, and the input load presented by an IC. No matter what type of logic is involved, the designer must know how many inputs can be driven from one IC output (without amplifiers, buffers, and so on). It is equally important to know that kind of load is presented by the input of an IC on the output of the previous stage (either IC or discrete component). As discussed in Sec. 8.2.5, fan-out is a simple, but not necessarily accurate, term to describe drive and load capabilities of an IC.

A more accurate system is where actual input and output currents are given. There are four terms of particular importance:

Output logic 1-state source current I_{OH}

Output logic 0-state sink current I_{OL}

Input forward current I_F

Input reverse current I_R

The main concern is that the datasheet value of I_{OL} must be equal to or greater than the combined I_F value of all gates (or other circuits) connected to an IC output. Likewise, I_{OH} must be equal to or greater than total I_R.

Unfortunately, the same condition exists for datasheet load and drive factors, as exists for other electrical characteristics; the values are not consistent from family to family, and for different manufacturers.

8.3.4.   Interfacing logic ICs

No matter what load and drive characteristics are given on the datasheet, it may be necessary to include some form of interfacing between logic ICs to provide the necessary drive current. This is especially true when the datasheet shows a very close tolerance. For example, assume that an IC with a rated fan-out of 3 is used to drive three gates. If the fan-out rating is typical or average, and the three gates are operating in their worst-case condition, the IC may not be able to supply (and dissipate) the necessary current. Because of its importance to proper IC logic design, the remainder of this chapter is devoted to interfacing.

No matter what drive and load characteristics are involved for a logic IC, it may be necessary to provide the necessary drive current, to change the logic levels, and so on with an interfacing circuit or device. It is not practical to have a universal interfacing circuit for all logic families of all IC manufacturers. Likewise, it is not even possible to have a universal circuit for interface between a given logic family of one manufacturer, and all other logic families of the same manufacturers. Nor is it possible to have a single interface circuit that will accommodate the same logic family of different manufacturers. For one thing, the logic voltage levels (or 0 and 1), the supply voltages, and the temperature ranges vary with manufacturers, and with logic families. For that reason, most logic IC manufacturers publish interfacing data for their particular lines.

We will make no attempt to duplicate all this data here, but instead, concentrate on the interface requirements for the most popular lines, and the most used logic families. A careful study of this information should provide the designer with sufficient background to understand the basic problems involved with interfacing logic systems, and to interpret interfacing data that appears on logic IC datasheets.

8.4. BASIC INTERFACING CIRCUIT

Although it is not practical to have a universal interfacing circuit for all logic ICs, the basic circuits of Fig. 8-34 should provide enough information to design interfacing between some IC logic elements. The equations shown in Fig. 8-34 are used to find the approximate or trial value of pull-up resistor R_1. The following is an example of how to use the equations.

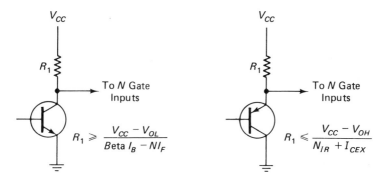

Fig. 8-34. Basic interfacing circuit for logic IC elements

Assume that the common-collector circuit is used, that V_{CC} is 5 V, that V_{OH} (high or logic-1 state voltage) is 3 V, that there are 7 gate inputs, and each has a forward current of 1 mA, and that I_{CEX} is 3 mA. Note that I_{CEX} is the input leakage current as tested with a voltage between base and emitter. I_{CEX} is a worst-case or active datasheet value, and is not to be confused with I_{CER}, which is input leakage current with no signal (tested with a fixed resistance between emitter and base).

Using the equation of Fig. 8-34, the value of R_1 is:

$$R_1 = \frac{5 - 3 \text{ V}}{(7 \times 1 \text{ mA}) + 3 \text{ mA}} = 200 \ \Omega$$

The next lowest standard value is 180 Ω.

Two cautions must be observed when using the circuits of Fig. 8-34. First, do not try to increase the output voltage of an IC (with an interface) to a level higher than the V_{CC} of either IC (input or output). For example, if both ICs have a V_{CC} of 5 V, keep the output of the interface transistor below 5 V, even if the IC datasheet may show that inputs greater than 5 V are safe. This is because many ICs have a diode between resistors in the circuit and the power supply terminal (with the cathode of the diode connected to the power supply side).

Also, do not pull the interface circuit output below ground. In many ICs, each internal transistor has a diode connected between the collector and ground terminal. The diode polarity is such that the diode is normally reverse-biased (anode connected to ground). However, the diode will be forward-biased if the output is pulled below ground. Unfortunately, the datasheet schematics do not always show these diodes, even though they exist in the circuit. To be safe, *keep the interface input and outputs above ground and below V_{CC}.*

Sometimes, there are signal problems with interface circuits, particularly when high speeds are involved. Interface circuits slow down rise and fall times of the logic pulses. This slowdown can cause saturated logic elements (which affects just about every family but ECL) to break into oscillation on the leading and trailing edges of the logic pulses.

The basic rule to follow is that the *rise and fall times of an input to any logic IC must be shorter than the typical propagation delay time of the IC.* No oscillation should occur if this rule is followed. Note that the terms rise and fall time refer to the time required to go from a 1 state to a 0 state, and vice versa. These rise and fall times are somewhat longer than the conventional 10 and 90 percent values for pulse measurement.

8.5. MOS INTERFACE

Many designs presently under consideration use both MOS and two-junction technologies in order to take full advantage of the low-cost and high-packaging density of MOS (such as MOS/LSI), as well as the flexibility of two-junction techniques for low-complexity functions. The interface of different logic families requires that the circuits operate at a common-supply voltage, and have logic-level compatibility. In addition, the devices must maintain safe power dissipation levels, and good noise immunity over the operating temperature range. In short, there must be system compatibility with all logic families used in the overall system.

8.5.1. MOS/LSI system compatibility

The following notes apply to the MOS/LSI line of Texas Instruments.

Power supplies. Two manufacturing technologies are common in MOS/LSI, and common throughout the industry: high-threshold and low-threshold MOS. The power supply requirements are:

	V_{SS}	V_{DD}	V_{GG}
high threshold	0	-12 V	-24 V
low threshold	0	-5 V	-17 V

where V_{SS} is the substrate supply,

V_{DD} is the drain supply,

and V_{GG} is the gate supply.

The drain supply will draw most of the current. Some circuits are designed to use only one power supply (saturated logic). V_{DD} and V_{GG} are then common.

To use MOS in a system, it is often convenient to translate all of the power supply voltages to a certain voltage. A common arrangement is:

	V_{SS}	V_{DD}	V_{GG}
high threshold	$+12$ V	0	-12 V
low threshold	$+5$ V	0	-12 V

Some high-threshold devices are specified at $V_{GG} = -28$ V and $V_{DD} = -14$ V.

Input compatibility.   Referencing all voltages to V_{SS}, the input swing on most MOS circuits is as follows:

	HIGH LEVEL	LOW LEVEL
High threshold	0 to -3 V	-9 to -24 V
Low threshold	0 to -1.5 V	-4.2 to -17 V

In relation to the translated power supplies, the input swing becomes:

	HIGH THRESHOLD	LOW THRESHOLD
V_{SS}	$+12$ V	$+5$ V
V_{DD}	0 V	0 V
V_{GG}	-12 V	-12 V
High level	$+9$ to $+12$ V	$+3.5$ to $+5$ V
Low level	$+3$ to -12 V	$+0.8$ to -12 V

In all cases, the input of the MOS circuit will look like a very high impedance, and input compatibility is achieved by the circuits of Fig. 8-35.

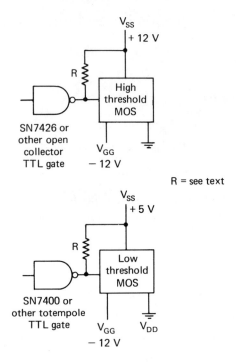

Fig. 8-35. MOS/LSI system input compatibility

The value of resistor R varies, depending on speed-power requirements. In many cases, this resistor R is diffused on the MOS chip. For low-threshold MOS, the resistor assures that the worst-case TTL output is pulled up to at least 3.5 V for proper MOS circuit operation.

Output compatibility. Three types of buffers are commonly used on MOS devices: open-drain, internal pull-up, and push-pull. These buffer arrangements are shown in Fig. 8-36.

With *open-drain* and *internal pull-up*, the buffer is simply a current switch. In the OFF state, the impedance of the buffer is extremely large, whereas in the ON state, impedance is typically under 1 kΩ. A discrete resistor or MOS transistor may be used as a load with an open-drain buffer. This resistor or the transistor may be internal to the MOS IC. When the load transistor is internal to the MOS, the buffer is called an *internal pull-up buffer*.

If the MOS is high-threshold with an open-drain buffer, the output can be made compatible with TTL, as shown in Fig. 8-37. Resistor R_2 provides the necessary current sink for the TTL input. Resistor R_1 limits the positive swing to +5 V.

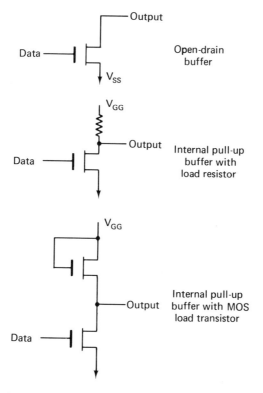

Fig. 8-36. MOS/LSI system output compatibility

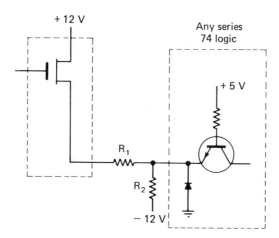

Fig. 8-37. MOS/LSI high-threshold (with an open-drain buffer) to TTL interface

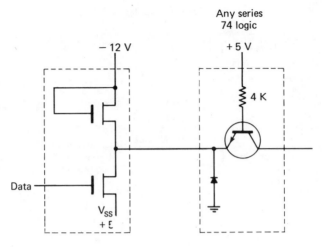

Fig. 8-38. MOS/LSI low-threshold to TTL interface

If the MOS is low threshold, V_{SS} is translated up to $+5$ V instead of to $+12$ V, eliminating the need for R_1. Further, if R_2 is on the chip (in resistor form or a MOS load resistor), no external components are necessary, permitting direct coupling of the MOS output to the TTL input, as shown in Fig. 8-38.

There are two common types of push-pull buffer, as seen in Fig. 8-39. The unsaturated push-pull buffer is most commonly used for low-threshold circuits, and permits direct TTL compatibility without external components (as well as direct compatibility with other low-threshold MOS circuits).

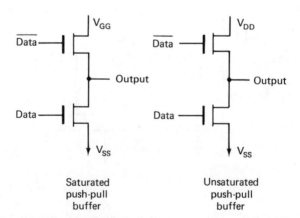

Fig. 8-39. Two common types of push-pull buffers used with MOS/LSI

Clocks.   MOS clock requirements depend on the circuit, as is the case with other logic families. For example, no clocks are required for static RAMs, ROMs, and so on. Some MOS devices require only one clock, with all other clocks generated internally. Most shift registers require two clocks. Highspeed, low-power dissipation shift registers may require four clocks.

For one-clock operation, an internal circuit generates the clocks from a single outside clock. This external-clock signal has the same swing as the data input signal, and the compatibility is identical. Generally, single-clock low-threshold MOS circuits will accept a TTL clock without adding components.

When two or four clocks are required, the clock signals must swing between V_{SS} and V_{GG}. To go from a single TTL-level clock to a multiple MOS-level clock, two circuits are required. First, a *clock generator* is necessary to generate the basic clock pulses. Second, a *clock driver* is necessary to bring the clock levels to the required values. In most cases, only one basic clock circuit is needed for an entire MOS/LSI system.

8.5.2. COS/MOS–TTL/DTL interface

The following notes apply to RCA COS/MOS. When interfacing one logic IC family with another, attention must be given to logic swing, output drive, dc input current, noise immunity, and speed of each family. Figure 8-40 shows a comparison of these for COS/MOS, medium power TTL, and medium power DTL. The supply voltage column of Fig. 8-40 shows that both saturated bipolar and COS/MOS devices may be operated at a supply voltage of 5 V. Both logic forms are directly compatible at this supply voltage (with certain restrictions).

Figure 8-40 also shows the voltage characteristics required at the output and input terminals of saturated logic devices, as well as the COS/MOS input and output characteristics at $V_{DD} = 5$ V. The COS/MOS devices are designed to switch at a voltage level about one-half of the power supply voltage. However, TTL/DTL devices are designed to switch at about $+1.5$ V, which is not one-half of the supply voltage.

COS/MOS driven by bipolar.   When a bipolar device is used to drive a COS/MOS device, the output drive capability of the driving device, as well as the switching levels and input currents of the driven device, are important considerations. Figure 8-40 shows that only 10 pA of dc input current are required by a COS/MOS device in either the 1 or 0 state. The input thresholds (for the driven COS/MOS device) are 1.5 and 3.5 V. Thus, the output of the TTL/DTL driver must be no more than 1.5 V (0 logic voltage) and no less than 3.5 V (1 logic voltage) in order to obtain some noise immunity.

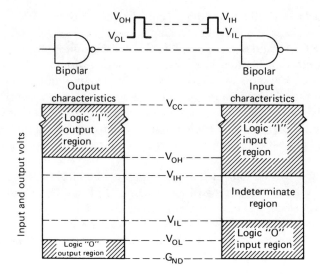

Interface voltage characteristics required at the
output and input terminals of saturated logic
devices.

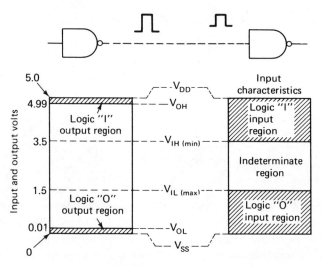

COS/MOS input and output characteristics
at a power-supply voltage of 5 volts

Fig. 8-40. COS/MOS to DTL/TTL interface characteristics

Family	Sup, voltage (volts)	Logic swing/output drive capability	DC input current	Noise immunity	Propagation delays
COS/MOS	3.0 to 15	V_{SS} to V_{DD} (driving COS/MOS) Output drive is type dependent (see text)	10 pA (typical) 1 and 0 state	1.49 at V_{DD} = 5 V The switching point occurs from 30% to 70% of V_{DD} which is 1.5 V to 3.5 V at V_{DD} = 5 V	35 ns (typical) for inverter C_L = 15 pF
DTL and TTL	5	0 state: 0.4 V max. at I_{sink} = 16 mA 1 state: 2.4 V min. at I_{load} = − 400 μA	0 state: − 1.6 mA max. 1 state: 40 μA max.	at V_{CC} = 5 V 0.4 V guaranteed The switching point occurs from 0.8 V to 2 V	20 ns (typical) for inverter C_L = 15 pF

Comparison of COS/MOS, TTL, DTL interfacing parameters

Logic voltage	Description	Voltage (volts)
V_{OL}	Maximum output level in low-level output state	0.4
V_{OH}	Minimum output level in high-level output state	2.4
V_{IL}	Maximum input level in low-level input state	0.8
V_{IH}	Minimum input level in high-level input state	2.0
V_{CC}	Positive supply voltage	5.0 ± 0.5
	Operating temperature range: − 55° to + 125°C−full temperature range product 0° to + 85°C−limited temperature range product	

Common logic voltages, supply voltage, and operating temperature range required to interface with DTL/TTL circuits

Fig. 8-40 (continued)

Current sinking. Figure 8-41 shows the low-state operation of a loaded bipolar driver stage. When the output drive circuit of the bipolar stage is in the low state, the collector is essentially at ground potential. The ON transistor must go into saturation in order to assure a reliable logic 0 level (0 to 0.4 V). To attain this voltage level, there should be a high impedance path from the output to the power supply. Current sinking capability is not a problem in this configuration because the COS/MOS devices have extremely high input impedances (typically 10^{11} ohms). Neither is the voltage level a problem; the COS/MOS devices have high noise immunity (greater than 1 V).

Current sourcing. Current flows from the V_{CC} terminal of a saturated logic (two-junction) output device into the input stages of the load. That is, the output device acts as a current source for the load. Figure 8-42 shows

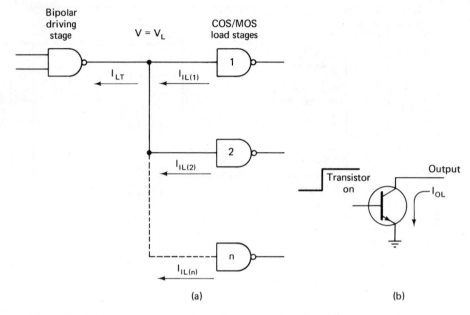

Fig. 8-41. (a) Low-state operation of a loaded bipolar driver stage; (b) typical bipolar output drive circuit in the low state

high-state operation of a loaded bipolar driver stage. Whenever a typical bipolar circuit is in the high state, a pull-up configuration (resistor or transistor) ties V_{CC} to the output pin. The total load configuration should not draw sufficient current to reduce the output voltage level below the V_{IH} required by the COS/MOS devices. (V_{IH} is the maximum acceptable input level for the device in the high-level input state, as shown in Fig. 8-43.)

There are three bipolar output configurations to consider: resistor pull-up, open collector, and active pull-up.

Resistor pull-up. Devices with resistor pull-ups, as shown in Fig. 8-42, present no problem in interface with COS/MOS.

Open collector. Devices with open collectors require an *external pull-up resistor*, as shown in Fig. 8-44. The selection of external pull-up resistors requires consideration of fan-out, maximum allowable collector current in the low state (I_{OL} max), collector-emitter leakage current in the high state (I_{CEX}), power consumption, power-supply voltage and propagation delay times.

The equations of Fig. 8-45 provide guidelines for selection of pull-up resistor maximum and minimum values. These equations neglect the values

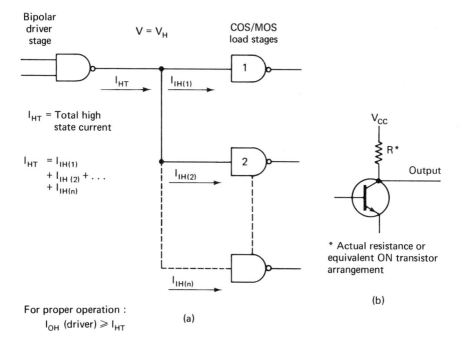

I_{HT} = Total high
state current

I_{HT} = $I_{IH(1)}$
+ $I_{IH (2)}$ + ...
+ $I_{IH(n)}$

* Actual resistance or
equivalent ON transistor
arrangement

(b)

For proper operation :
I_{OH} (driver) ⩾ I_{HT}

(a)

I_{OH} = Maximum permissible output driver current in high
state ("1") (driver leakage)
I_{OL} = Maximum output driver sinking current in low state ("0")
I_{IL} = Low state input current drawn from the load stage
(to the driver)
I_{IH} = High state input current flowing into the load stage
from driver

Fig. 8-42. (a) High-state operation of a loaded bipolar driver stage; (b) typical
bipolar output drive circuit in the high state

of I_{IH} and I_{IL} for the MOS device because such values (typically 10 pA) are
insignificant when compared with the value of the bipolar currents.

Assume that the equations of Fig. 8-45 are used to find the value of R_X
where conditions are V_{DD} = 5 V, V_{OL} = 0.4 V max, I_{OL} = 20 mA, V_{CC} = 5 V
± 0.5 V, V_{IH} (for the MOS) = 3.5 V, and I_{CEX} = 100 μA max.

For one driver (bipolar) and one load (MOS), the values of R_X are:

$$R_{X(min)} = \frac{(5.0 - 0.4)}{0.020} = 230 \ \Omega$$

$$R_{X(max)} = \frac{(4.5 - 3.5)}{0.0001} = 10 \ k\Omega$$

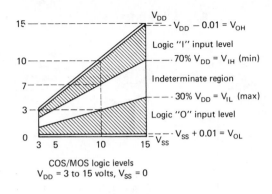

COS/MOS logic levels
V_{DD} = 3 to 15 volts, V_{SS} = 0

Legend within figure:
V_{DD} − 0.01 = V_{OH}
Logic "1" input level
70% V_{DD} = V_{IH} (min)
Indeterminate region
30% V_{DD} = V_{IL} (max)
Logic "0" input level
V_{SS} + 0.01 = V_{OL}

Logic voltage symbol	Description
V_{OH}	Minimum guaranteed noise free output level of device in high-level output state
V_{IH}	Minimum acceptable input level for device in high-level input state
V_{IL}	Maximum acceptable input level for device in low-level input state
V_{OL}	Maximum guaranteed noise free output level of device in low-level output state
V_{NL}	Maximum (positive) noise level tolerated at low level state
V_{NH}	Maximum (negative) noise level tolerated at high level state
$V_{OL} + V_{NL} \geqslant V_{IL}$	
$V_{OH} + V_{NH} \leqslant V_{IH}$	Operating temperature range
V_{DD}	Positive supply voltage — 55°C to + 125°C (full temperature prod).
V_{SS}	Negative supply voltage — 40°C to + 85°C (limited temperature prod).

Fig. 8-43. Common logic voltages; supply voltages and temperature range of COS/MOS

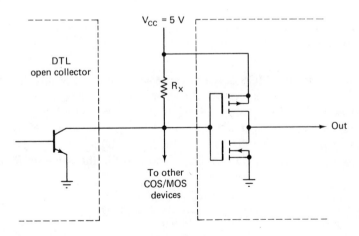

Fig. 8-44. Examples of DTL/TTL circuit with open collectors that require a resistor between the output and V_{cc}

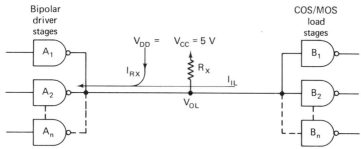

Conditions: Only one driver stage (A_2) in low state "O".
All other drivers in high state "I".

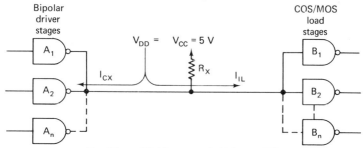

Conditions: All driver stages in high state "I".

$$R_X \text{ (min)} = \frac{V_{DD} - V_{OL \text{ (max)}}}{I_{OL}}$$

$$R_X \text{ (max)} = \frac{V_{CC \text{ (min)}} - V^*_{IH \text{ (min)}}}{(N)\,I_{CEX \text{ (max)}}}$$

*V_{IH} is the value for the COS/MOS device

Fig. 8-45. Bipolar output (with pull-up resistor) driving COS/MOS in low-state and high-state operation

If more than one driver is used with the open collector drive arrangement, the values of I_{OL} and I_{CEX} must be increased accordingly (I_{OL} and I_{CEX} multiplied by the number of drivers). However, the values of R_X will not be affected by an increase in the number of MOS loads. Of course, if an infinite number of MOS loads are added so that I_{IL} and I_{IH} become significant compared to I_{OL} (for the minimum value of R_X) or to I_{CEX} (for the maximum value of R_X), then the MOS input currents must be included in the calculations as follows:

$$R_{X(\text{min})} = \frac{V_{DD} - V_{OL(\text{max})}}{I_{OL} = I_{IL(\text{MOS})}}$$

$$R_{X(\text{max})} = \frac{V_{CC(\text{min})} - V_{IH(\text{min})(\text{MOS})}}{I_{CEX} + I_{IH}}$$

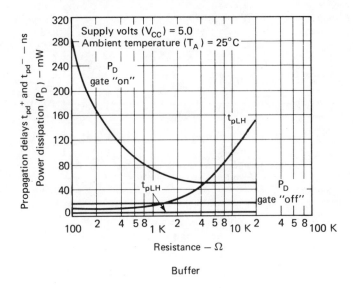

Buffer

Gate

Fig. 8-46. Typical speed-power trade-off of open collector TTL buffer or gate, and pull-up resistor

For short propagation delay times with an external pull-up resistor, it is best to keep R_X small. That is, use the $R_{X(min)}$ value, or the next higher standard resistor value. However, power consumption increases rapidly at values below about 1 kΩ so some compromise is usually necessary. Of course, final selection of the pull-up resistor depends on what is most important for the intended application: high speed or low power.

Figure 8-46 shows typical speed-power relationships as a function of R_X for two popular bipolar open collector drivers and illustrates the trade-off between speed and power. The power figure shown is the power dissipated in both the bipolar driver and the pull-up resistor.

Active pull-ups.   When an active pull-up is used, such as the transistor-plus-diode arrangement shown in Fig. 8-47, there can be a problem in the 1 state. This is because the minimum output level (2.4 V) cannot assure an acceptable 1 state input for the COS/MOS device. For example, assume that the 2.4 V minimum TTL/DTL output level is specified for a load current of 400 μA (which is typical). This would require approximately 40 MOS devices (with a typical input current of 10 pA).

If only one, or a few, MOS devices are used as a load, the minimum TTL/DTL high-output level will rise to about 3.4 to 3.6 V. There is no noise immunity in such a configuration. It is recommended that a pull-up

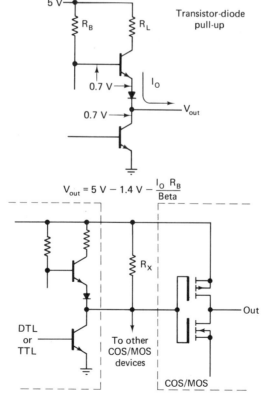

Fig. 8-47. Active pull-up (transistor-diode) to COS/MOS interface

resistor be added to V_{CC} from the output terminal of the bipolar device. The selection of this resistor should be based on the calculations of Fig. 8-45.

When driving a COS/MOS device from an output arrangement as in Fig. 8-47, the driver should not fan-out to TTL/DTL circuits: but only to COS/MOS devices. This is because it is not accepted practice to tie devices with active pull-ups together (if one device is at 0 and the other is at 1, the output in unpredictable).

COS/MOS driving bipolar. Figure 8-48 shows COS/MOS devices driving bipolar devices. The current sinking capacity of the COS/MOS device must be considered when the device is a medium-power DTL or TTL circuit. (Figure 8-40 illustrates that the TTL/DTL device requires no more than 1.6 mA in the 0 input state, and a maximum of 40 µA in the 1 state.)

The COS/MOS device must be capable of sinking and sourcing the currents while maintaining voltage output levels required by the TTL/DTL gate. Any given TTL/DTL gate will switch state at a voltage that ranges

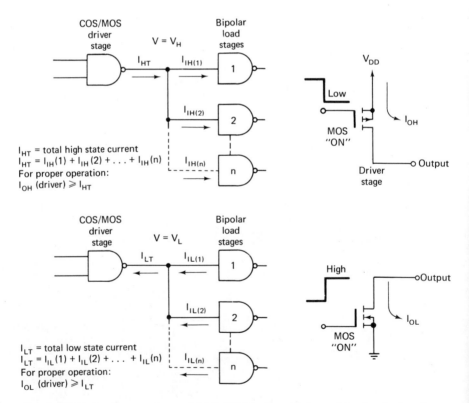

Fig. 8-48. Logic diagram for a COS/MOS device driving a bipolar device, low state and high state

from 0.8 to 2.0 V. Thus, the output drive capability of the COS/MOS driver must be at least 40 μA for a given 1-state output voltage of 2 V, and at least 1.6 mA for a given 0-state output voltage of 0.8 V. In order to provide a noise margin of 400 mV for the driven bipolar device, the COS/MOS device must sink 1.6 mA at a 0-state voltage of 0.4 V, and 40 μA at a logic 1 level of 2.4 V.

Current sourcing. In the high-state operation (Fig. 8-48), V_{DD} is normally connected to the driver output through one or more ON p-channel devices which must be able to source the *total leakage current* of the bipolar load stages. The published information for the particular COS/MOS and bipolar devices must be consulted in order to determine the leakage currents (for the logic 1 state), and drive fan-out to be used in the equations shown in Fig. 8-48. Saturated logic devices *will not reach their required switching levels* unless this equation has been satisfied.

Current sinking. When the output of a COS/MOS driver is in the low-state, an n-channel device is ON, and the output is approximately at ground. The COS/MOS device sinks the current flowing from the bipolar input-load stage. The published data for the COS/MOS device must be consulted to determine the maximum output low-level sinking current, and the published data for the bipolar device must be consulted to determine its input low-level current.

Not all COS/MOS devices can sink the required current for all bipolar logic families. This problem can be overcome by connecting several COS/MOS devices in parallel. Likewise, some COS/MOS MSI devices (such as counters and registers) have limited drive capability. Their outputs may require buffering if these COS/MOS devices are to drive TTL/DTL. The COS/MOS line contains several buffers. In some cases, COS/MOS drive and current sinking capability can be increased if the devices are operated at higher supply voltage. These capabilities are described in COS/MOS literature.

8.5.3. Level shifters (level translators)

When interfacing DTL and TTL devices with COS/MOS devices that are operated at a higher voltage supply, the same resistor interface shown in Fig. 8-44 can be used. The resistor is tied to the higher level (V_{DD}). The maximum supply voltage for the DTL and TTL gates is generally specified at about 8 V. Thus, not all DTL/TTL gates may be used for interface applications that require higher supply voltages (V_{DD}).

Guaranteed operation at these higher supply voltages can be accomplished by selection of DTL/TTL units with breakdown voltages $V_{(BR)CER}$

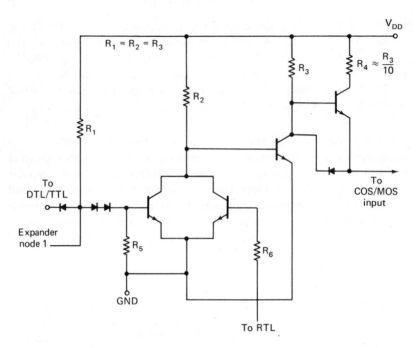

Fig. 8-49. Level translator used to convert DTL, TTL and RTL input logic levels and voltages compatible with COS/MOS circuitry

exceeding the COS/MOS operating voltage, or by using a level shifter (also known as a level translator) shown in Fig. 8-49. This circuit converts DTL, TTL, and even RTL input logic levels to voltages compatible with COS/MOS circuitry. In interface application, the supply voltage for the translator should be equal to the supply voltage required by the COS/MOS circuitry.

The *speed consideration* is not important when a separate interface circuit is used. It is desirable (unless high noise immunity is a prime consideration) for the speed of an interfacing circuit to be maximum (or at least no slower) than either type of logic joined by the interface. No interfacing device other than a pull-up resistor is required, however, between the COS/MOS and TTL logic at a supply voltage of 5 V. Speeds involved when COS/MOS drives TTL (which can be found in the published data for COS/MOS devices) are comparable to the COS/MOS propagation delays. Speeds involved when COS/MOS is driven by TTL, even with a large external resistor, are no slower than delay times for COS/MOS logic circuits. As a result, speed is not a problem in COS/MOS-TTL interfacing, provided *clock rates* are within the COS/MOS range.

8.5.4.   COS/MOS-HTL interface

HTL circuits operate at voltage levels between 14 and 16 V. COS/MOS circuits can operate at these voltages as well, but generally are limited to voltages no higher than 15 V. HTL circuits have more limited temperature range, and dissipate much more power than COS/MOS circuits. Thus, care should be exercised when using the combinations in extreme temperature environments.

Typically, HTL resistance values vary by about 20 percent from one end of their temperature range to the other. In addition, the transistors used in HTL are sensitive to temperature, and are subject to thermal runaway.

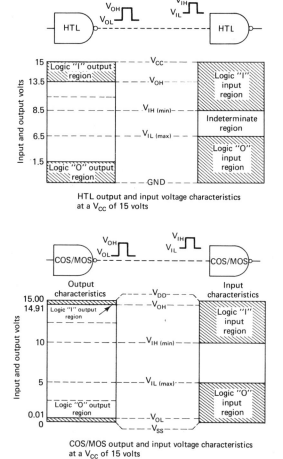

Fig. 8-50. COS/MOS and HTL interface characteristics

HTL common logic voltages, supply voltage
and operating temperature ranges

Logic voltage symbol	Description	Voltage
V_{OL}	Maximum output level in low-level output state	1.5 V
V_{OH}	Minimum output level in high-level output state	13.5 V
V_{IL}	Maximum input level in low-level input state	6.5 V
V_{IH}	Minimum input level in high-level input state	8.5 V
V_{NL}	Worst case positive noise level tolerated at low level state	5.0 V
V_{NH}	Worst case negative noise level tolerated at high level state	5.0 V
V_{CC}	Positive supply voltage	15.0 ± 1 V
Operating temperature range − 30°C to + 75°C		

Fig. 8-50. (*continued*)

The V_{OL} level, propagation delay, and noise immunity of HTL circuits vary widely across the temperature range. COS/MOS circuits show almost negligible variation for these same parameters over a temperature range that is approximately 75 percent wider than that of HTL. In COS/MOS-HTL interface, the main concern with temperature is the HTL parameters, not the COS/MOS parameters.

The same general rules described for COS/MOS-TTL/DTL (Sec. 8.5.2) apply to COS/MOS-HTL. Figure 8-50 shows the voltage characteristics required at the output and input of an HTL device for a V_{CC} 15 V, as well as the same characteristics for a COS/MOS device.

The HTL devices either have a built-in pull-up resistor (typically about 15 kΩ) or an active pull-up. An external pull-up resistor is unnecessary when COS/MOS devices are being driven by HTL. The dc noise immunity in the high state (logic 1) is 3.5 V for an active pull-up and 5 V for a resistor pull-up.

The published data should be consulted to be sure that the rise and fall times and the pulse widths of the HTL output are compatible with the required pulse width and input rise and fall times of the COS/MOS circuits. The rules for selection of external pull-up resistors are the same as described in Sec. 8.5.2, except that the values are different. For example, a typical HTL will have values such as: $V_{IH} = 8.5$ V (min), $V_{IL} = 6.5$ V (max), and $I_{IL(max)} = 1.2$ mA, and $I_{IH(max)} = 2$ μA.

8.5.5. COS/MOS-ECL interface

Figure 8-51 shows the interface of COS/MOS devices with ECL devices. The V_{CC} to V_{EE} voltage range is fixed from a ground level to −5 or −5.2 V. Logic 1 to logic 0 values are separated by only 0.3 to 0.5 V, depending on

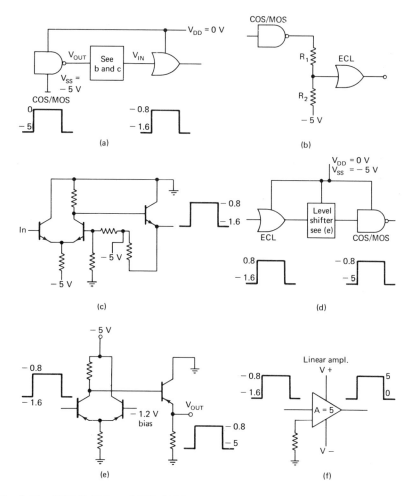

Fig. 8-51. COS/MOS and ECL/ECCSL interface characteristics

the particular type of ECL family used. Figure 8-52 shows some typical ECL values. However, since each manufacturer shows different logic levels for a number of ECL families, care should be taken to use only the applicable value obtained directly from the published data.

A logic 1 is the most positive frame of reference, and a logic 0 is the most negative. For example, for positive logic, an RCA type CS2150 OR/NOR gate is at a logic 1 when its voltage is -0.8 V, and at a logic 0 when its voltage is -1.6 V (more negative value).

The interfacing of COS/MOS devices driving ECL devices requires a method to reduce the output voltage swing to 0.9 V. This can be done with a precision resistor-divider network arrangement (Fig. 8-51b), an emitter-

Logic voltage symbol	Description	Voltage range***	
		From	To
V_{OL}	Maximum output level low-level state	− 1.6	− 1.45*
V_{OH}	Minimum output level high-level state	− 0.8	− 0.795*
V_{IL}	Minimum input level low-level state	− 1.4	− 1.7*
V_{IH}	Maximum input level high-level state	− 0.75	− 1.1
V_{NL}	Worst case positive noise level tolerated at low-level state	0.20*	0.35*
V_{NH}	Worst case negative noise level tolerated at high-level state	− 0.235	− 0.305*
V_{CC}	Positive supply voltage	0	0
V_{EE}	Negative supply voltage	− 5.5	− 5.0
	Temperature range + 10 to + 60°C		

* At T = + 25°C
** These values are representative of the range for several ECL families

Fig. 8-52. ECL common logic voltages; supply voltages, and operating temperature range

follower (Fig. 8-51c), or numerous combinations of resistor, diode and transistor configurations. For example, if the COS/MOS output for a logic 1 is 5 V, and the input for an ECL logic is 1 V, the resistor-divider network of Fig. 8-51b can be chosen to provide a five-to-one voltage division.

The interfacing of COS/MOS devices that are being driven by ECL generally include an amplifier. Amplification is necessary because the COS/MOS requires a greater voltage swing. That is, the ECL −0.8 to −1.6 V swing must be amplified to the COS/MOS 0 to 5 V swing. The use of a separate transistor, such as that shown in Fig. 8-51e, is recommended. Proper biasing of the transistor is essential. It is suggested that the V_{SS} level for the COS/MOS circuit be the same as the V_{EE} level of the ECL circuit. This will minimize the number of power supplies as well as provide better interface conditions.

8.5.6. COS/MOS to MOS interface

There are a number of MOS devices which function at the same V_{DD} and V_{SS} ranges of COS/MOS. These devices can be interfaced directly, provided V_{DD} and V_{SS} are the same. Direct interface applies (generally) to n-channel MOS devices. If p-channel devices are involved, there may be a problem of logic polarity. If the PMOS device uses negative logic, the COS/MOS positive logic must be converted.

8.5.7. MOS to RTL interface

Since RTL is most similar to discrete transistor logic, discrete interface circuits are generally required when translating between RTL and MOS. Figure 8-53 gives three discrete circuits that can be used in converting RTL

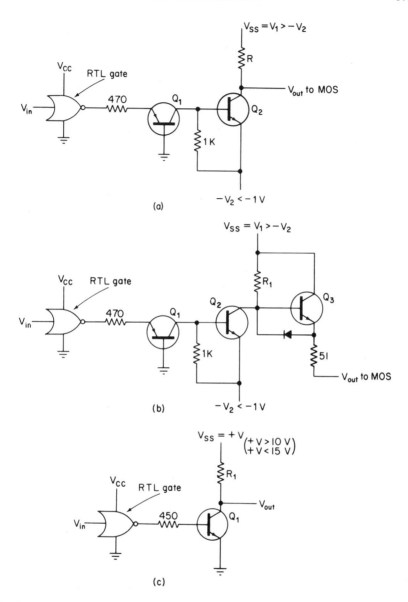

Fig. 8-53. RTL to MOS interface circuits

380 DIGITAL IC BASICS

to MOS. The circuits of Fig. 8-53a and 8-53b are particularly useful when the MOS substrate is at the RTL power supply voltage, or at ground.

The circuit of Fig. 8-53b has an active pull-up, and should be used when driving large capacitive loads. Resistor R_1 should be selected on the basis of power dissipation and the desired rise and fall time at the output. If the MOS substrate is at a voltage in the range of $+10$ to $+15$ V, the circuit of Fig. 8-53c provides an efficient interface. Again, the value of R_1 is selected on the basis of available drive from the RTL gate, desired rise and fall times, and power dissipation.

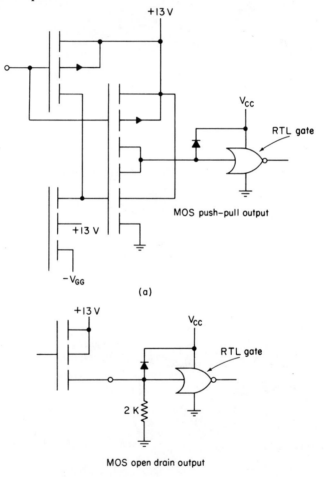

Note: Use germanium diode when $V_{CC} = 3.6$ V

(b)

Fig. 8-54. MOS to RTL interface (with MOS substrate at $+13$ V)

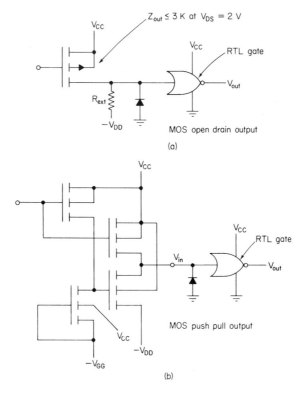

Fig. 8-55. MOS to RTL interface (with MOS substrate at V_{cc})

The circuit of Fig. 8-54 illustrates conversion from MOS to RTL when the MOS substrate voltage is at $+13$ V. The diode clamps the gate input to the RTL power supply voltage, which prevents the MOS output driver from pulling the gate input above the maximum specified rating at $+4$ V. When V_{CC} is 3.6 V, the clamp diode should be germanium.

In the circuit of Fig. 8-55, the MOS substrate is at the RTL power supply voltage. In this mode, the clamp diode is used to prevent the voltage on the RTL gate input from falling below the -4 V rating.

8.6. ECL AND HTL LEVEL TRANSLATORS

In general, level translators (or level shifters) are used when ECL and HTL must be interfaced with any other logic family. This is because of the special nature of ECL and HTL (ECL is nonsaturated logic; HTL operates with large logic swings).

8.6.1. HTL interface

Most HTL logic lines include at least two translators; one for interfacing from HTL to RTL, DTL, and TTL; and another for interfacing from these families to HTL. This is the case with the Motorola HTL line (designated as MHTL).

The translator for interface from MHTL to the other three logic families is shown in Fig. 8-56. For conversion to DTL and TTL, a 5 V supply is connected to a 2 kΩ pull-up resistor through pin 13. (Note that only one-third of the schematic is shown. However, all three logic elements are contained in one IC package.) For interface with RTL, pins 4, 9 and 12 are connected to the RTL supply voltage (nominally 3.6 V). Expander points (pins 2 and 5) without diodes are present at the inputs of the two units, but not on the third unit. This is because of the need for additional leads for the RTL power supply, and pin limitation of the 14-lead IC package.

The translator for interface from RTL, DTL, and TTL to MHTL is shown in Fig. 8-57. This translator is also a triple unit. Signals from

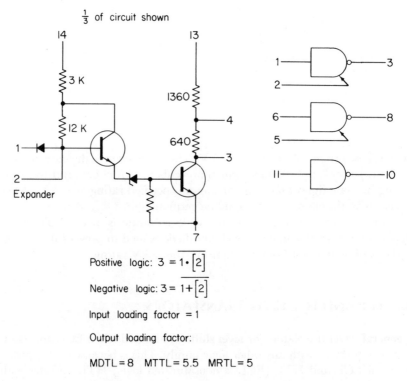

$\frac{1}{3}$ of circuit shown

Positive logic: $3 = \overline{1 \cdot \boxed{2}}$

Negative logic: $3 = \overline{1 + \boxed{2}}$

Input loading factor $= 1$

Output loading factor:

MDTL $= 8$ MTTL $= 5.5$ MRTL $= 5$

Fig. 8-56. Motorola translator from MHTL to MRTL, MDTL or MTTL

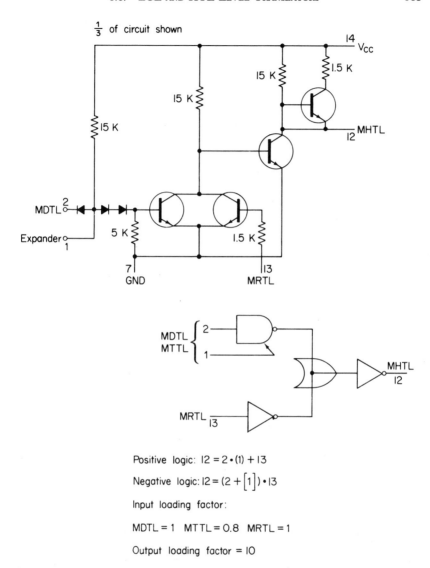

Fig. 8-57. Motorola translator from MRTL, MDTL, or MTTL to MHTL

DTL/TTL sources are applied to one set of input terminals, while signals from RTL sources are applied to another terminal. The different inputs provide threshold levels and characteristics compatible with MDTL/MTTL and MRTL (Motorola) families. Each DTL/TTL section is also an input expander terminal (pins 1, 6 and 9) without diodes. These terminals may be used to expand input logic capability or to use high-voltage diodes to readily interface high-voltage relay or switch circuits to HTL levels. Both types of inputs may be applied simultaneously, with one output going high if the logic function or either input goes high.

If the RTL input is used by itself, the DTL/TTL input must be grounded for proper operation. This is not necessary if the DTL/TTL input is being used by itself, but is advisable under this condition to ground the RTL input to reduce any possible noise pickup.

Driving discrete components from HTL. In some cases, HTL must be used to drive discrete transistors. This creates a problem because of the high output voltage from HTL. The output in the high state is not as much of a problem as the output of an HTL in the low state. Assuming positive logic, the output of an HTL in the low state (or 0) is about 1 V. (This is generally listed as V_{OL}.) If 1 V is applied to an *npn* in the low state, the *npn* will probably never turn off. However, an HTL with an active pull-up output can be used for driving *npn* transistors if the circuit of Fig. 8-58 is used.

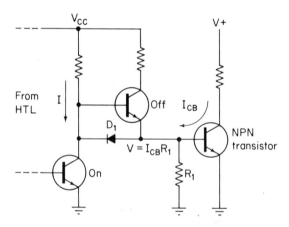

Fig. 8-58. Driving discrete transistors from HTL logic IC outputs

The higher V_{OL} voltage of the HTL is partially caused by the extra diode D_1 on the output. However, if the gate is not required to sink current, then the voltage on the base of the *npn* transistor is equal to the leakage current of the collector-base junction times the value of R_1. Thus, resistance R_1 should be chosen to provide a base voltage of about 0.2 V (or less, using I_{CB} leakage current as the factor). With such a circuit, the HTL should not be required to sink any current when the output is in the low state.

8.6.2. ECL interface

As in the case of HTL, ECL level translators are available for interfacing with the other logic lines. However, if the only problem is one of interfacing from an input to ECL, it is possible that the circuit of Fig. 8-59 will solve the

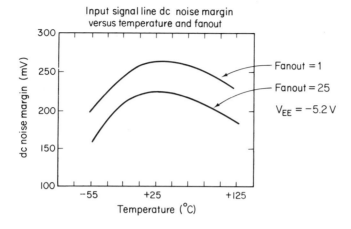

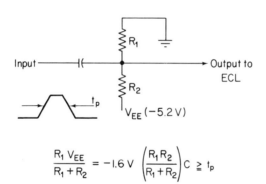

$$\frac{R_1 V_{EE}}{R_1 + R_2} = -1.6\ V \quad \left(\frac{R_1 R_2}{R_1 + R_2}\right) C \geq t_p$$

Fig. 8-59. Interfacing from non-ECL input to ECL

problem. The input (either discrete component or IC) must be on the order of 0 V. As shown by the equations, the values of R_1 and R_2 are selected to give a logic 0 (-1.6 V) at the output (to the ECL input). The value of C can be determined by the fact that the RC time constant should be several times greater than the duration of the input pulse.

8.7. RTL INTERFACE

The input signals to a typical RTL element should meet the following:

high level: $E_{in} \geqslant +1$ V

$0.5\ mA \leqslant I_{in} \leqslant 0.5$ mA per load

low level: $E_{in} \leqslant 0.3$ V

When RTL is interfaced with MOS, follow the recommendations of Sec. 8.5.6. When an RTL drives a DTL or TTL type input, there is a current sink requirement on the RTL output. To ensure operation under all conditions, a maximum sink current of 5 mA is permissible. When interfacing RTL at the input, the high-level output current of the driver is important. For a DTL driver with an output of 6 kΩ pull-up resistor to a +5 V supply, one RTL load is allowed. Five gate loads can be driven if the DTL output resistor is reduced to 1.5 kΩ, for a typical RTL IC.

The specifications of most RTL and DTL/TTL ICs are not compatible, so worst-case interfacing cannot be guaranteed between the two families without performing special tests and/or adding discrete components. However, by using conservative design principles and making one relatively simple test, interfacing can be accomplished.

The recommended circuitry for driving RTL from DTL outputs is shown in Fig. 8-60a. In order to ensure that this technique is valid for operation under worst-case conditions, it is necessary to test V_{OL} of the DTL circuit at an I_{OL} of 6.5 mA. Use only those DTL components with a V_{OL} less than 0.4 V for this application.

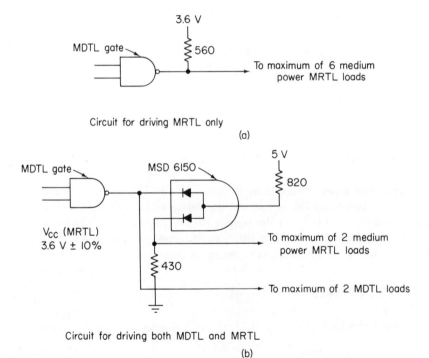

Fig. 8-60. DTL to MRTL interface

The circuit of Fig. 8-60a should not be used to drive both RTL and DTL from a common output. If both must be driven from a common point, use the circuit of Fig. 8-60b. The load current seen by the output transistor of the DTL gate in Fig. 8-60b is greater than that of Fig. 8-60a. The leakage current of the RTL gate may increase slightly also. Neither of these factors should cause significant problems.

Driving DTL from RTL outputs is not a situation where conservative design can result in no-test circuit limits. The primary reason for this is that the high-level output of RTL is tested only to the level of V_{ON} (about 0.67 to 1 V, depending on type and temperature). Although a typical RTL gate shows a high-level output well above the DTL threshold, a test is always recommended.

The test required is relatively simple. With the datasheet value of V_{OFF} applied to all points of the driving RTL gate, the unloaded output voltage should be a minimum of 2.5 V at the maximum operating temperature. If so, RTL can be used to drive DTL directly without special interface.

In cases where driving both RTL and DTL from an RTL output is required (including the case of an unbuffered flip-flop output driving a DTL

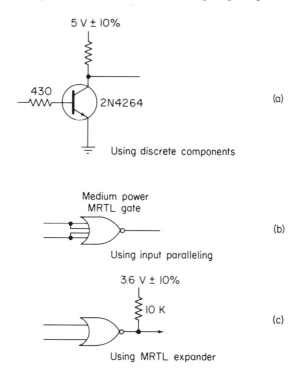

Fig. 8-61. MRTL to DTL interface

8.8 Summary of Interfacing Problems

component), use the circuit of Fig. 8-61. This is also the recommended alternative when the RTL fails the 2.5 V output test.

There are several techniques for ensuring that an RTL gate can sink enough current to drive a number of DTL inputs. The most common method is to double the input loading factors of the driving (RTL) gate (which is the equivalent to allowing up to a typical 5 mA of additional sink current, or enough for about 3 DTL gates). This is not a guaranteed condition, of course, but the incidence of devices that will not perform in this manner is very small. Even so, the conservative designer will drive at least two inputs of the RTL gate as shown in Fig. 8-61b, thus ensuring low-level drive capability under worst-case conditions.

Another worst-case design technique is shown in Fig. 8-61c. This circuit requires an RTL expander, rather than an RTL gate, as an external resistor.

8.8. SUMMARY OF INTERFACING PROBLEMS

As can be seen, each logic IC line or family has its own set of design or usage problems. Most of these problems can be resolved using the datasheets and brochures supplied with the IC. One area where the datasheets leave a gap is the interfacing between IC logic forms. For that reason, we summarize the interfacing problems here. If MOS is involved, follow the recommendations described in Sec. 8.5.

If either ECL or HTL is involved, the best bet is to use the various level translators supplied with most ECL and HTL lines, or use the specific discrete translator circuits shown in the datasheets and brochures. In the absence of such information, use the circuits described in Sec. 8.6.

DTL and TTL are compatible with each other, thus eliminating the need for special translators or interface circuits. However, the load and drive characteristics may be altered somewhat. These changes are generally noted on the datasheets. In the absence of any DTL/TTL interface data, use the information of Sec. 8.4 through 8.7.

8.8.1. Typical interface ICs

As discussed, most major IC manufacturers provide a complete line of interface ICs for their popular logic families. Such ICs provide for interfacing to and from other families, and other devices. For example, Motorola produces ICs that provide for interfacing with computer buses (including minicomputer and microcomputer), instrumentation buses, analog-to-digital and digital-to-analog conversion, memory interface (including registers and magnetic memories), terminal, peripheral, and numeric-display interface, voltage comparison, and communications interface (including crosspoint switches, and voice encoding/decoding).

9

SOLVING DESIGN
PROBLEMS WITH
INTEGRATED CIRCUITS

In this chapter, we discuss how ICs can be used to solve specific design problems. Most of the ICs covered are digital (or logic). However, both linear and special-purpose ICs are also used. The same basic format is used throughout. First, the design problems are stated, and the basic considerations are discussed. This is followed by state-of-the-art solutions to the problem. Each section is rounded off with a specific design solution using currently available ICs.

The design problems and solutions covered here include: pulse-width modulation for dc motor control, variable-speed control systems for induction motors, digitally controlled power supplies, recovery of recorded digital information for disc and tape systems, power control using the zero voltage switch, a 3-1/2-digit dual-ramp DVM, industrial clock/timer, and a battery-powered 5 MHz frequency counter.

9.1. DC MOTOR CONTROL WITH
PULSE-WIDTH MODULATION

Speed control of small dc motors is generally done using one of two methods. The simplest method is the insertion of resistance in series with the motor and its power supply. This method, although inexpensive and easily applied, is inefficient and generally undesirable in battery-driven systems where battery life is a serious consideration.

The second method is pulse-width modulation (PWM). With PWM, a full voltage is applied to the motor in a series of short pulses. The duty cycle of the pulses is varied to vary the *average motor voltage*. In addition to being more efficient, PWM has the advantage of generating high torque pulses, allowing easy starting of the motor even when set for low-speed operation. This section shows several circuits for developing PWM drive signals, and describes how they may be applied to a dc motor. Although any type of dc motor, series, shunt, or permanent magnet, may be controlled using PWM, it is most efficient when applied to a permanent magnet motor. Thus, all circuits in this note are based on a permanent magnet motor, although the circuits are applicable to other motor types.

9.1.1. Basic considerations for PWM motor control

Two basic factors must be considered in design of a PWM motor control system. Assuming that the motor size, voltage, and current have been determined by the load requirements, one must consider the *range of control necessary* and the *frequency of the control system*.

Control system range.   Some types of motor controls are limited to a range of approximately 5 to 95% of full speed. Such systems do not have the ability to turn on or off completely. Other motor control systems do have the ability to turn on and off with a "jump." This jump may be less than 1% of full speed, or as high as 3%, depending on the control system.

Control system frequency.   If it is necessary for the control to be free of audible noise, the frequency of operation should be in the 18 to 20 kHz range, so as to be above audible frequencies, and yet still be as low as possible to allow the use of the most economical power transistors. If noise is not a serious factor, lower frequencies may be used and limited only by the mechanical response of the motor and load. Since motor drive torque is a direct function of the armature current, it may be desirable to operate with constant rather than pulsed-drive currents. In this case, the frequency should be high enough so that the armature current never reduces below a predetermined level during the off time of the output power transistor (motor current will then be flowing through the free-wheeling diode). Once these two considerations have been resolved, then the type of drive system may be chosen.

9.1.2. PWM motor control with op-amp astable multivibrator

The classic approach to PWM motor control involves an astable multivibrator formed with discrete components. An IC version of this motor control circuit is shown in Fig. 9-1, where an astable multivibrator is formed with an op-amp. Range of the circuit is typically 5 to 95%. Frequency is limited by the op-amp switching characteristics.

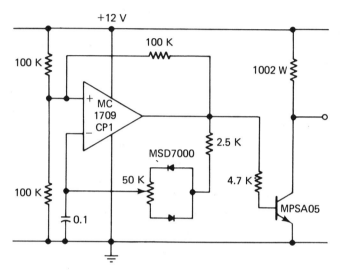

Fig. 9-1. Op-amp astable multivibrator for PWM motor control

In operation, the voltage divider consisting of two 100 kΩ resistors, establishes a voltage reference point on the noninverting amplifier input. The capacitor is charged through the timing resistor until the voltage exceeds the reference level. At this time, the amplifier switches and the capacitor is then discharged. Variable charge and discharge rates are obtained by using the MSD7000 dual diode to feed the timing potentiometer so that the time constant during charge and discharge may be different.

Using the same potentiometer and capacitor in both halves of the cycle maintains a constant frequency regardless of the control duty cycle. During the time the capacitor voltage is low compared to the reference voltage, the output of the amplifier is at positive saturation, effectively placing the 100 kΩ positive feedback resistor in parallel with the upper resistor of the voltage divider, raising the reference voltage. When the capacitor charges from the high output value of the amplifier and the voltage just exceeds the reference voltage, the amplifier output switches to negative saturation. Following this action, the positive-feedback resistor is essentially in parallel with the lower voltage divider resistor, lowering the reference voltage point and causing the capacitor to discharge toward the zero voltage level.

When the capacitor voltage drops just below the reference voltage, the amplifier switches and the cycle is repeated. Since the output of the amplifier is either a positive or negative saturation, the period in each position is controlled by the length of time necessary to charge or discharge the capacitor, the system provides a convenient and inexpensive PWM generator.

Using equal values for the voltage divider and positive feedback resistors allows the capacitors to charge from one-third to two-thirds of the power supply voltage. The frequency is unaffected by voltage variations within the limits of operation of the op-amp. The range of control is dependent on the ratio of series resistor value (2 kΩ) to the sum of the control potentiometer resistance and the series resistor. Again, 5 to 95% is a practical limit for this circuit.

9.1.3. PWM motor control with digital astable multivibrator

Figure 9-2 shows another PWM motor control circuit using the astable multivibrator principle. However, the circuit of Fig. 9-2 uses a digital IC. The IC is a Motorola MC14011 CMOS quad NAND gate. In the Fig. 9-2 circuit, the range of operation is approximately 3 to 97%, since the CMOS output is internally limited and a series resistor is unnecessary. Smaller ranges of operation may be obtained by inserting a resistance in series with

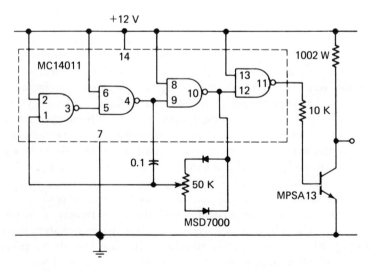

Fig. 9-2. Digital astable multivibrator for PWM motor control

the center point of the dual diode as shown in Fig. 9-1. The voltage range on the capacitor is slightly greater than the full power supply voltage.

An obvious advantage of the Fig. 9-2 circuit is that a minimum of external components is required. The disadvantage is the low output current available (about 1 mA) such that a Darlington type output buffer transistor is required. As in the circuit of Fig. 9-1, either *npn* or *pnp* output buffers may be used.

Note that both the Fig. 9-1 and 9-2 circuits require a power output stage to handle the current required by the motor. Power output stage design is discussed in Sec. 9.1.6.

9.1.4. PWM motor control with op-amp comparator

Both the Fig. 9-1 and 9-2 circuits are manually controlled, with no provision for any electrical input signal to control pulse width. If electrical control is required, say from a tachometer to feed back motor speed, or from a current sensor to feed back motor load current, a comparator type circuit is preferred. The basic comparator circuit is shown in Fig. 9-3. Note that the

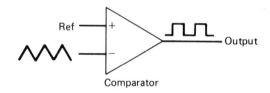

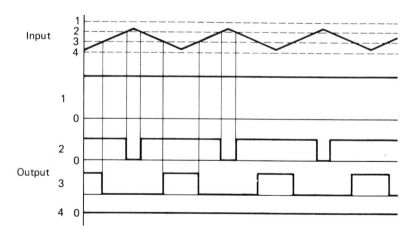

Fig. 9-3. Basic comparator circuit for PWM motor control

comparator can be either an op-amp (or OTA) or an IC voltage comparator. Both options are discussed in the following paragraphs.

With either option, the comparator requires two input signals. One
signal is a triangular wave of fixed frequency and amplitude. The other
signal is a variable dc voltage that can come from a potentiometer, tachometer, load current sensor, etc. Operation is shown in Fig. 9-3 for four different
situations.

In situation 1, the direct voltage is higher at all times than the
triangular wave level, and the output is a continuously high signal. In
situation 2, the triangular wave exceeds the dc input for only a very short
period, and thus produces a narrow, low-voltage pulse on the output
waveform. In situation 3, the dc level is below the average triangular wave
level, thus, the low period of the output signal is greater than the high
period. In situation 4, the reference level is below the triangular wave level
at all times, thus giving low comparator output.

The comparator circuit can control pulse widths from full-on to
full-off. However, there is a slight jump in the transition from pulsing control
to steady output caused by rounding at the peaks and valleys of the
triangular waves. In most cases, it is not at all difficult to limit this jump to
less than 1% of the on or off time.

Practical op-amp comparator motor control circuit.   The circuit of Fig.
9-4 uses a single IC to provide the motor control function. The Motorola
MC1458 dual op-amp allows the use of the same IC to generate the
triangular wave, and to provide the comparator function. The triangular
wave generator is the same basic circuit as described in Fig. 9-1. The

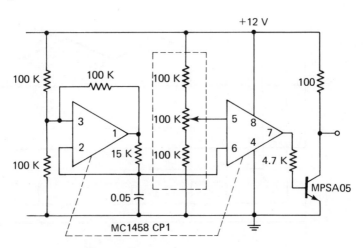

Fig. 9-4. Op-amp comparator PWM motor control circuit

adjustable duty cycle circuit is replaced by a single 15 kΩ resistor that provides a 50% duty cycle. The triangular wave is picked off the timing capacitor, resulting in a slightly curved triangular pattern. The peak and valley rounding is limited by the slew rate of this first amplifier. At frequencies up to about 20 kHz, jump-on and jump-off pulse widths of less than 1% are easily obtained.

The reference input to the comparator section in Fig. 9-4 is shown as a potentiometer in a circuit designed to allow the voltage variation to be one-third to two-thirds of the supply voltage. This allows manual control with full-range operation. If the resistance of the end resistors on the potentiometer is made slightly smaller than that of the potentiometer, there will be a slight amount of dead space at either end of the control. This is desirable to ensure that the output voltage will vary from full-on to full-off under all combinations of resistor tolerance values.

Although the circuit of Fig. 9-4 uses a manual control, the input need not be from a potentiometer. Any voltage signal of the appropriate level may be used, such as might be derived from a motor tachometer, or a positioning feedback system.

9.1.5. PWM motor control with IC comparator

The circuit of Fig. 9-5 uses an IC quad comparator for PWM motor control. This IC, the Motorola MC3302, has four identical voltage comparator stages. Two of the comparators are used for sensing and control, in a circuit

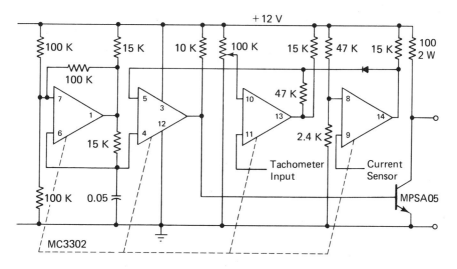

Fig. 9-5. IC comparator PWM motor control circuit

similar to that of Fig. 9-4. The third comparator of the Fig. 9-5 circuit is used as a feedback device, allowing manual control of motor speed with tachometer feedback. The fourth comparator is connected as a current-limiting circuit.

The MC3302 has the unique ability of controlling its output in response to control signals very close to the zero voltage level even when using a single power supply. A small sensing resistor between the load and ground developing a signal of as low as 0.4 V will cause the output of the current-limiter comparator to switch to a high state, raising the reference input to the second comparator, and reducing the output pulse width if the maximum allowable current is exceeded. The diode in the feedback line prevents the current limit circuitry from reducing the reference voltage when the current is below the set point.

9.1.6.   Power output stages

Note that none of the circuits shown in Figs. 9-1 through 9-5 are connected directly to the motor. A power output stage is required in order to apply the PWM signal to a motor. The circuit of Fig. 9-6 is the simplest possible power output stage, and allows a maximum motor stall current of about 15 A. The 2N6286 Darlington transistor offers the simplest and most direct means of

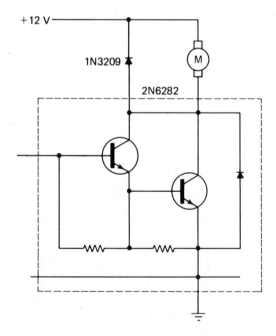

Fig. 9-6. Power output stage for small dc motor control

coupling the pulse energy to the motor. In addition to the transistor, a free-wheeling diode is required. This diode must be able to carry approximately 0.4 of the maximum current expected in the motor (which could be easily larger than the normal operating current if the motor is expected to be stalled for periods longer than a second or two).

9.1.7. Other PWM control circuits

The circuits described thus far provide the designer with a starting point for design of PWM control systems. The circuits are by no means limited to motors. The circuits may be used with minor modifications in such applications as switching power supplies, battery chargers, and other uses requiring a controlled direct current that is reduced from higher supply voltages at high efficiencies.

9.2. INDUCTION MOTOR VARIABLE SPEED CONTROL

The speed of an induction motor is determined, in general, by four factors: the number of poles, the frequency and amplitude of the supply voltage, and the magnitude of the load. When a fixed-frequency source such as the 60 Hz power system is used, the speed may be controlled by varying the applied voltage. This technique has a disadvantage in that torque of the motor is proportional to the square of the applied voltage. Thus, at slow speeds, the load capacity is reduced. If, instead, the voltage is held constant while the frequency is varied, the maximum torque will not decrease as the speed is reduced. The circuits described in this section provide a means of controlling the speed of a permanent-split, capacitor type, induction motor by the variable frequency technique.

9.2.1. General circuit discussion

A permanent-split, capacitor induction motor requires two drive signals 90° apart. A capacitor is normally used in series with one winding to obtain the necessary phase shift when the motor is operated from a single-phase source. Since the capacitive reactance is inversely proportional to frequency, the capacitor cannot maintain the proper phase shift when operation over a range of frequencies is desired. Thus, the capacitor must be eliminated in a suitable variable speed control system.

Figure 9-7 shows the block diagram of a variable-speed control system for induction motors. A pair of flip-flops, operated in *time-quadrature*, can provide the same function as the phase-shifting capacitor. An oscillator is used to set the frequency of operation. A shaping circuit converts the

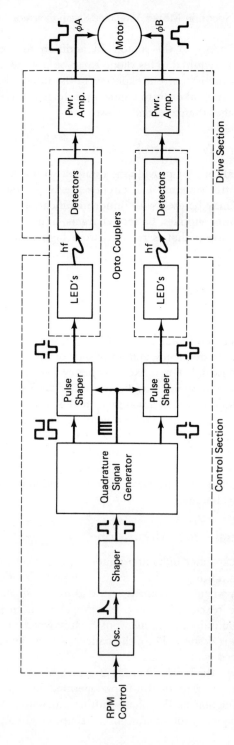

Fig. 9-7. Block diagram of variable speed control for induction motors

oscillator signal to MHTL (Motorola HTL) logic levels. The quadrature generator that follows the shaper performs two functions. First, the generator provides two complementary pairs of quadrature-phased signals. Second, the generator distributes a pulse of fixed width to the pulse shapers. The pulse-shaping circuits combine the square wave with the fixed-width pulse to produce asymmetrical drive signals for the LEDs (light-emitting diodes).

Because of the asymmetry of the drive signals, the duty cycle of each LED is less than 50%. The drive signals are coupled via the optical path (hf) to the power amplifiers. The power stages have a finite switching speed that is dependent upon the type of power transistor used and the load current. The nonsymmetrical drive allows each transistor to turn off before its complement is turned on.

9.2.2. Detailed circuit discussion

Figure 9-8 shows the complete schematic of a variable speed control for induction motors. Note that both digital ICs and discrete components, including unijunction transistors (2N4870) and optical couplers (4N26), are used. Design techniques for both unijunction transistors and optical couplers are covered in the author's *Handbook of Electronic Circuit Designs* (Prentice-Hall Inc., Englewood Cliffs, New Jersey, 07632, 1976).

The unijunction transistor (UJT) is connected as a free-running oscillator in the circuit of Fig. 9-8. The frequency range of this oscillator is 40 Hz to 1200 Hz. Because the logic that follows is in a divide-by-four configuration, the actual drive frequency range is 10 Hz to 300 Hz. This implies a speed range for an induction motor with two pairs of poles, of 300 to 9000 rpm. In a practical circuit, the speed range is limited by a variety of losses.

The resistor in the Base 1 lead of the UJT controls the width of the oscillator pulse. This pulse is then shaped and used to control the duty cycle of the LED drive signals.

The MPS6515 and MPS6519 pulse amplifiers translate the oscillator signal to MHTL levels to drive the Set and Reset inputs of the X flip-flop(FF). Since this FF is of the RS type (set-reset), FF operation is dependent upon the input levels and duration, exclusive of the input rise and fall times.

The $\overline{Q}$ output of X, called $\overline{X}$ on the timing diagram of Fig. 9-9, clocks the A FF, that is used as a divide-by-two toggle, and provides out-of-phase clock signals for B and C FFs. The B and C outputs are shifted 90° with respect to each other. (The MC688 FFs toggle on the negative transition of their clock signals. Since their clock inputs are 180° apart, the B and C outputs are 90° apart, because of the divide-by-two operation.) In this manner, the need for a phase-shifting capacitor in series with one winding of the motor is eliminated.

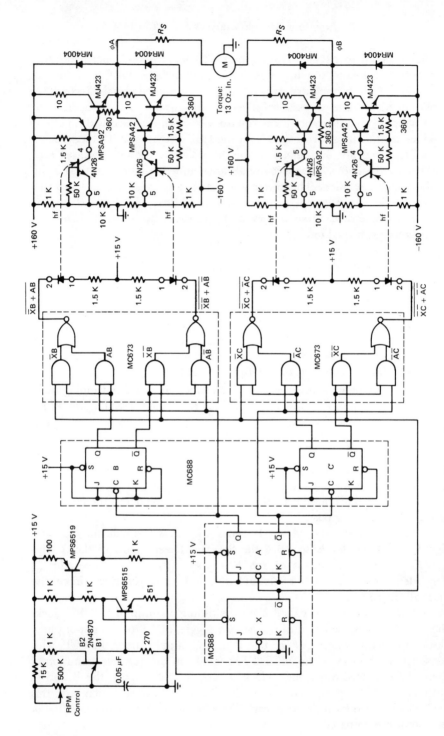

Fig. 9-8. Variable speed control circuit for induction motors

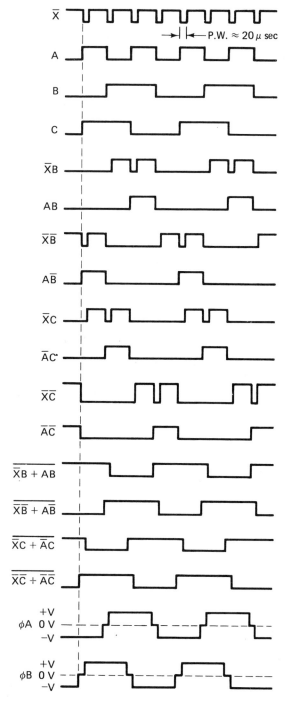

Fig. 9-9. Timing diagram of variable speed control circuit for induction motors

The MC673 AND-OR-INVERT gates combine the B and C square waves with the $\overline{X}$ pulse. This combination results in a zero voltage step in the phase A and phase B drive signals. The MJ423s require about 1 μs to shut off. If the complementary MJ423 was turned on during this interval, current will flow between the transistors without benefit of any load-limiting impedance. This would result in device overstress. To ensure protection, $\overline{X}$ has been set (by the delay through the logic circuits) to produce a 20μs dead time during the switching crossover.

The A *FF*, in addition to providing clock signals for B and C, is also ANDed with the B and C outputs to prevent dead time from occurring in the middle of the drive signals.

The MHTL logic operates from +15 V, whereas the motor drive is ±160 V. Motorola 4N26 optical couplers are used to provide bilateral drive signals from the unilateral control signals. With this technique, the drive circuits can be operated at frequencies down to, and including, direct current without the problems found in typical broadband *RC* coupling (attenuation at low frequencies, etc.). Since photons are the coupling mechanism, and typical input-output isolation of the 4N26 is 1500 V, transients from the motor drive section up to 1500 V will not be transferred to the control section via the 4N26.

A drive circuit, consisting of an MJ423, MPSA92 and a 4N26, is normally OFF, and is turned ON only when the corresponding LED is ON. In this manner, if anything happens to the logic power, the drive circuits are disabled, turning off the motor and providing a fail-safe feature.

The parameters of a high voltage supply are dependent upon the motor and its load. The control section requirements are +15 V and about 50 mA. If a high degree of noise in the motor is anticipated, the control logic power should be suppled via a transformer to prevent noise coupling along an otherwise common voltage return bus.

9.2.3. Circuit application

Figure 9-10 shows a torque-speed diagram of a Class F, 60 Hz, 13 ounce-inch, permanent-split, capacitor-induction motor controlled by the circuit of Fig. 9-8. Note that the load capacity (torque) actually increases as the speed is decreased. This is in contrast to a motor controlled by a variable voltage scheme, where output torque decreases with voltage and speed.

To obtain the curves of Fig. 9-10, resistors were added in series with the windings to limit the current, since the motor reactance is proportional to drive frequency. The value of the series resistor for the motor shown in Fig. 9-10 is 25 Ω at 50 W.

Heat problems. At 1700 rpm and with a 13 ounce-inch load, the motor temperature was approximately 40 °C when operated by the circuit of Fig. 9-8. A 60 Hz sinewave produces a 32 °C case temperature at the

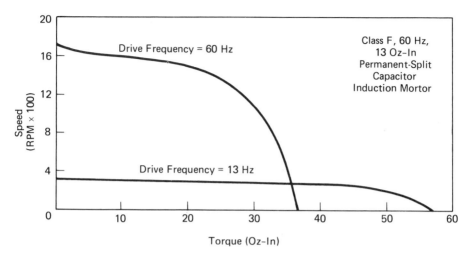

Fig. 9-10. Torque-speed diagram

same speed-load values. The difference in temperature is because of the harmonic components of the squarewave drive signals. Because of the additional heat, care should be taken to ensure that adequate ventilation and heat sinking are provided for the motor when operated by the circuit of Fig. 9-8.

9.3. DIGITALLY CONTROLLED POWER SUPPLIES

This section describes design of two digitally programmable power supplies using Motorola digital-to-analog (DAC) ICs, and Motorola IC voltage regulators. The power supplies can be used for a programmable laboratory power supply, a computer-controlled power supply for automated test equipment, or in industrial control systems. IC voltage regulators are described in Sec. 7.7. DACs (as well as analog-to-digital converters, A/D) are described in the author's *Logic Designer's Manual* (Reston Publishing Company, Reston, Virginia, 22090, 1977).

One of these power supplies uses an MC1408L 8-bit DAC coupled to an MC1723 regulator. This supply has a voltage range from 0 to 25.5 V in 0.1 V increments, and is capable of supplying currents in excess of 100 mA. This value can be extended with the addition of an external current-boost transistor.

The second power supply uses the MC1406L 6-bit DAC and the MC1466L floating voltage regulator to overcome the voltage and current limitations of the standard DACs. The MC1406L DAC floats on the output voltage with the MC1466L regulator. The digital word is coupled into the

DAC with Motorola 4N28 optoelectronic couplers, that allow the DAC to float as high as 500 V. The output voltage of the circuit, designed to supply up to 1 A and 63 V, may be incremented in 63 steps of 1 V each.

9.3.1.   DAC operating principles

Before going into design of the power supplies, let us review the operating principles of DACs. The MC1406 and MC1408 are digital-to-analog converters that have 6- and 8-bit digital inputs, respectively. The inputs are TTL compatible and are converted to an analog output by means of an *R-2R ladder network*, reference current, and current switches.

An *R-2R* ladder network is shown in Fig. 9-11. At each node of the *R-2R* ladder the reference current is split in half. Fifty percent of the current continues on through the ladder, while the other half is shunted into the ladder termination circuits. The currents that flow into the ladder termination circuits are summed together in the output, or shunted to the positive supply, depending upon the digital inputs. The input is a digital word consisting of ones and zeros that turn the appropriate switches on or off. The output quantity is a unilateral current sink with the full-scale current being set by an external resistor and reference voltage. Typically, the reference current is set at about 2 mA, which means that the output will sink current from 0 to $(63/64) \times 2$ mA, with 64 discrete levels for the MC1406, and from 0 to $(255/256) \times 2$ mA, with 256 discrete levels for the MC1408. A single capacitor is used for compensation of the DAC.

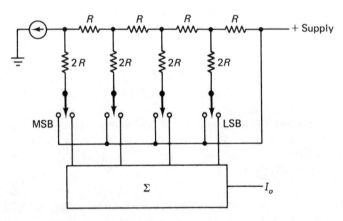

Fig. 9-11. R-2R ladder network

9.3.2.   25 V power supply

The output of the MC1408L is a current sink, and an op-amp is needed to convert from current to voltage, for the 25 V supply. In this particular application, the MC1723 voltage regulator was chosen because it contains the needed op-amp with a current capability in excess of 100 mA. In addition, the MC1723 has an internal 7 V reference which can be used as the reference supply for the MC1408. (The MC1723 regulator is also quite inexpensive.)

Negative supplies. The MC1408 and MC1723 combination, shown in Fig. 9-12, requires two negative power supply voltages. One is a −5 V supply which is used for the MC1723, while the other negative supply is for the MC1408. Minus 15 V is a typical supply value for the MC1408, although this can be less if convenient. However, it should be at least 3 V greater in magnitude than the negative supply for the regulator. The −5 V supply does not require a high degree of regulation, and it can be obtained from a resistor network off the − 15 V supply. The 15 kΩ trimpot is used for calibration of the output voltage. The calibration should be done for the maximum output voltage.

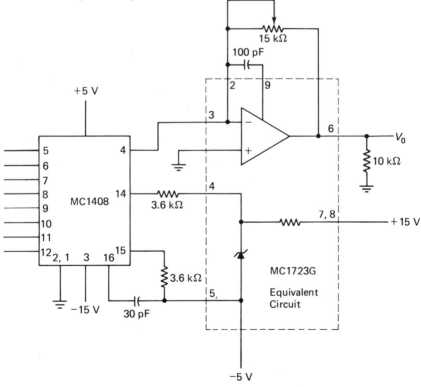

Fig. 9-12. 8-bit DAC with voltage regulator

Positive supplies. Two positive voltages are also required for operation. One is a 5 V supply needed for the MC1408, that is usually available from other TTL power supplies. The second is the voltage to be regulated by

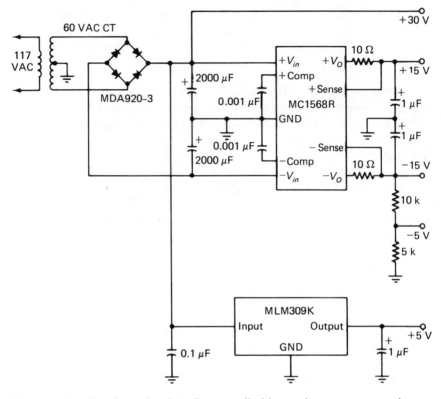

Fig. 9-13. Supply voltage for digitally controlled low voltage power supply

the MC1723, that should be at least 3 V above the maximum output voltage required. To take advantage of the entire range of the MC1408, to 25.5 V with 0.1 V increments, 28.5 V would be required for this supply. One method for obtaining the required voltages for the supply is shown in Fig. 9-13. This technique uses the MC1468 ±15-V regulator for the ±15 V, and an MLM109 5-V regulator for the logic and DAC supply.

Current boost.   The output V_o of the circuit shown in Fig. 9-12 is about 100 mA. The output of the MC1723 can be current-boosted to provide the desired output current range. Figure 9-14 shows a current-boost technique for a 2 A programmable power supply. Other current-boost configurations can be found on the Motorola MC1723 datasheet.

Accuracy, regulation, and settling time.   The MC1408 is accurate to one-half of the least significant bit (LSB). The power supply is set up so that the LSB is 0.1 V, which gives an accuracy of ±0.05 V. The amplifier in the MC1723 is essentially being operated in the unity-gain mode over the entire

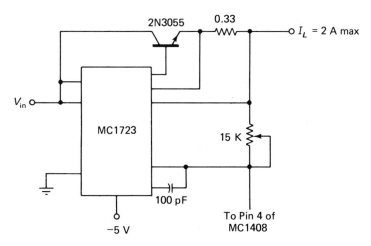

Fig. 9-14. Current-boosted MC1723 voltage regulator

output range. This implies a constant output impedance over the entire voltage range. The output impedance is about 30 milliohm. Line regulation remains about the same as with the MC1723 in a normal configuration. Settling time for a 20 V switch for the MC1723, without current boost, is less than 5 μs.

9.3.3. 63 V supply

This supply uses the MC1466 as the regulator, and the MC1406 for the DAC. The MC1408 can also be used if more resolution is needed. (The MC1406 is a 6-bit DAC, whereas the MC1408 is an 8-bit DAC. Thus, the MC1408 can provide greater resolution.)

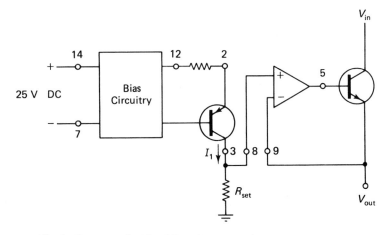

Fig. 9-15. Block diagram of MC1466L voltage regulator

MC1466L operation. The MC1466L regulator is unique in that it operates from a separate 25 V supply, floating on the output voltage which may be as high as desired. The output voltage range is limited only by the capability of the external pass transistors and, in this application, by the voltage rating of the optoelectronic couplers.

The regulated output voltage is established as shown in Fig. 9-15. A reference current I_1, normally 1 mA, is passed to ground through a voltage setting resistor R_{SET}. This technique establishes a reference voltage which is applied to the noninverting input of a differential amplifier. The inverting input of the amplifier is connected to the regulated output in a voltage follower configuration, thus assuring that the output voltage is equal to the reference voltage. If the reference voltage is varied by changing either I_1 or R_{SET}, the output voltage changes accordingly.

Optoelectronic couplers. The 4N28 optoelectronic coupler (also known as an optical coupler) is an *npn* phototransistor and a gallium-arsenide infrared LED packaged together so they are optically coupled, but electrically isolated. Breakdown voltage between the phototransistor and the LED is 500 V minimum. Current-transfer ratio between the LED current and the transistor collector current is typically 30%, with a 10% minimum. The maximum current for the MC1406L digital input logic 0 level is 1.5 mA. Thus, 15 mA of current through the LED assures a logic 0 at the MC1406L input, even at the minimum transfer ratio. The 15 mA diode current is also within the current-sink capability of standard TTL gates, that may be used to drive the diodes.

If higher breakdown voltages or current-transfer ratios are needed in a particular application, other couplers are available. For example, the Motorola 4N25 has a breakdown voltage rating of 2500 V and a 20% minimum transfer efficiency.

Circuit operation. The MC1466L output voltage is set by means of a current through a resistor, and the MC1406L output sinks current in proportion to a digital word input. To take advantage of these conditions, the MC1406L must float with the MC1466L on the output voltage. This is done by powering the MC1406L by the same 25 V that powers the MC1466L. A block diagram of circuit operation is shown in Fig. 9-16. The complete schematic is given in Fig. 9-17.

As shown in Fig. 9-16, I_1 is established at 1 mA by a current source in the MC1466L. The output of the MC1406L, current I_2, is programmed from 0 to 1 mA by the digital word coupled into the MC1406L through the 4N28 optical couplers. When I_2 is at 0, I_3 is 1 mA, and the voltage developed across R_{SET} is at maximum. As I_2 is increased, the current available for I_3 decreases. When I_2 equals 1 mA, the voltage across R_{SET} is zero. Since the MC1466L holds the output voltage V_{out} equal to the voltage developed

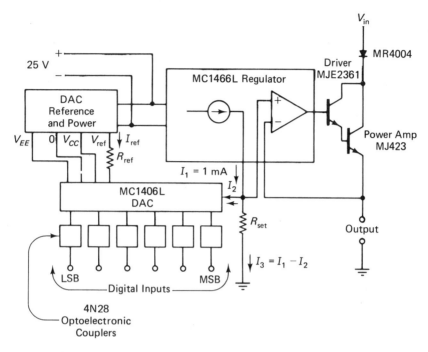

Fig. 9-16. High-voltage supply block diagram

across R_{SET}, V_{out} is programmable by programming I_2.

Figure 9-17 shows how the zero voltage reference for the MC1406L is established by the 1N5235, a 6.8 V Zener diode. This approach places the MC1406L zero reference at approximately the same voltage level as established at pins 3, 8, and 9 on the MC1466L by an internal Zener. Since the output voltage at pin 4 of the MC1406L must stay within 1 diode drop of its zero reference, and since pin 4 is connected to pins 3 and 8 of the MC1466L, the Zener that establishes the MC1406L zero reference must match the MC1466L internal Zener to within about ±0.5 V. This internal zener is CR_1 on the MC1466L datasheet, and is nominally 7.25 V as measured between pins 7 and 9.

The positive supply voltage for the MC1406L is provided by the 1N5231, a 5.1 V Zener. Reference current is established by the 1N827, a temperature-compensated 6.2 V Zener reference diode, and a 5.76 kΩ resistor in series with a 1 kΩ trimpot. For less critical applications, the reference current could be derived from the 5.1 V positive supply Zener or from pin 12 of the MC1466L.

The output voltage is established by the current through the 57.6 kΩ 1% resistor in series with the 10 kΩ potentiometer R_1. This resistor and potentiometer should both be temperature-stable because of the wide

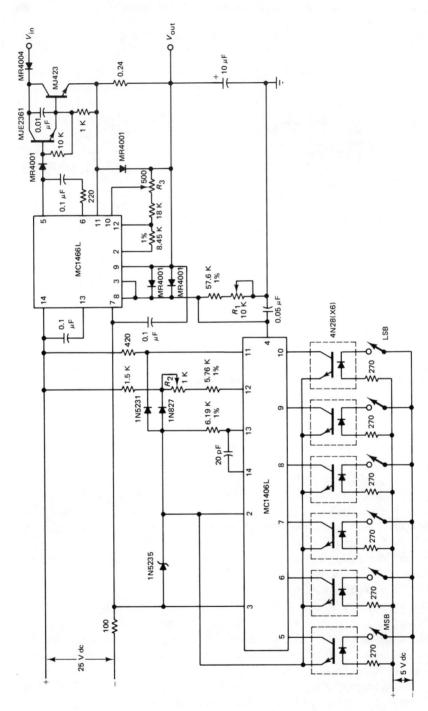

Fig. 9-17. High-voltage power supply schematic

410

changes in power dissipation in these components as the voltage is programmed over its range. Maximum output current is adjustable by means of R_3 and may be set from near zero to over 1 A.

The digital inputs of the MC1406L have internal pull-up resistors to the V_{CC} supply. This technique allows the collectors of the phototransistors in the 4N28 coupler to be connected directly to the digital inputs with no external pull-up resistors required. The emitters all are returned to the MC1406L zero reference at pin 2. The infrared LED current is about 15 mA when operated from a +5 V power supply, assuring that the phototransistor will be well saturated.

The current-amplifier output stage, consisting of the MJE2361 and the MJ423 in a Darlington connection, is designed to operate with a +70 V supply and deliver up to 1 A of load current over the full output voltage range of 0 to +63 V. The MJ423 must have a heat sink adequate to maintain the case temperature at +80 °C while dissipating 70 W if operation at or near 0 V and 1 A output is expected, such as under short circuit conditions. Operation at higher voltages is possible simply by increasing the size of resistor R_{REF}, and increasing V_{in} accordingly. However, when this is done care must be taken not to exceed the power dissipation ratings or the safe operating area curves of the MJE2361 and MJ423 transistors.

Adjustment and performance. The digitally controlled output voltage is initially set to +63.00 V by setting the digital inputs to all zeros (switches closed) and adjusting the 10 kΩ trimpot R_1. Set all the digital inputs to 1 (switches open) and adjust the output to zero volts with resistor R_2. This adjustment should be done with the output loaded rather than open. Without a load, leakage currents may affect the accuracy of the zero setting. The output voltage should now remain within ±0.5 V of its programmed value over the entire operating range of 0 to +63 V. Best overall accuracy may be obtained by checking each of the 6-bits to see which has the greatest error, and then trimming resistor R_2 to split this error with the other bits.

An output filter capacitor is necessary for stability and good load transient response. Minimum value should not be less than 1 µF. Large capacitors will begin to degrade the output voltage slew rate, as shown in the following:

OUTPUT VOLTAGE RISE AND FALL TIMES

OUTPUT CAPACITOR VALUE WITH $R_L = 100 \ \Omega$

	1 µF	10 µF	100 µF
Risetime (t_r)			
(3.2 V to 28.8 V)	7 ms	8 ms	10 ms
Falltime (t_f)			
(28.8 V to 3.2 V)	8 ms	9 ms	30 ms

The above table shows the output voltage rise and fall times when the most significant bit is switched with a 100 Ω load resistor and three different output filter capacitors.

The regulation of V_{in} has virtually no effect on output voltage. However, regulation of the floating 25 V does have an effect, as shown in the following:

OUTPUT VOLTAGES FOR $\pm 10\%$ VARIATIONS
IN ISOLATED SUPPLY VOLTAGE

22.5 V	25.0 V	27.5 V
63.00	63.00	63.00
31.10	31.00	30.93
3.37	3.17	3.02
1.41	1.20	1.05

Alternate voltages. By changing R_{SET}, the maximum output voltage may be set anywhere desired, up to the limits of the available power supply, series-pass transistors, and the optical couplers used. Another possible variation is to set the reference current I_1 from the MC1466L to twice the maximum I_2 of the MC1406, either by doubling the current from the MC1466L or halving the MC1406L reference current into pin 12 of the MC1406L. This would allow twice the resolution over half the output voltage range, for example, from $+31.5$ V to $+63$ V in 0.5 V steps.

The MC1408L, an 8-bit monolithic DAC, offers four times the resolution of the 6-bit MC1406L. Using the MC1408L, the output voltage could be programmed to $+255$ V in 1 V steps. The MC1408L functions in the same manner as the MC1406L except that the input logic is inverted. With the MC1408L, maximum output voltage occurs with a logic 1 (switches open) applied to the 4N28.

9.3.4. Circuit options

The basic circuit of Fig. 9-17 has several options for the input instead of the simple switches shown. For example, the inputs can be from a BCD-to-binary converter so that thumbwheel switches can be used, or the inputs can be from a memory where the digital inputs are momentary. Likewise, the inputs can come from some remote location. The following paragraphs describe these options.

BCD-to-binary conversion. There are a number of thumbwheel switches available for digital applications. These switches read in decimal numbers suitable for human operators, and generally have a BCD (binary-coded-decimal) output. Since the input of the Fig. 9-17 circuit is pure binary, a BCD-to-binary interface is required.

Figure 9-18 shows a BCD-to-binary converter using CMOS MC14008 binary adders. The range of the converter is that of an 8-bit DAC which is from 0 to 255 V. Each BCD input code is converted to the corresponding binary number by means of the binary adders. For example, the BCD input with a value of 40 is converted to a binary 40 which consists of 32, plus 8. Likewise, the BCD number 80 is converted to 64, plus 16. The MC14008 is a 4-bit binary adder which has carry-in and carry-out terminals. A separate adder is needed for a third level of addition. The last ladder in IC_4 is used for the third level since there can be no carry-out from the previous stage.

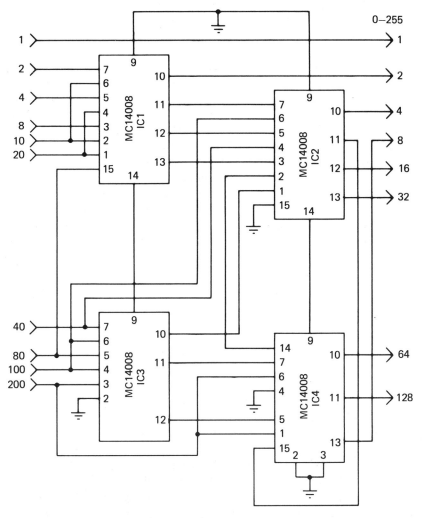

Fig. 9-18. CMOS BCD-to-binary converter

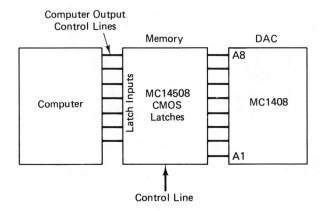

Fig. 9-19. CMOS memory system for programmable power supply

Memory option.   A memory can be used to store the controlling information when the digital inputs to the power supply are momentary, or where the inputs are to be changed while the output is to remain constant. Figure 9-19 shows a block diagram of the MC1408 DAC receiving data from a CMOS memory, the MC14508, that consists of dual 4-bit latches. Each of the inputs to the DAC is connected to the input signal (from a computer or similar source) through the latches. The strobe or control line for the latches controls the flow of data to the DAC inputs. With a logic 1 on the strobe line, the output follows the inputs. When the strobe goes to logic 0, the output retains the state of the data inputs at the time of transition.

This memory option can be used for both the computer-controlled power supply and the laboratory supply. For a laboratory supply, the advantages of the method are apparent when the voltage is set to 9.9 V and the operator wishes to go directly to 10.0 V. Without the series of latches, the output voltage would have to be returned to zero and then changed to the desired 10.0 V, and then this final value transferred to the DAC.

System isolation.   Another requirement for the computer-controlled or remotely controlled power supply is that of isolation between the systems. This can be done with optical couplers to eliminate unwanted ground paths between the computer or controller and the system being controlled.

Motorola's 4N25-4N28 series of optoelectronic couplers may be used for this purpose, and can provide up to 2500 V of isolation between the programmable supply and the remote controlling equipment. The optoelectronic couplers must be used with the high voltage supply, since the inputs to the MC1406 are floating on the output voltage.

The MC1408 DAC is TTL/CMOS compatible. When CMOS ICs are used with the MC1406, or with the MC1408 through optoelectronic couplers, a buffer must be used to supply or sink the required current. This

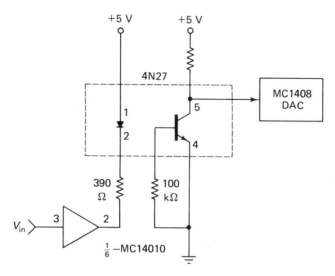

Fig. 9-20. Conversion of CMOS to-opto-to-DAC

buffer can be either a CMOS MC14000, an inverting buffer, or an MC14010 noninverting buffer. Figure 9-20 shows an interface circuit for CMOS ICs through a 4N27 coupler. This arrangement provides 1500 V isolation through the coupler. Note that there are 6 buffers in the MC14010; each buffer is capable of sinking 8 mA at $V_{OL} = 0.5$ V.

9.4. RECOVERING RECORDED DIGITAL INFORMATION WITH INTEGRATED CIRCUITS

This section describes the general problem of recovering digital information stored on a recording medium such as drum, disc, or tape. General circuit approaches, as well as specific circuit examples, are covered. Two design examples of read circuitry, based on the maximum use of ICs, are included here.

9.4.1. General problems in the recovery of digital information.

The design parameters of the read circuitry are determined to a large extent of equipment characteristics and recording method. A review of terminology (expressed in terms of a tape system) serves to show the relationship. Of particular interest are: tape speed, transfer rate, packing density, and encoding format.

Tape speed.   The velocity at which the tape moves past the head is called tape speed. Generally, tape speed is expressed in inches per second (ips). Typical values range from a few ips to several hundred ips, depending on the recording medium.

Transfer rate.   The specification of how fast data is being handled by the recording system is called transfer rate. Generally, transfer rate is expressed in bits per second (or, in the case of parallel recording, bytes per second).

Packing density.   The number of bits per inch (bpi) stored on the tape is called packing density. Generally, packing density varies from a few hundred bpi to several thousand bpi. Density is determined by the tape speed, data transfer rate, and the physical limitation of the tape. Typical rates for various recording systems are as follows:

SYSTEM	TRANSFER RATE (IN BITS/SECOND)
Cassette	2–20 k
Cartridge	50–100 k
	100–500 k
Tape reel	100–200 k
	200–800 k
Disc	1.5–20 M

Encoding format.   The specific manner in which a data bit is represented is called the encoding format. With most encoding formats in common use, data is written by passing current through an inductive recording head as the tape moves past the head, causing a flux transition to be recorded. The polarity of the flux reversal is determined by the direction of the current through the write head. As the tape is read, each flux reversal is sensed by the read head, and this causes peaks to occur in the read signal.

There are numerous encoding methods or formats presently in use. However, *phase encoding* is widely used, and all of the design approaches described in this section are based on phase encoding. Figure 9-21 shows a typical data block to be encoded using the phase-encoding method. As shown, the phase-encoded format records a 1 as a flux transition at the midpoint of the data cell, toward the magnetization level representing erased tape. Logic 0 is recorded as a flux reversal in the opposite direction.

Phase transitions are introduced as needed at the end of each data cell. For example, if a 1 is to be recorded immediately following another 1, there must be a phase transition so that another transition toward erase polarity

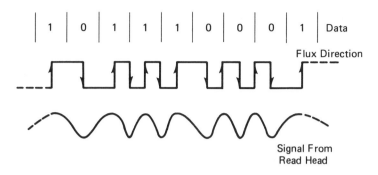

Fig. 9-21. Phase encoding example

can occur at the midpoint of the second data cell. The various possibilities of phase transition, as well as the corresponding analog signal that results from reading such a format are shown in Fig. 9-21.

9.4.2. Basic read circuit without threshold

Figure 9-22 shows a block diagram of a simple read circuit that does not have a threshold function. The *amplification* block increases the level of the signal from the read head to a level adequate to drive the *peak detector*. Read head signal levels vary with such factors as tape speed, whether the system is disc or tape, type of head, and circuitry. Typically, read head signals are in the order of 7 to 25 mV, whereas the peak detector requires about 2 V. (Unless otherwise specified, all signal voltages referred to in this section are peak-to-peak.) Voltage amplification between the read head and peak detector must be in the order of 38 to 50 dB. Usually, two stages of amplification are used. For example, a typical tape cassette system with variable speed tape will use a first stage for most of the gain (say about 35 dB), and the second stage for some gain (about 15 dB) and for gain control.

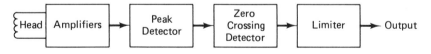

Fig. 9-22. Basic read circuit without threshold

Either op-amp or wideband amplifier ICs can be used for the amplification stages. Op-amps are generally used for low speed equipment with low transfer rates. Examples of op-amps suitable for read head signal amplification include the MC1741, MC1458, MC1709, and MLM301. Wideband amplifiers such as the MC1733 are generally used for high transfer rates, such as disc systems.

As a comparison, the MLM301 has slightly less than a 40 dB open-loop gain at 100 kHz; the MC1741 (a compensated op-amp) has approximately 20 dB open-loop gain at 100 kHz; the MC1733 (a video wideband amplifier) has approximately 33 dB gain up to 100 MHz (depending on gain option and loading).

Peak detector.   There are a number of peak detector configurations. The simplest and most widely used peak detector is a passive differentiator that generates *zero-crossings* for each of the data peaks in the read signal. The circuits used to differentiate the read signal vary from a differential *LC* type in disc systems to a simple *RC* type in reel and cassette systems. Either type attenuates the signal by an amount depending on the circuit used and system specifications. Typically, the *RC* type differentiator attenuates the amplified read signal of about 2 V by 20 dB, resulting in an output from the differentiator of about 200 mV.

Zero-crossing detector.   In most cases, detection of the zero-crossings is combined with the limiter. These functions serve to generate a TTL-compatible pulse waveform with edges corresponding to zero-crossings. An op-amp with series or shunt limiting is often used for low transfer rates. Comparators are used for transfer rates greater than 100 k B/s. An example of an op-amp suitable for comparator applications is the MC1709. When operated open-loop with no compensation, the MC1709 is considerably faster then compensated counterparts such as the MC1714. Suitable comparators for higher transfer rates are the MC1710, MC1514, MC1711 and MC3302.

9.4.3. Read circuits with threshold detectors

The basic read circuit shown in Fig. 9-22 is often modified to include *threshold detection* where there is an output only when the read signal peaks reach a predetermined threshold or amplitude level. This eliminates the possiblity of an erroneous output being produced by minor variations (such as noise) in read signal amplitude. The following paragraphs describe three typical read circuits with threshold detection.

Threshold detection with gate output.   Figure 9-23 is the block diagram of a read circuit that uses a double-ended limit-detector to enable an output NAND gate when either the positive or negative data peaks of the read signal exceed a given threshold. There are two basic methods of implementing this function. With one method, the signal is rectified before the signal is applied to a comparator with a given threshold. The other method uses two

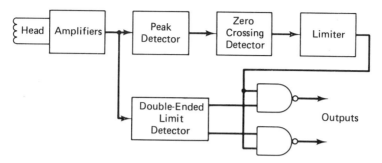

Fig. 9-23. Threshold detection with gate output

comparators, one comparator for positive-going peaks and the other comparator for negative-going peaks. The comparator outputs are combined in the output logic gates.

Threshold detection with a single flip-flop output. In the circuit of Fig. 9-24, the rectifier, peak detector, and zero-crossing blocks provide a clock transition to the flip-flop each time a negative or positive data peak occurs. (This branch of the circuit may or may not include threshold circuitry prior to the peak detector.) The detector in the lower signal path detects whether the signal peaks are positive or negative, and feeds this data to the flip-flop. (This detector can be implemented using a comparator with preset threshold.) The circuit of Fig. 9-24 produces an output only when a positive or a negative (but not both) read signal peak occurs.

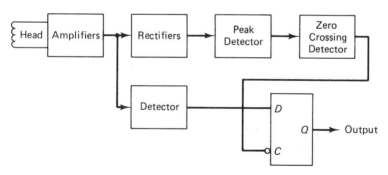

Fig. 9-24. Threshold detection with single flip-flop output

Threshold detection with double flip-flop output. Figure 9-25 shows a read circuit that uses separate circuits with threshold provisions for both negative and positive peaks. That is, both negative and positive peaks produce an output. The peak detectors and threshold detectors may be implemented with two comparators and two passive differentiators.

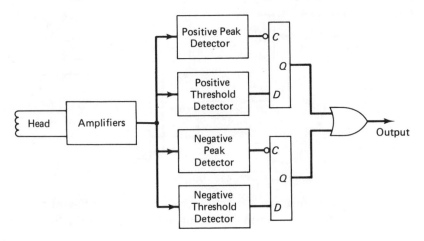

Fig. 9-25. Threshold detection with double flip-flop output

9.4.4. Choosing a read circuit design

Each of the approaches shown in Figs. 9-22 through 9-25 has certain advantages and disadvantages. For example, if cost is all important, the approach shown in Fig. 9-22 is probably the best of the four designs. On the other hand, if threshold sensing is required, and it is desired to have such sensing for both positive and negative peaks, the approach in Fig. 9-25 is best. The final choice is up to the designer, and must be based on such factors as cost, system requirements (circuits driven by the read circuit, etc.).

Section 9.4.5 describes complete circuits for each of the four approaches shown in Figs. 9-22 through 9-25. Section 9.4.6 describes the step-by-step design of a read circuit using the simple approach of Fig. 9-22. Section 9.4.7 covers design of a circuit using the double flip-flop output of Fig. 9-25.

9.4.5   Read circuit theory of operation

The following paragraphs discuss specific circuitry for the approaches shown in Figs. 9-22 through 9-25.

Read circuit without threshold.   Figure 9-26 shows a method of implementing the approach of Fig. 9-22, using two ICs, 8 resistors and 3 capacitors. In the Fig. 9-26 circuit, the read amplification is provided by op-amps IC_1 and IC_2 (both ICs are on a single MC1458 package). The peak detector differentiator is formed by C_2R_7. The zero-crossing and limiter functions are performed by IC_3, an MC1710 comparator.

Typical waveforms are also included on Fig. 9-26. Note that the zero-crossing (differentiated output from C_2R_7) occurs at the peaks of the read signal. Also, the rising edges of the output pulses from IC_3 correspond

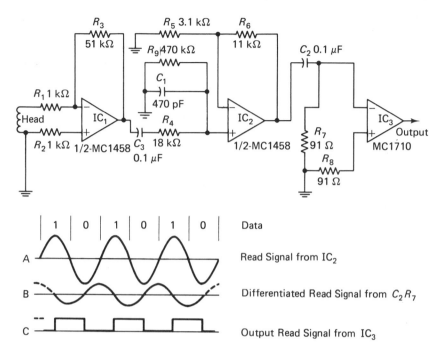

Fig. 9-26. Read circuit without threshold, schematic diagram

to positive peaks of the read signal, whereas falling (negative) edges of the pulse train output correspond to read signal negative peaks.

Threshold detector with double-ended limit detector output. Figure 9-27 shows a method of implementing the approach of Fig. 9-23. In the Fig. 9-27 circuit, the first three stages IC_1, IC_2 and IC_3 amplify, differentiate, and limit the read signal in a fashion similar to that shown in Fig. 9-26. The resultant output is fed to two AND gates, IC_6 and IC_7. The output signals from IC_6 and IC_7 appear only when the gates are enabled by IC_4 and IC_5. Comparators IC_4 and IC_5 provided the double-ended limiter function. IC_6 and IC_7 are enabled only at the positive and negative peaks, respectively, of the read signal. Signal magnitude required to trigger the comparator is set by the positive and negative threshold levels.

Typical waveforms are also included on Fig. 9-27. Note that the leading edge of the pulse train from IC_6 corresponds to the positive peaks of the read signal. The trailing edge of the pulse train from IC_7 corresponds to the negative peaks of the read signals. The output of IC_7 drives a one-shot flip-flop IC_8, which triggers on the trailing edge of the output pulse train from IC_7. The desired result, with the leading edge of the signal from IC_8 corresponding to the negative signal peaks is shown on the bottom waveform of Fig. 9-27.

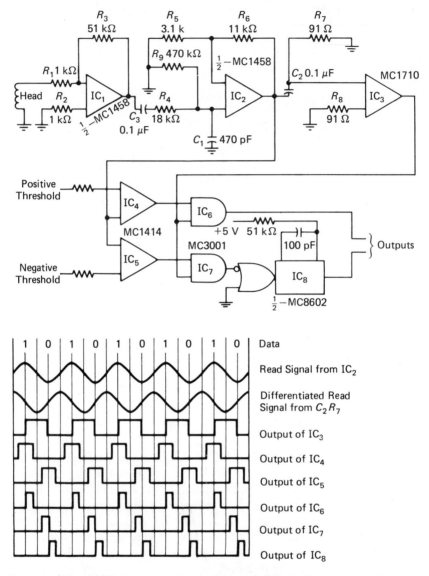

Fig. 9-27. Threshold detector with double-ended limit detector (gate) output, schematic, and waveforms

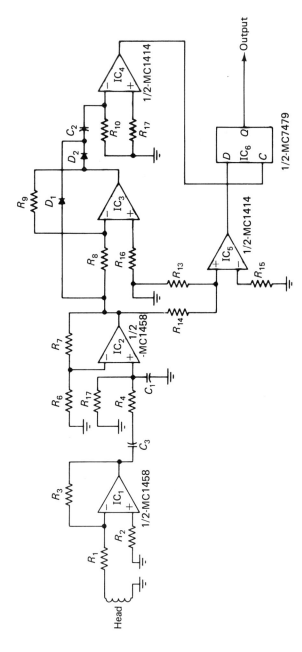

Fig. 9-28. Threshold detector with single flip-flop output, schematic diagram

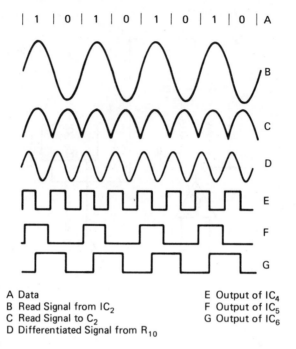

A Data

B Read Signal from IC_2

C Read Signal to C_2

D Differentiated Signal from R_{10}

E Output of IC_4

F Output of IC_5

G Output of IC_6

Fig. 9-28. (*continued*)

Threshold detector with single flip-flop output. Figure 9-28 shows a method of implementing the approach of Fig. 9-24. In the Fig. 9-28 circuit, the flip-flop IC_6 provides an output when clocked by IC_4 which in turn produces a positive-going pulse for each positive and negative peak in the read signal (as shown in the waveforms of Fig. 9-28). IC_1 and IC_2 provide the gain. The amplified read signal drives D_1 directly, and D_2 indirectly (through inverter IC_3, which functions a unity-gain amplifier). The resultant rectified read signal is fed to the differentiator $R_{10}C_2$ and to the zero-crossing detector/limiter IC_4 (a comparator).

A pulse train is fed to IC_6 with each negative transition corresponding to the positive and negative peaks of the read signal. Comparator IC_5 also receives the amplified read signal. IC_5 supplies a data pulse to IC_6 when the positive peaks exceed a threshold established by R_{13}, R_{14}, and R_{15}.

Threshold detector with double flip-flop output. Figure 9-29 shows a method of implementing the approach of Fig. 9-25. In the Fig. 9-29 circuit, the positive and negative peaks of the read signal are processed separately. IC_3 and IC_4 process the negative-going peaks. The positive-going peaks are processed by IC_5 and IC_6. The outputs of IC_4 and IC_6 drive flip-flops IC_8 and IC_9, which serve as data inputs for IC_7 and IC_{10}. The timing is shown by the waveforms of Fig. 9-29.

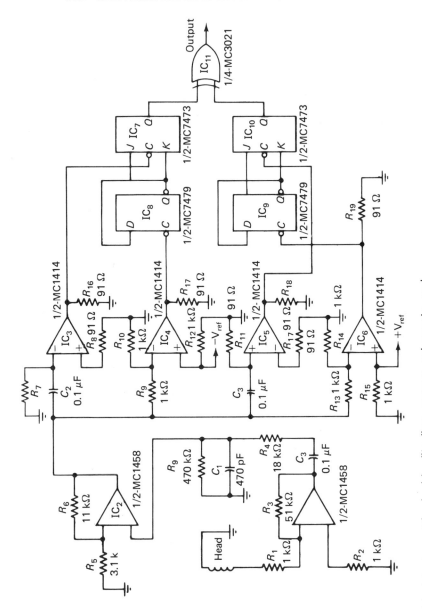

Fig. 9-29. Threshold detector with double flip-flop output, schematic and waveforms

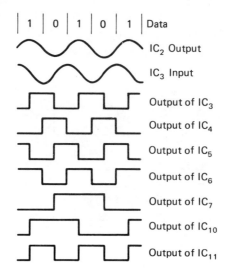

Fig. 9-29. (*continued*)

9.4.6. Design steps for the read circuit without threshold

The following paragraphs describe design of a read circuit without threshold such as shown in Figs. 9-22 and 9-26. (Most of the design steps also apply to the read circuit with double flip-flop output, as discussed in Sec. 9.4.7.) Either design is suitable for a cassette recorder with the following specifications:

FRPI*	1600
Tape speed	15 ips
Head signal	10 mV at 15 ips
Encoding method	phase encoding
Logic	TTL

The complete circuit, with all values, is shown in Fig. 9-26. As discussed, this circuit uses two stages of gain (MC1458), a passive differentiator to generate the zero-crossing, and a comparator as a detector-limiter (MC1710).

Signal level. The minimum signal level to comparator IC_3 was chosen to be 70 mV (140 mV peak-to-peak), since signal levels less than 50 mV can

---

*Flux reversals per inch with a maximum of two bits per time.

cause undesirable oscillation during the transition period. This is a common problem with comparators, and is usually the result of noise and a slowly changing input waveform. The problem can be overcome by providing adequate input signal levels.

Comparator accuracy. An inadequate signal level (below 50 mV) to the comparator can result in a greater difference between where the zero-crossing points occur, and where the MC1710 actually triggers. The resultant error of less than 5 percent (using the 70 mV input) is adequate for this design.

Differentiator design. The differentiator network consists of C_2R_7 which provides the phase shift necessary to generate the zero-crossing at both positive and negative peaks of the read signal. The differentiator network also attenuates the signal. Figure 9-30 shows the phase shift and attenuation (as a function of normalized frequency) for this type of differentiator. As shown in Fig. 9-30, a phase shift and attenuation of 86° and 23 dB, respectively, results from a value of $F = 0.07\,f_c$. This difference of 4° from the

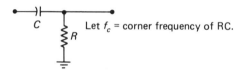

Let f_c = corner frequency of RC.

$\dfrac{*f}{f_c}$	Attenuation, dB	Phase Shift
1.0	3.014	44.97
0.9	3.496	47.98
0.8	4.090	51.31
0.7	4.832	54.98
0.6	5.774	59.01
0.5	6.991	63.41
0.4	8.604	68.18
0.3	10.832	73.28
0.2	14.15	78.68
0.1	20.043	84.28
0.09	20.95	84.85
0.08	21.965	85.42
0.07	23.119	85.99
0.06	24.452	86.56
0.05	26.031	87.13
0.04	27.965	87.71
0.03	30.461	88.28
0.02	33.981	88.85
0.01	40.000	89.43

*Where f_c is the -3.0 dB corner frequency

Fig. 9-30. Differentiator network with corresponding phase shift and attenuation

desired 90° results in an error of less than 5% in regards to actual peak location.

For a value of 0.07 f_c and a minimum frequency f_{min} the corner frequency f_c is:

$$f_c = \frac{f_{min}}{0.07}$$

The minimum frequency f_{min} corresponds to one-half the value of f_{max}, and occurs when alternate 1 and 0 logic states are recorded, and no phase transitions are required at the end of the data cells. The maximum frequency can be calculated from tape speed, and the maximum flux reversals per inch as:

$$f_{max} = \frac{1600 \text{ FRPI}}{}$$

$$f_{max} = \frac{1600 \text{ flux reversals}}{\text{inch}} \times \frac{15 \text{ inches}}{\text{second}} \times \frac{\text{cycle}}{2 \text{ flux reversals}}$$

$$= 12 \text{ kHz}$$

Therefore, $f_{min} = \dfrac{12 \text{ kHz}}{2} = 6 \text{ kHz}.$

The corner frequency that provides less than 5% phase error is:

$$f_c = \frac{6 \text{ kHz}}{0.07} = 86 \text{ kHz}.$$

Choosing a value of R_7, much less than the parallel input of the IC_3 comparator, will assure that the corner frequency is determined by $C_2 R_7$ only. With this in mind, R_7 is chosen to be 91 Ω, and C_2 is calculated to be 0.1 μF.

The 23 dB of attenuation, in conjunction with the desired level of 140 mV (peak-to-peak) into IC_3, requires a minimum drive level to the differentiator of 2 V at 6 kHz. At 12 kHz, the attenuation is 17 dB for the given values of C_2 and R_7. Thus, the minimum drive level to the differentiator at 12 kHz is 1 V.

Read amplifier requirements. Design of the read amplifier involves such considerations as how much gain is required, what is the bandwidth requirement, and how much outband rolloff is required.

The *overall gain* requirement is straightforward in that the 10 mV signal available from the read head must be amplified to the previously determined level of 2 V (at 6 kHz) to drive the differentiator. This requires an overall voltage gain A of: $A = 2$ V/10 mV = 46 dB. Correspondingly, the required gain at 12 kHz is 40 dB.

The required *bandwidth*, based on the maximum frequency of interest, is 12 kHz. The *outband rolloff* includes among other factors the response of the differentiator. In this example, the differentiator is actually a high-pass filter that readily passes noise higher than 12 kHz (at 6 dB per octave). (That is, noise figure increases directly with attenuation. The severity of this problem depends on the signal level going into the differentiator, and the filter characteristics.) Therefore, it is highly desirable to incorporate at least 6 dB per octave of outband rolloff in the read amplifier. The actual filter used in practical circuits varies with the system and manufacturer. For example, it is not uncommon to find read amplifiers with 18 dB per octave of outband rolloff in a reel tape system.

The values shown in Fig. 9-26 provide 12 dB per octave of outband rolloff using the normal 6 dB per octave of IC_1 (with the corner frequency properly chosen), and the 6 dB per octave rolloff of a simple RC low-pass filter. Both corner frequencies should be at 19 kHz to achieve an overall 3 dB bandwidth of 12 kHz. Placing the corner frequency of IC_1 at 19 kHz results in a closed-loop gain of 32.5 dB at 12 kHz. The value of R_3, with R_1 assumed to be 1 kΩ, is approximately 52 kΩ. This is obtained from $A = R_3/R_1$, where $A = 34$ dB (which is the gain if there were no rolloff at 12 kHz).

The second network R_4C_1 is also selected to have a corner frequency of 19 kHz. Choosing a convenient value of $C_1 = 470$ pF results in a value of 18 kΩ for R_4.

As discussed, the overall gain of the read amplifier (IC_1 and IC_2) must be 46 dB at 6 kHz and 40 dB at 12 kHz. The gain of IC_1 is set at 34 dB (excluding the 1.5 dB rolloff at 12 kHz) by the ratio of R_3/R_1. To ensure that adequate overall gain is available, the second stage is designed for 6 kHz, rather than 12 kHz. The gain of IC_2 must be sufficient to provide the difference between the total required gain (46 dB), the first stage IC_1 gain of 34 dB, and the rolloff. Allow a total of 1 dB for rolloff (0.5 dB for IC_1 and 0.5 dB for the low-pass filter). Thus, the required gain of IC_2 is 13 dB ($46 + 1 = 47$; $47 - 34 = 13$).

The gain of a noninverting amplifier (Chapter 3) is: $A = 1 +$ (feedback resistance/input resistance). Assuming a value 11 kΩ for R_6, R_5 should be approximately 3.1 kΩ.

13 dB is an approximate voltage gain of 4.5;

$$4.5 \approx 1 + \frac{11 \text{ k}\Omega}{3.1 \text{ k}\Omega}$$

Using this design, the overall gain of the read amplifier at 12 kHz is 34 dB + 13 dB − 3 dB, of 44 dB. This value is 4 dB higher than the minimum required value of 40 dB. Also, the overall gain at 6 kHz is 46 dB, there is a 3 dB bandwidth of 12 kHz, and an outband rolloff of 12 dB/octave.

9.4.7. Design steps for the read circuit with threshold

The following describes design of a read circuit with a double flip-flop output such as shown in Figs. 9-25 and 9-29. Most of the design steps of Sec. 9.4.6 apply. Except for IC_7 through IC_{11}, and IC_4 and IC_6, the design is identical. For example, although two differentiators (C_2R_7 and C_3R_{11}) are used, the corner frequency for each is again 86 kHz. Also, the required signal level to IC_3 and IC_5 is approximately 70 mV. The read amplifier design is also identical to Fig. 9-26.

The threshold circuitry for IC_4 and IC_6 is unique for the circuit of Fig. 9-29. The threshold of either IC can be varied by changing $+ V_{ref}$ and $- V_{ref}$. For example, increasing $+ V_{ref}$ will require a large signal to the noninverting terminal of IC_6, thus providing a higher degree of noise immunity. Typical values of V_{ref} (both positive and negative) were experimentally determined to be about 1 V for a typical cassette.

9.5. POWER CONTROL USING AN IC ZERO-VOLTAGE SWITCH

An IC zero-voltage switch is used to trigger triacs, SCRs, and other power-control devices, during the time when the power-line potential is passing through the zero voltage region. The advantages of a zero-voltage switch are to minimize radiated interference, arcing in mechanical switches, and stress on the power-control components.

When a dc power source is switched to a load, there is a step change in load voltage. If there are reactive components in the load, this voltage step may generate ringing in the load current as shown in Fig. 9-31. The voltage step generates two undesirable current components; the high rate of current change during the *inrush* period, and *ringing* which is a high-frequency oscillation.

If an ac voltage is switched to a load at any time other than when the voltage is near zero potential, a similar condition occurs, as shown in Fig. 9-32. Both the inrush current and the ringing currents are capable of radiating strong interference into other electrical control and communications systems. Also, the instantaneous voltage steps at turn-on and turn-off with reactive loads can be sources of arcing in mechanical switches, and cause severe stress to semiconductor thyristors.

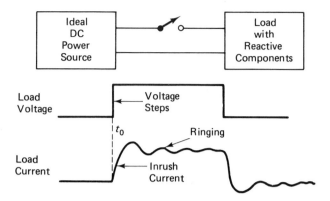

Fig. 9-31. DC turn-on disturbance

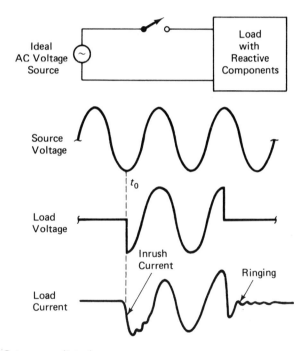

Fig. 9-32. AC turn-on disturbance

This section discusses the Motorola MCF8070 zero-crossing switch. The MCF8070 will turn on a thyristor at the beginning of a half-cycle of load voltage (near the zero-voltage point). The thyristor then conducts until the zero point of load current. Such action eliminates voltage steps in loads, and minimizes problems of interference and stress on switching devices.

9.5.1. Operation of the MCF8070 zero-voltage switch

A block diagram of the MCF8070 is shown in Fig. 9-33. Note that the circuit is composed of a zero-voltage detector, a pulse gate, an input-control section, and a charging circuit.

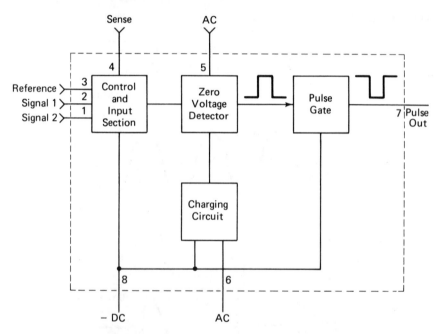

Fig. 9-33. Zero-voltage switch

Zero-voltage detector. As indicated by the current arrows in Fig. 9-34, base current flows in the *pnp* transistor for both polarities of the line voltage. During the intervals when the line voltage is near zero potential, current stops and the pulse gate inhibiting current ceases, allowing the pulse gate to conduct. Output pulses can occur only when the *pnp* transistor in the voltage detector loses base current during line voltage zero-crossings. However, the input-control section can also furnish base current to the voltage detector transistor. When this happens, the voltage detector delivers the inhibiting

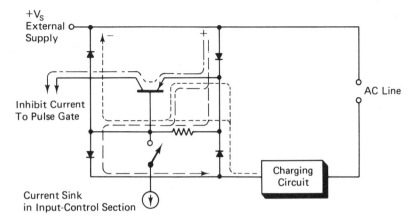

Fig. 9-34. Zero-voltage detector

signal to the pulse gate, even during zero-crossings, and no output pulses are produced.

Pulse gate. As shown in Fig. 9-35, the pulse gate is a Darlington amplifier switch with a current limiting resistor in series with the pin 7 output. A saturated switch is connected between the Darlington input and the negative supply. This switch saturates whenever inhibit current flows from the zero-voltage detector. When the switch is saturated, the Darlington input current is shunted through the saturated switch and the Darlington does not conduct.

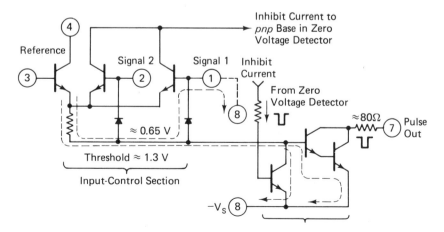

Fig. 9-35. Input-control section and pulse gate

Input-control section. The input-control section is basically a simple differential amplifier. There are two identical control signal input transistors in parallel, and the circuit may be used or described with respect to either one alone. The input-control circuit voltages are measured and described with respect to pin 8, the negative side of the dc supply.

Assume both signal inputs (pins 1 and 2) are open. With the dc supply in the normal region of approximately 10 V, current flows into the pulse gate from the input-control section when the reference input (pin 3) reaches approximately 2.5 V.

Assume the reference potential is set to 5 V. With the input signal terminals open, a constant current flows into the pulse gate. The pulse gate then conducts, or not, depending upon the inhibit current from the zero voltage detector.

If pin 1 or 2 is now connected to pin 8, as might happen with a short-circuited sensor, the pulse gate input current is diverted to pin 8 through the short circuit protection diodes on the two signal input terminals. Thus, with a short-circuited sensor at either input, the pulse gate cannot operate because its drive has been diverted through a protection diode.

If either of the signal inputs is connected to a higher potential than the reference, that input transistor then conducts, and the reference transistor becomes biased off by the increase in its emitter potential. In this mode, the input transistor collector current is drawn from the zero-voltage detector, maintaining the flow of inhibit current, even through the zero-crossing region.

The following conditions must be satisfied to maintain operation of the pulse gate.

1. The reference voltage must be in the 2.5 to 10 V range.
2. Both input signals must be higher in potential than 1.5 V.
3. Both inputs signal must be 0.7 V (or more) below the reference voltage.

In turn, the pulse gate can be inhibited in any one of the following ways:

1. By dropping the reference 0.7 V below one of the input potentials,
2. By raising either of the inputs 0.7 V above the reference,
3. By shorting either signal input to pin 8.

Combinations of these operating modes give a wide range of possibilities to the circuit designer, both in signal states, and types of external sensors and input devices.

Charging circuit. The charging circuit is shown in Fig. 9-36. Current flow through the dc supply external capacitor C_S and through the MCF8070 is limited by an external resistor R_S. The drop across the internal Zener

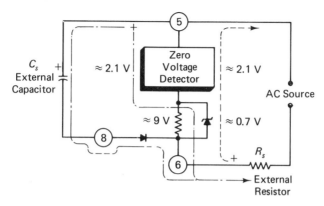

Fig. 9-36. Charging circuit

diode is approximately 9 V when pin 5 is positive with respect to pin 6. This limits the maximum voltage between 5 and 6 to the sum of the Zener IR drop (9 V), and the approximately 2.1 V drop of the detector, or to about 11.1 V. The peak charging voltage across C_S is then approximately 11.1 V, less one diode drop (0.7 V) or about 10.4 V.

For most circuit operations, this approximately 10 V charge on C_S serves as a voltage source to power the pulse gate and input-control section, plus any external sensors (bridge circuits, etc.). During the half-cycle of line voltage when pin 5 is negative with respect to pin 6, the Zener conducts as a forward-biased diode, and no charge is added to C_S.

9.5.2. Basic zero-voltage switch circuits

The following are some typical circuits in which the MCF8070 can be used.

Zero-voltage switch with external sensor. Figure 9-37 shows the MCF8070 operated by an external bridge-type sensor. Any one of the resistors R_1 through R_6 can be used as a control sensor. For example, R_1 could be a temperature-sensing thermistor, with the remaining resistors at some fixed value. Since pins 1 and 2 are redundant, two resistors (R_1R_2 or R_3R_4) may be eliminated, and the unused pin left open, unless the control is to be a function of two variables. Pins 1 and 2 may also be connected together for single sensor input.

Zero-voltage switch with mechanical on–off. Figure 9-38 shows the MCF8070 operated by a mechanical switch. For this function, only three resistors need be used to bias the input-control section away from the inhibit mode. The unused input is then controlled with the switch. Shorting pin 1 to pin 8 with the switch inhibits output pulses, and inhibits power flow in the triac.

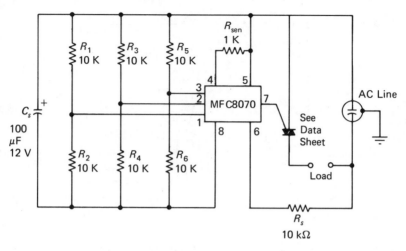

Fig. 9-37. Basic zero-voltage switch circuit with external sensor

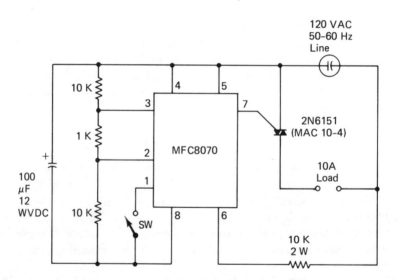

Fig. 9-38. Zero-voltage switch with mechanical on–off

Zero voltage switch with optical coupler control.   Figure 9-39 shows the MCF8070 operated by an electrical signal through an optical coupler. In this circuit, the triac is pulsed on at each zero crossing, if there is no current flow into the optical coupler. When the optical coupler phototransistor is turned on by the LED, pin 2 potential rises above that of pin 3, inhibiting output pulses. To invert the mode of operations, interchange pins 2 and 3 which makes the switch work in the normal open or off mode.

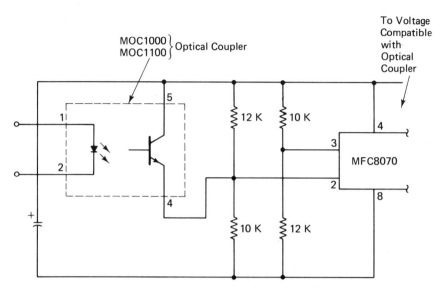

Fig. 9-39. Zero-voltage switch with optical coupler control

9.5.3. Using zero-voltage switches to drive SCRs

When the zero voltage switch is used to drive an SCR, additional circuitry is required. This is because the MCF8070 sinks current through its pulse gate, rather than sourcing current. Figure 9-40 shows one method of driving an SCR with the MCF8070. This method is for a single SCR driving a half-wave load. Figure 9-41 shows the same devices driving a full-wave load through a full-wave bridge.

In both circuits, an external rectifier is used for charging the supply capacitor. The tables in Figs. 9-40 and 9-41 list some possible combinations of devices and components for a 120 V ac service. The values for Q_1, R_S, R_1, R_2, and R_3 in Fig. 9-40 also apply to corresponding current values in the table of Fig. 9-41.

9.5.4. Using the dc supply with zero-voltage switches

The external dc supply capacitor is limited in its charging rate by the external resistor R_S. An approximate expression for the maximum amount of charge increase on each full cycle is:

$$\Delta E_{CS} \approx \frac{2}{6.28 \text{ V } R_S C_S} \approx \frac{0.45}{R_S C_S} \text{ volts/cycle}$$

where R is in ohms and C is in farads. For $R = 10$ kΩ and
$\quad\quad\quad C = 100$ μF:

$$\Delta E_{CS} \approx 0.45 \times 10^{-4} \times 10^4 \approx 0.45 \text{ V/cycle}$$

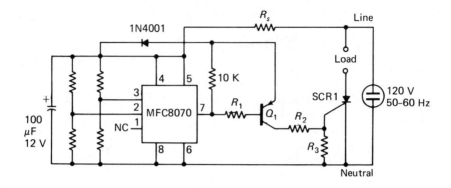

Load Amperes	Plastic SCR	Metal SCR	Q_1	R_1	R_2	R_s	R_3
0.5	—	MCR206	MM869B	10 K	2.7 K	10 K 2W	1 K
0.8	MCR120	—	MPS6517	10 K	2.7 K	10 K 2W	1 K
	2N5064		MPS6517	10 K	2.7 K	10 K 2W	1 K
1.6	—	2N1597	MM869B	10 K	10 K	10 K 2W	1 K
4.0	MCR107.4	—	2M3905	10 K	470	10 K 2W	10 K
	MCR407.4		MPS6517	10 K	4.7 K	10 K 2W	1 K
8.0	2N4443	—	2N3905	270	120	10 K 2 W	10 K
	—	2N4154 2N4170 2N4186	2N2907A	270	120	10 K 2W	10 K
16.0	—	2N1846	2N2907A	270	120	10 K 2W	10 K
20.0	—	2N6168	2N2907	270	120	10 K 2W	10 K
35.0	—	2N3871 2N3879 2N6172	2N2907	270	120	10 K 2W	10 K

Fig. 9-40. Half-wave SCR drive

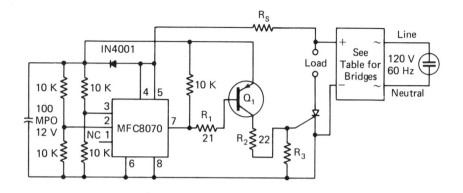

Load Current	Bridge Type
1.0 A @ 70°C	MDS920-4
1.8 A @ 40°C	MDA922-4
2.5 A @ 55°C	MDA960-3
4.0 A @ 55°C	MDA970-3
12 A @ 55°C	MDA980-3
27 A @ 55°C	MDA990-3

Fig. 9-41. Full-wave SCR bridge

To charge the nominal 10 V operating voltage it takes $10/0.45 = 22$ cycles of line current or 366 ms. This time can be decreased by decreasing C_S or R_S.

For 120 V ac operation we are limited as to the smallest value of R_S by the maximum dissipation ratings of the MCF8070 to about 5 kΩ.

With a 5 kΩ resistor for R_S, a 100 μF capacitor can be recharged approximately 0.9 V per cycle. This places a limit on the rate at which current can be drawn from the capacitor to that equal to its maximum charge rate.

With a worst-case pulse load (that is, with pins 5 and 7 shorted), the pulse gate sinks approximately 125 mA for pulse widths of approximately 150 μs twice per cycle.

A typical sensor bridge circuit of approximately 10 kΩ per element draws an additional 2 mA from the capacitor. Also, there is a nominal 1 mA drain through the input-control section for a total device-sensor drain of approximately 3 mA.

The discharge rate for the 100 μF capacitor under these worst-case conditions is approximately 0.875 volts per cycle. Thus, the normal capacitor charge rate barely exceeds the discharge rate.

If the loading on the capacitor supply is increased, the supply voltage will fall, reducing the load currents until equilibrium is reached between the capacitor's charge and discharge rates. As a guideline for new circuit designs, verify that the capacitor voltage is at least 7 V in all modes of operation.

The actual pulse load is less than the worst-case figures used here. Thus, there is a margin for low current interfacing devices to be used to control the power delivery modes of the MCF8070. Examples of these devices are Motorola McMOS logic devices (Chapter 8), and Motorola MOC1000-1100 optical couplers.

9.5.5. Temperature controller using the external supply

Figure 9-42 shows the MCF8070 used to provide temperature control of a load. Circuit waveforms are given in Fig. 9-43.

The temperature control operates on the principle of *dual ramp comparison*. Both sections of a CMOS dual-D flip-flop are clocked at the line frequency rate through the 1N4001 diode, and the 47 kΩ pull-up resistor. The first flip-flop drives the two integrating networks from the $\bar{Q}$ output. As long as $\bar{Q}$ is in the high state, current flows through the networks, charging the two 10 μF timing capacitors. When the first capacitor has charged to the CMOS logic threshold, the logic value at the first D input is taken as a high at the time of the next clock, changing the state of F_{F1}.

When Q_1 goes low, both 10 μF capacitors discharge, the first through the 1N4001 and the second through the gate junction of the 2N5457 FET. At the time of the next clock pulse, Q_1 goes high again, which starts another ramp. Because the capacitor was discharged during the previous cycle, the D input is now under its threshold voltage.

The time it takes (number of cycles) for the first ramp to reach threshold is controlled by the resistance setting of the 50 kΩ variable resistor. For a given setting, the ramp may build up for, say, 15 cycles and be discharged for one cycle, giving a 16 cycle sequence length. This is the reference against which the second ramp works.

The second ramp period is controlled by the resistance value of the thermistor which has a negative temperature coefficient (NTC). When the ramps start up, the MCF8070 pulses the triac across the threshold. At this point, by inhibiting the MCF8070, power is removed from the load, until both ramps are restarted by the first flip-flop. Thus, the percentage of time that the load received power is a ratio of the thermistor resistance to the variable resistance. As the thermistor heats and its resistance drops, less power is delivered to the load until equilibrium is reached. The point of equilibrium is a function of the variable resistance.

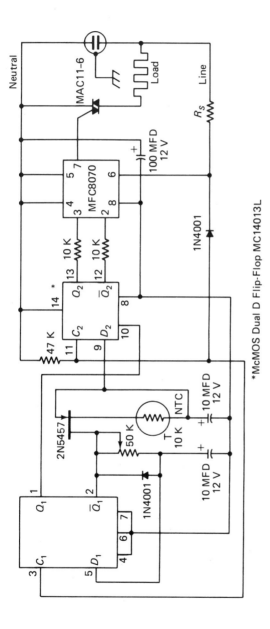

Line Voltage	R_S
120 VAC	10 K 2 W
240 VAC	20 K 4 W

*McMOS Dual D Flip-Flop MC14013L

Fig. 9-42. Temperature controller using external supply

441

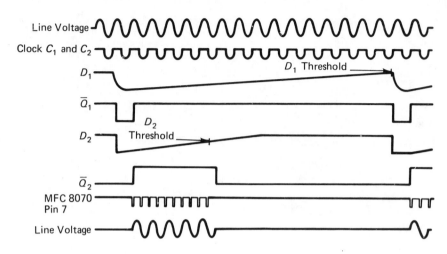

Fig. 9-43. Temperature controller waveforms

A nominal 10 Ω thermistor was used for the design of Fig. 9-42. If a 50 Ω thermistor had been used, the 2N5457 would not have been used, and the second capacitor would have been discharged through a 1N4001.

9.6. DIGITAL VOLTMETER USING AN IC DUAL RAMP SYSTEM

This section describes design of a $3\frac{1}{2}$-digit digital voltmeter (DVM) using Motorola MC1405 and MC14435 ICs. These two ICs combine to provide an analog-to-digital (A/D) package using the dual ramp technique.

9.6.1. Dual ramp system

There are many techniques of A/D conversion, each having different characteristics and each favoring different applications. The dual ramp technique of A/D conversion provides an inexpensive method of obtaining high accuracy, which makes it ideal for DVM type applications. A complete discussion of other A/D techniques is found in the author's *Logic Designer's Manual* (Reston Publishing Company, Reston, Virginia, 22090, 1977). Here, we discuss the dual ramp system briefly, to help understand DVM circuit operation.

Figure 9-44 shows a block diagram and corresponding waveforms of the basic dual ramp A/D converter. The dual ramp conversion cycle consists of two basic time periods. Time period T_1 results from the input

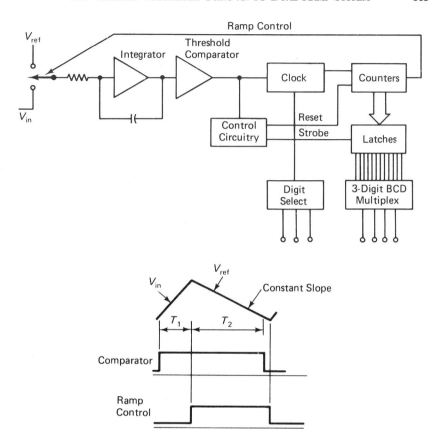

Fig. 9-44. Dual ramp system block diagram and waveforms

unknown voltage V_{in} being integrated for a fixed time interval. This integration results in the output voltage of the integrator being proportional to V_{in}. At the end of the time period T_1, the reference voltage V_{ref} is applied to the integrator, causing the integrator output voltage to decrease. This integration continues until the output voltage again reaches the zero reference level. This time interval T_2 is the down-ramp time period.

Time period T_1 is constant for each conversion cycle. The time interval T_2 is dependent upon the input unknown voltage. Referring to the dual ramp waveform, $T_2/T_1 = V_{in}/V_{ref}$.

Considering the four variables in this relationship, T_1 is a fixed time period, T_2 is measured from the start of the down-ramp time period until the zero level is reached, and V_{ref} is calibrated into the system. The only remaining variable in the equation is V_{in}, which is the analog voltage input to be determined. By counting out a time period T_1, measuring the down-ramp time interval T_2, and calibrating the reference voltage, the dual ramp A/D conversion technique determines the value of an analog input voltage.

The A/D conversion method also eliminates inaccuracies caused by integrator capacitor drift and time or temperature drift in clock frequencies that are common to other A/D systems. These advantages result from the fact that the same integrator capacitor and the same clock frequency are used during both the up-ramp time period T_1, and the down-ramp time period T_2 when using the dual-ramp system. For example, if the clock frequency has drifted up 10 percent over some time period, both T_1 and T_2 will be 10 percent less, thus cancelling the effect of the drift. Likewise, if the capacitor value drifts over time or temperature, the integrator voltage will be higher or lower at the end of T_1, and at the beginning of T_2. This eliminates the errors owed to a changing capacitor value.

9.6.2. MC1405 and MC14435 operating principles

The MC1405 and MC14435 together produce a complete dual-ramp A/D. The MC1405 is bipolar, whereas the MC14435 is CMOS. Thus, the system takes advantages of both technologies.

MC1405. Figure 9-45 is a block diagram of the MC1405. As shown, the MC1405 contains the integrator, as well as the analog switch required to switch between the unknown input and the reference voltage. The MC1405 also includes the reference voltage, reference voltage to current converter, unknown voltage to current converter, and a comparator for the integrator output.

The selection of clock frequency and power supply voltage determines the integrating capacitor value. A polar capacitor may be used with pin 7 of the MC1405 connected to the + terminal. However, setting time will be increased if electrolytics are used. Tantalum electrolytics are preferred.

MC14435. Figure 9-46 is a block diagram of the MC14435. As shown, the MC14435 contains the counters and control circuitry of the $3\frac{1}{2}$-digit dual-ramp converter. The MC14435 has a $3\frac{1}{2}$-digit BCD output with 3-digits multiplexed, and the half-digit unmultiplexed on a separate output pin. An overrange indication is provided, as well as a display update pin that allows the data rate into the latches to be slowed or stopped. The multiplexed BCD outputs are normally low, whereas the digit select lines are in a normally high state.

The clock capacitor across pins 3 and 4 of the MC14435 determines the frequency of the system clock. The oscillator frequency is dependent on the capacitor value, the operating voltage, and some process variations. The

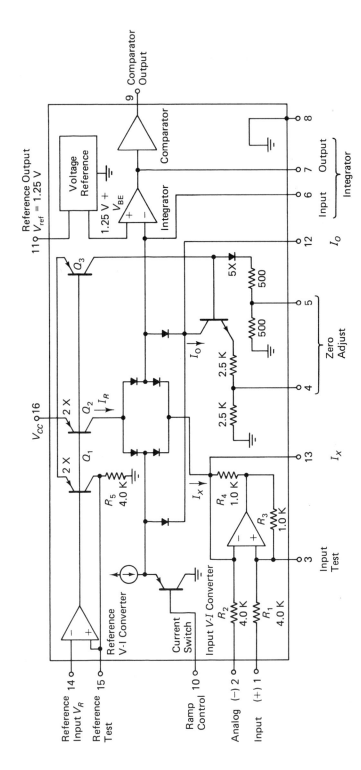

Fig. 9-45. MC1405 A/D converter analog subsystem

445

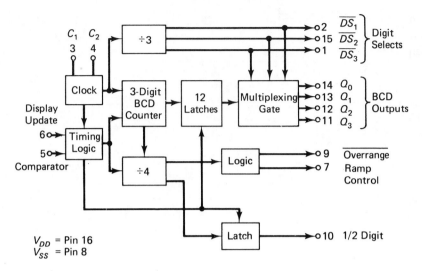

Fig. 9-46. MC14435 digital subsystem

dual-slope system may be operated from 1 kHz to 1 MHz, depending upon application. A full-scale conversion requires 3100 counts. Thus, up to 300 conversions per second are possible. However, when used to drive a display, such update rates are annoying to a viewer. 10 kHz is often chosen to provide about three conversions per second in visual display instruments. For 15 V operation, a clock frequency of 10 kHz requires a 0.001 μF capacitor (as indicated by the IC datasheet).

Interconnections. There are two connections between the MC1405 and MC14435. The comparator output of the MC1405 indicates whether the integrator output is above or below the threshold level. The ramp control output from the MC14435 determines whether the input to the integrator is the unknown voltage or the reference voltage. At frequencies below 200 kHz, an *RC* delay may be required on the output of the MC14435 ramp control. This time delay indirectly assures that the oscillator restarts (controlled by the comparator) are synchronized with each new conversion. At 10 kHz, the delay should be sufficient to extend the comparator off time to 50 μs. That is one-half the period of the clock frequency.

Offset circuits. The MC1405 and MC14435 also include offset circuitry to stabilize voltage readings at or near zero. With the basic dual-ramp system, near zero inputs will only charge the integrator capacitor to very small voltage levels. Then, during the down-ramp period, noise can cause the comparator to trigger prematurely, resulting in instability of the output display. By adding a constant current to the input of the integrator, and subtracting out the number of clock pulses produced by this offset current, the Motorola dual-ramp system never reaches the undesired zero input conditions.

Power supplies. The two ICs will operate from a single power supply between +5 and +15 V. However, the DVM described here operates from 15 V, and requires a negative voltage in addition to the positive supply voltage. This is used for the input and autopolarity circuitry.

9.6.3. DVM circuit description

The complete DVM schematic is shown in Fig. 9-47, and includes the digital readout, autopolarity circuit with indicator, high impedance inputs, and overrange indicator. There are three input voltage ranges that allow the DVM to measure a dc voltage of 0–1.999 V, 0–19.99 V and 0–199.9 V.

The MC1405 has two calibration potentiometers R_5 and R_6. R_5 sets the full-scale calibration and R_6 is used for zero adjust. Capacitor C_2 is the integration capacitor. For $3\frac{1}{2}$-digit continuous conversion, such as in this DVM, an electrolytic capacitor may be used. Capacitor C_3 is added to the MC14435 to set the frequency of the internal oscillator. With the 0.001 μF value shown, the clock frequency is about 10 kHz with a 15 V supply.

Output indicators. The output of the DVM consists of the $3\frac{1}{2}$-digit readout for voltage, polarity indication, and overrange indication. Seven-segment LED displays are used for the three least significant digits of the display, with a $\frac{1}{2}$-digit display used to indicate both the $\frac{1}{2}$-digit and the input polarity.

Output decoder. A single MC14543 seven-segment decoder is used to decode the multiplexed BCD digital output from the MC14435. By tying the phase control (pin 6) positive, the MC14543 will drive a common-anode LED display such as is used in this DVM design. The outputs of the MC14543 are buffered with MPS6517 *pnp* transistors. The digit-select outputs from the MC14435 are also buffered with *pnp* transistors. Each ON segment of the LED has an instantaneous current of about 30 mA, or an average current of 10 mA owed to the 30 percent duty cycle obtained when the LEDs are multiplexed.

The internal operation of the MC14435 causes the $\frac{1}{2}$-digit to be blanked off until a voltage of 1.000 is reached. An *npn* MPS6513 transistor is used to buffer the MC14435 output for the $\frac{1}{2}$-digit display. The plus and minus indicators are driven with *pnp* transistors from the CMOS autopolarity (discussed in later paragraphs).

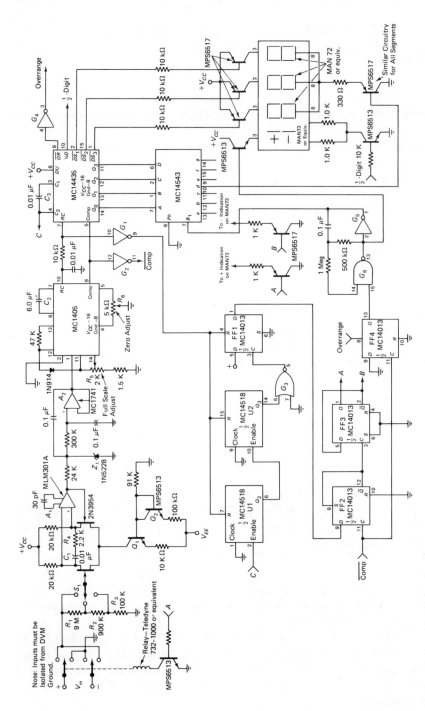

Fig. 9-47. 3½-digit DVM schematic

Overvoltage indication. Although there are many methods to indicate that the input voltage has exceeded the allowable maximum, the method used here uses the overrange information from the MC14435 to blink the entire display ON and OFF at a 2 Hz rate. This condition occurs when the MC1405 input exceeds 1.999 V. Flip-flop FF_4 is used to latch the overrange information from the MC14435 and control a low-frequency oscillator formed by gates G_4 and G_5. The blanking control of the MC14543 is used for blinking the three full-digits of the display while the $\frac{1}{2}$-digit and polarity indicators are blinked via the MPS6513 *npn* transistor in the anode of the $\frac{1}{2}$-digit display.

Input buffer circuits. The MC1405 has a low impedance input (typically 4 kΩ) which in many applications must be buffered to minimize the loading effect on the circuit being evaluated. An op-amp may be used as a buffer. However, a typical bipolar op-amp requires a sizeable bias current that must be supplied by the voltage source under test. This fact makes a bipolar op-amp generally undesirable. Op-amps with FET inputs are generally more expensive. An inexpensive, yet practical, compromise is to use a discrete FET input in conjunction with a bipolar op-amp.

The input impedance buffer shown in Fig. 9-47 consists of a 2N3954 dual FET, and an MLM301A bipolar op-amp. This amplifier arrangement has a high input impedance, and operation as a voltage follower, thus assuring an accurate closed-loop gain. Input impedance of the amplifier (10 MΩ) is set by the input voltage divider R_1, R_2 and R_3. The FETs are connected as a differential amplifier with the source leads common, and returned to a constant-current generator. This constant-current source is bipolar transistor Q_1 with temperature compensation by transistor Q_2, used to match the V_{BE} drifts of Q_1. The drain current is about 350 uA, and the drain voltage about 7 V with a ±15 V power supply.

The MLM301A op-amp is compensated by the 30 pF capacitor. Additional compensation for loop stability is done with the resistor/capacitor network R_4 and C_1. The temperature drift of this amplifier is typically well under 1 mV for a temperature from 0 °C to 50 °C.

Range select. Switch S_1 is used to select the maximum input voltage range of the DVM. These ranges are 1.999 V in 1 mV steps, 19.99 V in 10 mV steps, and 199.9 V in 100 mV steps. Resistors R_1, R_2 and R_3 should be matched to better than 0.05%, or each could be trimmed with a small series trimpot. Zener diode Z_1 which follows amplifier A_1 (the input buffer) is used to protect the MC1405 if a voltage greater than 5.0 V is accidentally applied to the input of the 2N3954.

Low-pass filter. Amplifier A_2 and associated components produce a Sallen and Key complex pole, low-pass filter that helps reduce 60 Hz signals that are picked up when resistance is placed between the signal source and the DVM ground. With the component values shown in Fig. 9-47, the filter has a corner frequency of 6 Hz, a damping factor of 0.6, and a gain of 1. With this design, the 60 Hz amplitude is reduced by 40 dB at the MC1405 input. Small gain and offset errors can be calibrated out with the existing full-scale and zero adjustments of the MC1405.

Autopolarity circuits. Autopolarity, or the ability to handle both positive and negative input voltages, is accomplished by switching the inputs with a mechanical relay to always provide a positive input to the MC1405. This technique is simple but has the disadvantage of requiring either the inputs to the relay or the inputs to the MC1405 be isolated from the DVM ground reference. However, if the MC1405 inputs were allowed to float, and ground reference was established at the input of the relay, reversal of the polarity switch would result in a common-mode level shift at the MC1405 input that will affect calibration of the DVM.

A *digital polarity detection scheme* is used in order to detect the polarity of the input voltage and to control the relay so that positive polarity is always maintained at the MC1405 input. The digital polarity detector has the advantage over analog techniques of not requiring any additional offset adjustments or temperature tracking. Polarity detection is accomplished by using the comparator output of the MC1405 in conjunction with the MC14435 circuits.

The MC1405 adds an offset current during the reference integration. In this case, a 100-count offset is produced that is subtracted out in the digital subsystem. If the input polarity is positive or zero, there are at least 100 clock pulses coming from the MC14435 clock. However, if the input polarity is negative, the unknown current is subtracted from the offset current, and less than 100 clock pulses are gated into the counters. This fact can be used to determine the input polarity.

The number of counts may be determined by the first two stages of the counter chain in the digital subsystem. However, since these are not available externally to the MC14435, MC14518 dual BCD counters are used. The MC14518 (U_1 and U_2) counter output is fed into a D flip-flop FF_1, which toggles the Q output high when 100 counts have come from the clock. This information is latched into FF_2, holding $\overline{Q}$ low, when the comparator output from the MC1405 goes low, indicating the end of a conversion cycle. This output does not toggle FF_3, and the relay does not change position. However, when the input polarity is reversed, FF_1 will not have been toggled high when the comparator goes low. This causes FF_2 $\overline{Q}$ to momentarily go high, toggling FF_3 and reversing the position of the relay. Thus, after completing a conversion cycle with a negative input voltage to the MC1405, the relay is toggled to apply a positive voltage to the MC1405. The positive voltage can then be measured in the normal manner. Diode D_1 on the MC1405 is used for recovery from negative differential input voltages. A 47 kΩ resistor between pins 12 and 13 is used to ensure that negative inputs do not latch the comparator low before the relay is reversed.

9.6.4. DVM calibration

Only two potentiometers on the MC1405 need be adjusted for calibration of the MC1405. First, short the input leads together, and adjust R_6 until the display reads zero. If R_6 is adjusted slightly negative, the display will toggle between plus and minus polarity.

Next, apply an accurate reference to the DVM input and adjust R_5 until the display reads the corresponding value. For best accuracy, the value of the reference should be as close to full scale as possible on the direct input, or in this case 1.999 V. If the voltage divider resistors R_1, R_2, and R_3 consist of trimpots, then each of the three input voltage ranges must also be adjusted.

9.7. INDUSTRIAL CLOCK/TIMER WITH BACK-UP POWER SUPPLY OPERATION

The section describes the design and operation of a clock/timer for industrial control applications. A block diagram of the clock/timer is shown in

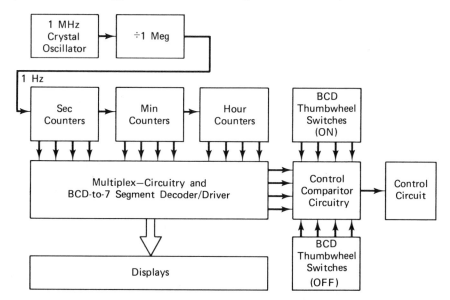

Fig. 9-48. Clock/timer block diagram

Fig. 9-48. The clock/timer reference is a crystal-controlled oscillator, with low-powered CMOS logic making up the clock reference divider chain and digital portions. A battery-powered back-up power supply is included to power the logic portion during power outages. To conserve battery power, digital displays are line-powered during normal operation, and are off during power failure. Digital comparators are used in the control section that is also designed to allow expansion for any size and type of device control.

9.7.1. Basic clock operation

The basic 24-hour clock circuit, shown in Fig. 9-49, uses Motorola McMOS dual BCD up-counters, transmission gates, and a Johnson counter to provide a multiplexed BCD output of seconds, minutes, and hours that drive the display decoder. The basic clock operates from a 1 Hz reference frequency. The first two MC14518 counters C_1 and C_2, and associated gates G_1 and G_2, are used to divide the 1 Hz signal to a frequency of one pulse-per-minute. The counters, which trigger on the negative-going transition of the enable E line, are connected to divide by 60, and thus 60 pulses on the input produce one pulse on the output.

An AND gate, made from a McMOS NAND and inverter ($\frac{1}{2}$ of MC14011), is used to decode the counter output corresponding to 60 input pulses, and resets counters C_1 and C_2. The second pair of counters C_3 and C_4, and gates G_3 and G_4, are connected similarly to produce one output pulse for 60 input pulses (for the minutes digit). Output Q_3 of counter C_4 provide a one pulse-per-hour signal. The next pair of counters C_5 and C_6, and associated gates G_5 and G_6, are arranged to divide the one pulse-per-hour frequency by 24, which will provide one output pulse per day.

A divide-by-12 logic section may be substituted for the 24-hour counter to provide a 12-hour readout as shown in Fig. 9-50. The McMOS MC14510 presettable BCD up/down counter C_1 and the MC17027 J-K flip-flop F_1 must be used to go directly from 12 hours to one hour. The three-input NAND gate G_1, and the following inverter G_2, are used to decode a number 13 at the outputs of counter C_1 and flip-flop F_1. At a count of 13, the J-K flip-flop is reset to zero and counter C_1 is preset to a number 1 so that the output goes directly from a readout of 12-to-1.

The multiplexed BCD outputs of the seconds, minutes, and hours counters can be decoded directly to drive the display as shown in Fig. 9-51. This method of multiplexing the digits uses five MC14016 transmission gates and only one MC14511 BCD-to-7 segment decoder/driver. The circuitry to perform this function is more complex, but less expensive, than individual decoder/drivers. Multiplexing of a digital display is a technique of time-sharing circuit components, and allows the eye to integrate the readout over the total time period to make a continuously appearing display. The

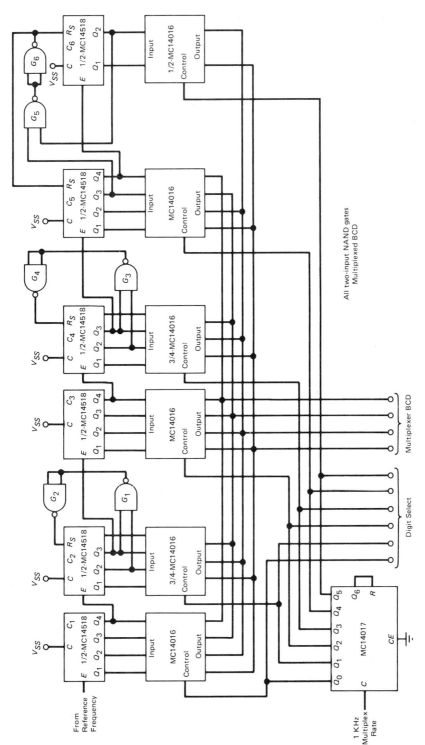

Fig. 9-49. Basic 24-hour clock circuit

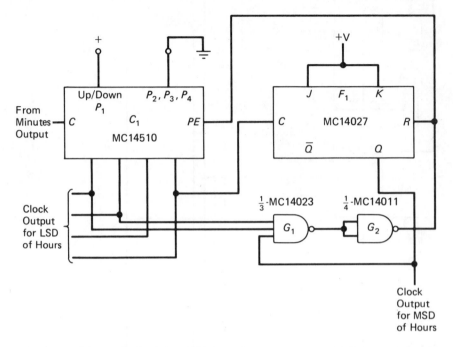

Fig. 9-50. 12-hour clock connection

multiplex rate must be fast enough so that the human eye cannot detect the individual enabling of each digit.

Digit selection is performed with the CMOS MC14017 Johnson counter, as shown in Fig. 9-49. Initially, the units seconds digits is enabled through the CMOS transmission gates, and is followed by the tens seconds digit, etc., until the tens hours digit is enabled. The cycle is then repeated. In this manner, a single BCD-to-7 segment decoder can be time-shared between all six counter digits. The Johnson counter selects the counter outputs to be applied to the decoder through the transmission gates, and also enables the correct display for this digit. Since the Johnson counter has 10 outputs, and only 6 are needed for the clock, the counter is reset after the ninth output goes to the logic 1 state (by connecting Q_6 to R). The 1 kHz multiplex rate ensures that the human eye does not detect a display flicker.

9.7.2.   Display and interface circuitry

The display is a planar gas-discharge seven-segment readout SP352, manufactured by Beckman Information Displays Division. The ionization potential for this type of orange-glow display is about 135 V, which requires a power supply of about 160 to 200 V. As shown in Fig. 9-51, the outputs of the BCD-to-7 segment decoder drive the MPS-A42 high-voltage transistors

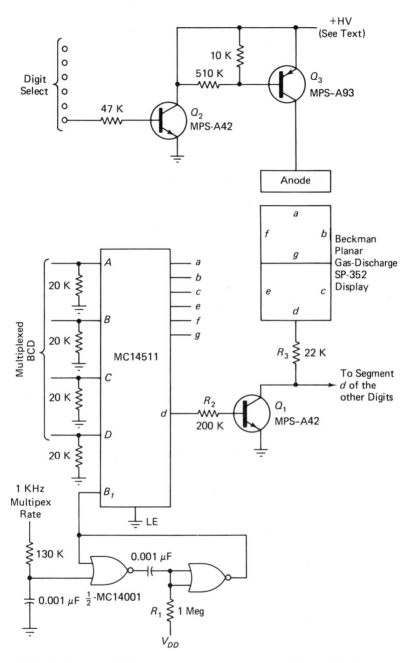

Fig. 9-51. Multiplexed BCD-to-seven segment decoder/driver/displays

to enable the proper cathode segments of the SP-352 display. Resistors R_2 and R_3, and transistor Q_1 are duplicated for each segment, with the segment of each digit connected in parallel. The anode-drive circuit is used to select the correct digit. In this case, the MPS-A42 transistor Q_2 is turned on, which activates the MPS-A93 transistor Q_3, and applies the high voltage to the anode of the display. The anode-driver circuitry is duplicated for each of the six digits of the clock.

When the transmission gates are disabled and the control lines are at a logic 0, the outputs are in a high-impedance state, and float to some undetermined voltage level. To prevent oscillation in the BCD-to-7 segment decoder, 20 kΩ pulldown resistors are necessary to maintain logic 0 states at the decoder inputs when the transmission gates are disabled.

The duty cycle of the cathode on-time of the display can also be varied, as shown in Fig. 9-51, to adjust the brightness of the digital readout. This feature is obtained by driving the blanking control pin B_1 of the BCD-to-7 segment decoder/driver (MC14511) from a monostable multivibrator made by using one-half of the MC14001 quad 2-input NOR gate. The multivibrator, or one-shot circuit, is triggered from the 1 kHz multiplex rate. The duty-cycle control from the one-shot may also be obtained with a photocell (used in place of resistor R_1) the resistance of which varies as a function of the light intensity. Thus, as the light level decreases, the brightness of the display also decreases, and prevents the readout light output from illuminating a dark room.

The Beckman SP-352 display has two 7-segment digits per package, with a character height of 0.55 inch. Thus, three packages are required for the 6-digit display. A multidigit readout of this type requires a delay time between the turn-on of the anode and cathode segments for each digit. This requirement prevents visible connecting paths, or "streamers," between digits in the same package. The delay time is set at 120 μs, with a cathode on-time of 550 μs. A 300 μs cathode blanking time remains before the next anode is activated.

Other displays such as LED, incandescent, or liquid crystal, can be substituted for the planar gas-discharge with minor changes in the drive circuitry and power requirements. For additional information on the Beckman SP-352 display or other types of digital readouts, the manufacturer of the desired display should be consulted.

9.7.3. Generation of reference and multiplexer frequencies

The 1 Hz reference frequency for the basic 24-hour clock is supplied from a 1 MHz crystal oscillator with a CMOS decade driver chain as shown in Fig. 9-52. The crystal oscillator includes gates G_1 and G_2 from the McMOS MC14001 NAND gate, with the crystal unit in the feedback circuit. The

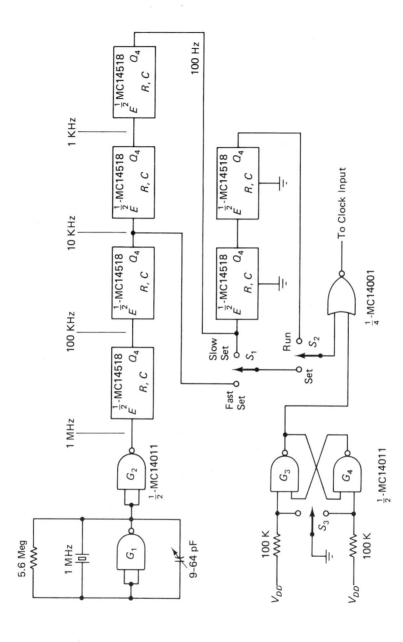

Fig. 9-52. Clock timebase with 1 MHz reference

457

MC14518 dual-decade counters are used for the divider chain and also generate the 1 kHz multiplex rate.

9.7.4. Control comparators and output control circuitry

In addition to providing a visual readout of the time-of-day, the basic clock can be combined with various digital comparator functions to produce an industrial clock/timer controller. This clock/timer can be used in a wide variety of applications, including the control of plant lighting, water sprinklers, bells, buzzers, etc.

The control comparators compare the time present on one set of input switches to the multiplexed BCD output of the basic clock. When the preset time and the clock time-of-day correspond, the output is set to turn a device on, or to ring a bell, for a preset period of time. A second set of comparators can be used to control the on/off operation of several different devices in a large control system.

The comparator circuits shown in Fig. 9-53 for the turn-off condition compare the BCD output of the thumbwheel switches to the multiplexed output of the basic clock. Even though the individual BCD outputs of the basic clock design are available, the use of the multiplexed BCD outputs was chosen to provide more versatility and eliminate extraneous wiring, especially if the comparators are not located on the same board with the clock. By using the multiplexed BCD outputs, each of the digits must be compared in sequence and the results stored. For the example shown in Fig. 9-53, only the hours and minutes digits are compared.

The digital comparison is performed with McMOS MC14519 4-bit AND/OR select ICs, which feature exclusive NOR operation when both control lines K_A and K_B are in the logic 1 state. When input $X_1 = Y_1$, $X_2 = Y_2$, etc., each output is at a logic 1. Correspondingly, for input $X_1 \neq Y_1$, etc., each output is in a logic 0 state. Also, all four outputs of the MC14519 are in the logic 0 state when both control lines are set at logic 0. The outputs of each comparator are ANDed with gates G_1 through G_4, the outputs of which are then ORed with a 4-input NAND gate. The output information of each comparator is clocked into the 4-bit shift register SR_1 ($\frac{1}{2}$ of MC14015) as the digits are multiplexed. When a comparison is made, all four outputs of the register are at a logic 1. This forces the output of the four-input AND gate to a logic 1 state. The output of this gate, derived from gate G_6 and inverter G_7, remains in the 1 state for one minute and is further used to trigger the S R flip-flop F_1. A second set of comparators are required to reset the flip-flop when the preset off-time is reached. In addition, manual setting and resetting F_1 is accomplished with switches S_1 and S_2, and gates G_8 through G_{11}. Comparators for the seconds digit can also be included if more accuracy is desired.

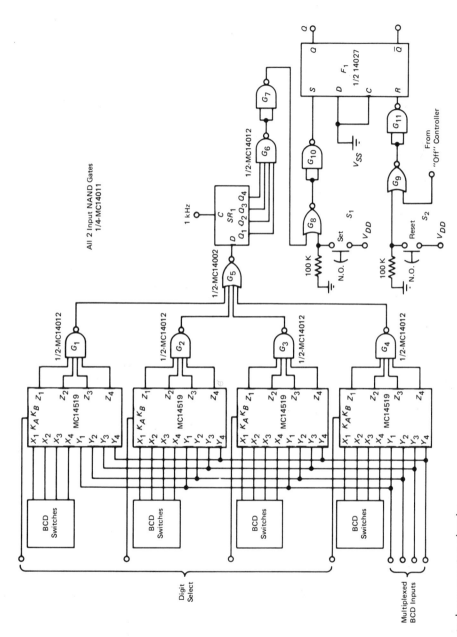

Fig. 9-53. Control comparator circuits

459

The Q output of F_1 can be used to control load power through a triac, as shown in Fig. 9-54. A 4N31 optoelectronic coupler isolates the 60 Hz line from the ground potential associated with the logic circuitry. When the Q output of the F_1 is at logic 1, the LED in the 4N31 is turned on and activates the phototransistor pair. This completes the gate-drive circuit through the MDA920-4 bridge, and triggers the 10 A MAC11-4 triac. The MDA920-4 also causes the triac trigger voltage to be positive for first-quadrant operation, and negative for third-quadrant operation. This ensures maximum sensitivity of triac control. In addition, the 10 kΩ resistor and 0.02 μF capacitor provide a voltage divider network from the power line so that the voltage rating of the phototransistor pair is not exceeded.

Output frequencies at 100 Hz and 10 kHz are used to set the correct clock time as follows. (See Fig. 9-52.) Initially, a 10 kHz signal is applied to the input of the basic clock, instead of the normal 1 kHz reference frequency. As the readout approaches the correct time of day, a 100 Hz signal is applied until the correct time is reached. The function is selected by either switch S_1 or S_2, and is gated on or off with switch S_3. Contact bounce elimination for switch S_3 is implemented with NAND gates G_3 and G_4. The clock can also be stopped to allow precise timing with the WWV radio station time/frequency standard as a reference.

If battery operation is not desired, the 1 kHz reference frequency can be obtained from the 60 Hz power line by dividing the 60 Hz signal by 60. While the short-term accuracy will not be as good as that of the 1 MHz crystal oscillator, long-term stability will be exceptional. In addition, the 1 kHz reference frequency will also have to be generated.

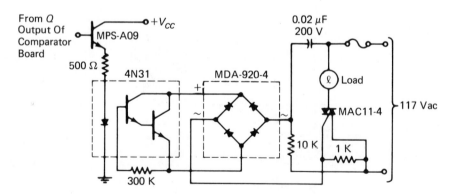

Fig. 9-54. Typical output control circuit for clock comparators

9.7.5. Power supply with back-up operation

The clock/timer controller requires two power supply voltages for normal line operation. A low-voltage supply is required to power the logic circuitry, and a high voltage (of between 160 and 200 V) is required for the gas-discharge display. The clock/timer also includes back-up battery operation so that proper time-of-day reference and comparison functions are maintained in the event of a sudden power outage. CMOS circuits are well suited for battery operation because of their inherent low power consumption and wide power supply operating range, specified as from about 3 to 16 V for industrial applications.

The entire clock/timer CMOS logic circuitry consumes only about 7 mW of power, whereas the major power dissipation of the clock/timer is in the digital readout. Thus, the display, which is normally powered from the 60 Hz line, is off during a power failure.

Since the low-voltage supply is to be battery-operated during power outage, a 5 V supply was chosen, since this is equivalent to four nickel-cadmium (Ni-Cd) batteries connected in series, as shown in Fig. 9-55. Under normal line-operated conditions, the battery is continually being recharged. Operating from a 6.3 V transformer winding, a half-wave rectified and

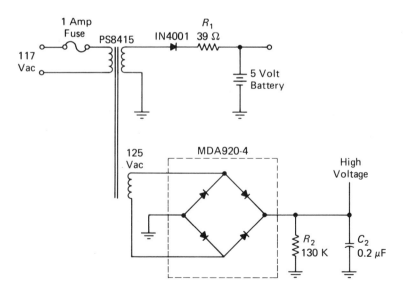

Fig. 9-55. Clock/timer power supply with battery back-up for logic

unfiltered supply provides pulsed energy to the battery. This power source is quite suitable for recharging the Ni-Cd battery, since it allows internal chemical reactions to settle before the next energy pulse (half-wave cycle) appears. The average input current to the battery is preset (with resistor R_1) at approximately $\frac{1}{20}$ of the mA/hour battery rating. Without any external power being applied, a 500 mA/hour battery is capable of providing nearly two weeks of continuous operation.

The high-voltage supply for the gas-discharge display is also illustrated in Fig. 9-55. The power source is obtained from a 125 V transformer winding, and uses a MDA920-4 bridge for full-wave rectification. The output waveform is filtered by C_1 and R_2.

9.8. BATTERY-POWERED FREQUENCY COUNTERS

This section describes design of two frequency counters. Both counters are battery operated. The first counter is for frequencies up to 5 MHz. The second counter operates up to 1 MHz, but at a reduced consumption of power, and lower sensitivity. Before going into design, let us review basic operation of electronic counters.

9.8.1. Basic frequency counter operation

Figure 9-56 is the block diagram of a basic frequency counter. The primary function of a frequency counter is to count the *number of events per unit of time*. An electronic frequency counter monitors the input events as transitions between two voltage levels. The actual counting operation takes place during a precise time period called the *gating* or *enable*. This time period is usually generated by a crystal oscillator and a series of decade dividers to produce the desired time period. Enable-time lengths may be changed by selecting the number of decade dividers. Input voltage transitions are counted by a series of decade counters. The input signal frequency is determined by the number of times the counters have been toggled during the enable time.

The counting sequence consists of three steps: *enable* or *gating* time, *strobing* the count into latches, and *resetting* of the counters. The waveforms of this sequence are shown in Fig. 9-56. The sequence is started with the enable pulse, which allows the counters to count. At the end of the enable pulse, the output of the counters is strobed or transferred into the latches and displayed on the digital readouts (R/O). After this step is completed, the counters are reset and the cycle is ready to repeat. The count is held by the latches while the counters are reset and the sequence is repeated.

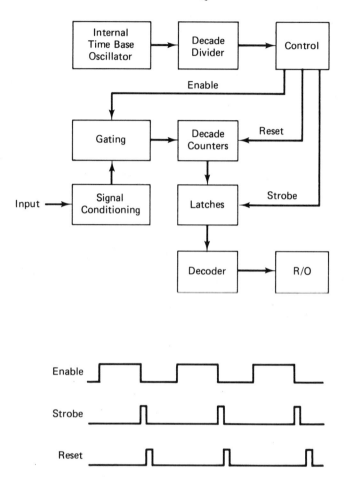

Fig. 9-56. Frequency counter block diagram and waveforms

9.8.2. 5 MHz frequency counter

Figure 9-57 is the schematic of the basic 5 MHz counter. The circuit is complete, except for a front end or signal conditioner (discussed in Sec. 9.8.3) and the displays (discussed in Sec 9.8.4). Note that Motorola McMOS is used through the circuit of Fig. 9-57.

Crystal oscillator and time base.   The time-base circuitry determines the length of the enable time. The reference frequency is obtained from a 1 MHz crystal oscillator using a CMOS gate G_1, with a crystal as a feedback element. Gate G_2 is used as a buffer for the 1 MHz oscillator. Provision for an external reference frequency is provided by switch S_1. The external

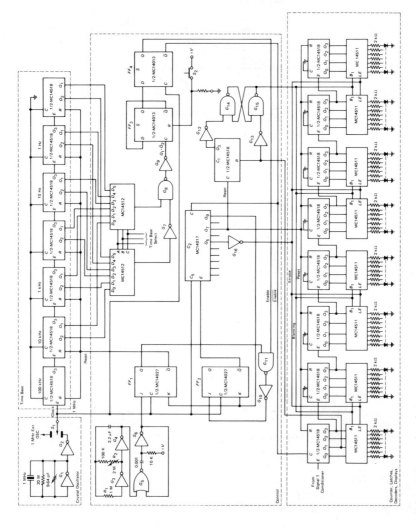

Fig. 9-57. Basic 5 MHz counter circuit

reference must have a 0–10 V swing to drive the time-base counters. This may be used when a more stable reference than the internal oscillator is required.

The time base divides the 1 MHz oscillator frequency down to the desired enable time by using MC14518 dual-decade counters. Three and one-half dual counters (seven decades) are used to provide a maximum enable period of 10 seconds. The counters are connected in a conventional ripple-through configuration. This type of counter operation has the advantage of simplicity, but the disadvantage of a time delay from the input of the first counter to the output of the last counter. The time delay is compensated by the counter control section as described in following paragraphs.

Counters. The MC14518 dual BCD counter is used for the actual counting of the input signal. Although the CMOS logic family is capable of operation over a wide voltage range, the highest frequency of operation is obtained at higher supply voltages. Thus, a 12 V operating voltage was chosen, which is also a convenient battery voltage for portable operation. The series of eight BCD counters (4 dual counters) are connected in a ripple mode.

Latches and decoders. MC14511 ICs are used for the latches and BCD-to-7 segment decoders. These CMOS ICs have both latches with a strobe control, and a decoder for the 7-segment display, with a blanking control pin for turning the displays off. This blanking feature can be used to conserve the power drain by turning the displays off when there is no input signal. This is accomplished with one-half dual BCD counter C_1, and a flip-flop made from 2-input NAND gates G_{12} through G_{15}. If four reset pulses are received on the BCD counter with no input signal, the displays are blanked out, thus conserving power. If there is an input signal, C_1 is constantly reset by Q_1 of the first MC14518 counter, thus keeping the displays in the on state.

Displays. The displays shown in Fig. 9-57 are direct driven. The displays can be multiplexed at a slightly lower system cost and lower power dissipation. However, package count is increased, along with increased circuit complexity. Displays are discussed further in Sec. 9.8.4.

Control section. The function of the control section is to process the precise timing signals from the time base to provide the proper logic sequences to the counters and latches. Figure 9-58 is the timing diagram for the basic counter. The control section has four outputs:

An *enable line* which turns the first counter on and off for the precise enable time period;

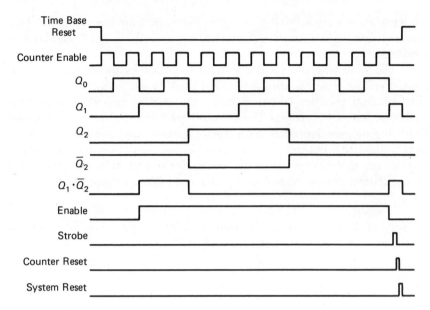

Fig. 9-58. Timing diagram for basic counter

The *strobe line* which transfers the count into the memory of the MC14511 latch decoder;

The *reset line* used for resetting the MC14518 decade counters to ready them for the next count cycle;

The *display blanking* output used to blank the display when four reset pulses are received and there is no input signal.

Counting cycle. If the counting cycle were run at the minimum enable time, the count would be updated into the latches at this cycle time. In the case of a 10 μs time base, the count would be updated at a 100 kHz rate. If the input frequency has any instability, the last few digits of the display would change at the 100 kHz rate, and would result in an unreadable flicker. This is eliminated by operating the count cycle at a much slower rate; in this case, 0.1 s to 10 s, depending on the setting of the potentiometer R_2.

Counting cycle sequence. Gates G_3 and G_4 form a low-frequency oscillator. Gates G_5 and G_6 make up a one-shot pulse generator. The sequence starts with a 5 μs pulse from the one-shot. This pulse toggles $\bar{Q}$ of FF_1 low, which removes the reset from the time-base counters, allowing the time-base counters to divide down the 1 MHz clock. The enable inputs of the selected time-base counter, and three of its outputs Q_0 through Q_2 are shown in Fig. 9-58. By ANDing Q_1 and $\bar{Q}_2$ together, two pulses with starting edges equal to the desired time base are formed. The gate delays through the time base are equal for both pulses, thus compensating for these delays. The pulses are applied to the input of FF_3.

Time-base selection. FF_3 is wired as a toggle flip-flop of which the first pulse sets the Q output high, and the second pulse resets the Q output low. The three NAND gates G_7, G_8, and G_9 are used to decode the $Q_1 \cdot \bar{Q}_2$ from the desired time-base counter. This counter can be selected electrically by the MC14512, or a manual 2-pole, 6-position switch. The MC14512 ICs are 8-channel data selectors used to select one of eight channels of information in response to electrical data from the 3-bit binary control lines (shown as time base select) on Fig. 9-57.

Enable pulse synchronization. FF_4 is used to ensure that the edges of the enable pulse occur exactly with the 1 MHz clock pulses. Since the delays through the time base and data selector are constant for both turn-on and turn-off pulses, these delays are cancelled and the waveform appearing at the $\bar{Q}$ output of FF_4 is a precisely controlled time period. This time period is the enable or gating time.

Strobe, counter reset, and control system reset. These functions are derived from the MC14017 Johnson counter C_2, FF_2, and the associated gates. Until the end of the enable pulse, the Johnson counter is disabled through either the clock line or the reset line. At the end of the enable pulse, the Johnson counter is enabled, and begins to count the 1 MHz clock frequency. Of the 10 outputs of the counter, Q_5 goes high for 1 μs after the sixth clock pulse. This is used for the *strobe pulse*. After the eighth clock pulse, Q_7 goes high, and the *counters are reset*. The tenth clock (Q_9) pulse *resets the control system* through FF_1 and FF_2. The control system then awaits another pulse from the low-frequency oscillator and one-shot generator.

9.8.3. Counter front end or signal conditioner

The input signal for the basic frequency counter must swing from a logic 0 (of zero volts) to a logic 1 (of about 10 V). A front-end or signal-conditioner circuit is needed to translate from any input waveform shape and level to a signal that is capable of toggling the CMOS counters. The front end should also have a high impedance so that the input waveform is left undisturbed by the frequency measurement. Also, the input should be protected from very large signals.

The front-end circuit of Fig. 9-59 includes all of these features. As shown, the circuit consists of a FET-bipolar buffer followed by the Schmitt trigger made from an MC75108 dual-line receiver. The circuit of Fig. 9-59 will operate up to 5 MHz with a 10 mV input signal.

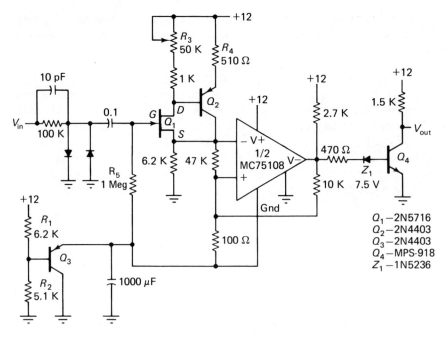

Fig. 9-59. High-frequency front end

Input protection. The circuit input is protected from large signals by the two diodes on the input line. Both positive and negative signal swings in excess of about 0.5-0.7 V are clipped by these diodes.

High-impedance buffer. The n-channel JFET Q_1 and *pnp* bipolar Q_2 provide a unity-gain, high-impedance buffer for the MC75108 line receiver. Potentiometer R_3 adjusts the buffer output voltage to be equal to the 6 V reference voltage developed by Q_3.

Line receiver. The MC75108 line receiver is designed to operate from a ±6 V power supply but can be biased up to operate from a single +12 V supply. In addition to both a positive and negative power supply, the line receiver requires a ground connection at the midpoint of the supply voltages. The MC75108 datasheet indicates a typical positive supply current of 18 mA, and a negative supply current of −8.4 mA. Thus, the ground pin returns about 10 mA to the positive supply. For single-supply operation, the positive supply pin is connected to the +12 V supply, with the negative supply pin connected to the counter circuit ground. The ground pin on the MC75108 must then be returned to a 6 V reference capable of sinking up to a maximum of 15 mA. The *pnp* transistor Q_3 and resistors R_1–R_2 operate as a simple voltage regulator for the 6 V reference.

The MC75108 is a dual-line receiver. The power consumption of the package is constant whether both line receivers are used or not. Both line receivers in the package can be used to provide two separate input channels with a switch to select between them, or they could be connected to the basic counter such that they are switched on and off, alternating with each count cycle. Thus, the first count cycle would look at channel A, with the second count cycle looking at channel B.

Voltage translation. Under normal split-supply operation, the output swing of the MC75108 is from zero to the positive supply. Since the MC75108 is biased up by 6 V in this case, the output swing is from 6 to 12 V. This voltage swing must be translated to voltage levels compatible with CMOS operating at $+12$ V. This translation is done with the 7.5 V Zener Z_1 and transistor Q_4. When the output of the MC75108 is at 6 V, Z_1 is off and Q_4 is turned off. When the output of the MC75108 is in the high state, Z_1 is conducting and Q_4 turns on. The output voltage swings from cutoff to saturation of Q_4, or from about 0.5 to 12 V.

9.8.4. Optimum power considerations

Although the counter of Fig. 9-57 is capable of low-power operation, it has been optimized for maximum performance. This optimization (for maximum frequency and sensitivity) increases the power dissipation. The circuit can be modified to decrease power consumption, but at the expense of input signal sensitivity, operating frequency, and component count. Before going into such modifications, let us consider where the power is being consumed. The following shows a comparison of the maximum performance Fig. 9-57 circuit and a modified minimum power counter.

	MAXIMUM PERFORMANCE	MIMIMUM POWER
Maximum frequency	5 MHz	1 MHz
Sensitivity at maximum frequency	10 mV	120 mV
Basic counter power	10 mA	2 mA
Front end power (no signal input)	25 mA	6 μA
Front end power (at 1 MHz)	25 mA	180 μA
Display power	225 mA	70 mA
Power supply	$+12$ V	$+6$ V
Total current	260 mA	72 mA
Total power	3.12 W	0.43 W

The largest portion of this power is being consumed by the digital displays. (Monsanto MAN-4 LEDs are used because of their wide adaptability and compatibility to the MC14511 7-segment decoder drivers.) In the Fig. 9-57 counter, the MAN-4 displays are operated at 4 mA per segment. While this current produces a bright and easily read display, a great deal of power is consumed in the displays. The same displays can be used, but with reduced power consumption, if the displays are multiplexed, as described in the following paragraphs.

9.8.5. 1 MHZ frequency counter with multiplexed displays

Figure 9-60 shows a technique that multiplexes a single decoder driver between all of the displays, and operates the displays at a lower level. Multiplexing of a digital display is a technique of time-sharing the circuit components and allowing the eye to integrate the readout over the total time period to make a display appear continuous. The multiplexing technique reduces the overall cost of the counter, while increasing the complexity of counter design. With the multiplex technique, the displays are operated at a peak current of 20 mA, but at a duty cycle of only 12.5 percent, which is a smaller power consumption.

Four MC14021 8-bit shift registers are used in the multiplexing technique. These shift registers have both parallel and serial inputs, and permit not only a unique technique of multiplexing, but also provide the latches that are required to store the count from the MC14518 counter chain.

The BCD outputs of the MC14518 counter chain are connected to the parallel inputs of the shift register. The least significant BCD bit of each of the eight counters is connected to each of the eight inputs of the first shift register. The next most significant bits of the eight counters are connected to the eight inputs of the next shift register. Thus, with eight 4-bit BCD words, four 8-bit shift registers are required. The eighth output of each shift register is connected to the series input of the same register so that the digital word in the shift register is shifted around in a loop from the first bit through the last bit.

When a logic 1 is applied to the parallel/series control line of the shift registers, the digital information on the parallel inputs is loaded in the register. This control signal is the strobe pulse from the counter-control section (Fig. 9-57). Then, after the register returns to series operation, the digital word is shifted around through the registers with each of the 8-bits of information appearing at the inputs to the single 7-segment decoder.

An MC14022 octal counter/divider is used to keep track of which counter digit is presently being applied to the 7-segment decoder input. The clock signal for both digit selection and the shift register is derived from an

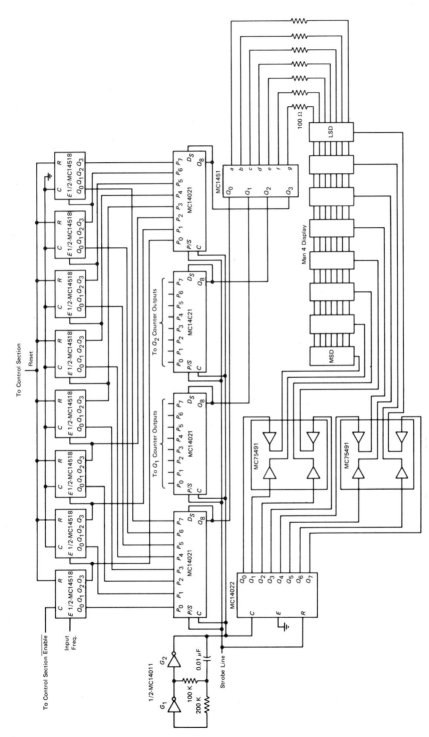

Fig. 9-60. 1 MHz frequency counter with multiplexed displays

471

oscillator constructed from gates G_1 and G_2 of Fig. 9-60. This oscillator controls the multiplex frequency. The values shown produce a frequency of about 1 kHz. The outputs of the MC1402? digit selector are buffered through the MC76491 LED drivers to the MAN-4 display cathodes. The anodes of the MAN-4s are driven directly from the MC14511 BCD-to-7 segment decoder. This technique represents an overall savings in both power and system costs, with an increase in system complexity and a decrease in LED brightness.

9.8.6. Low-power, low-frequency front end

Another area of possible power reduction is in the signal conditioning or front end. The front end of Fig. 9-59 is optimized for maximum performance, and not only requires about 25 mA of current from the 12 V supply, but this current is constant whether an input signal is present or not.

The circuit of Fig. 9-61 is a low-power front end for the counter. This design not only reduces the power consumption, but the power consumption is proportional to the input frequency. With no signal input to the Fig. 9-61 front end, current drain is only a few μA. The Fig. 9-61 front end uses one-half of an MC14583 CMOS Scmitt trigger, which operates from the single 6 V supply used in the Fig. 9-60 counter. A single resistor R_1 is used to set the threshold voltages from both the high and low logic states. The over-voltage input protection network $(D_1\text{-}D_2)$ used with the Fig. 9-57 circuit is also used with the Fig. 9-61 circuit.

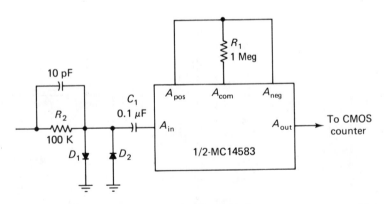

Fig. 9-61. Low-power, low-frequency front end

Sensitivity versus frequency. Both the input sensitivity and maximum frequency limit of the Fig. 9-60 circuit is reduced from that of the Fig. 9-57 circuit. Figures 9-62 and 9-63 show sensitivity versus frequency for the Fig. 9-57 and 9-60 counter circuits, respectively.

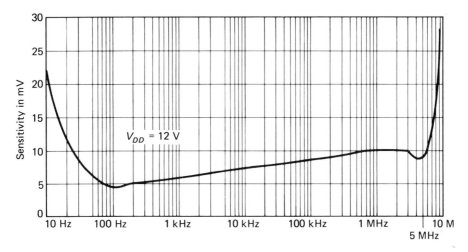

Fig. 9-62. Sensitivity versus frequency for 5 MHz counter circuit

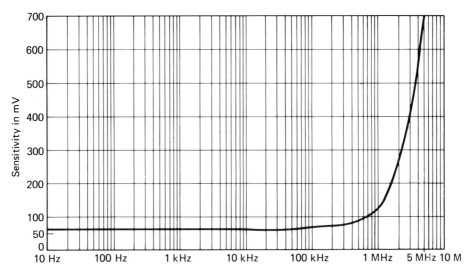

Fig. 9-63. Sensitivity versus frequency for 1 MHz counter circuit with multiplexed displays

9.8.7.   Battery power supplies

There are many types and sizes of batteries available for operation of these counters. The battery may be a "throw-away" non-rechargeable unit, or one of the many types of rechargeable units. In the case of the rechargeable battery, the method of battery charging depends upon the type of battery and may be a simple transformer-rectifier-current-limit-resistor, or a very complex unit with discharge indicators, timing circuitry or automatic recharge circuitry. The manufacturers of the desired battery should be consulted as to the method of charging compatible to that type of battery. Recommended batteries include Globe Battery GC1245 (12 V at 4.5 ampere-hours) and the GC626 (6 V at 2.6 ampere-hours.).

INDEX

AM detector, 297
AM/FM radio amplifiers, IC, 228–49
Amplifier
  audio, IC, 222
  bias current, OTA, 107
  power, IC, 223–28
  pulse, IC, 221
  video, 304
  wideband, IC, 212
  with AGC, 213
Amplitude modulator, 295
Arrays, IC, 189, 299
Audio amplifier, IC, 222
Audio circuits, op-amp, 121

Balanced modulator, 292
Bandwidth, op-amp, 73, 93
Battery-powered frequency counter, 462

Bias current, op-amp, 99
Bias, ECL, 331
BiMOS op-amp, 68
Bipolar construction, 3
Bipolar-COS/MOS interface, 372

Capacitor fabrication, IC, 6
Chip, IC, 8
Clock/timer, industrial, 451
Closed-loop, op-amp, 73
CMOS, 7, 11
CMOS logic, 333
Common-mode rejection (CMR), 62, 63, 99
  definitions, 65
Common-mode voltage swing, op-amp, 98
Communications ICs, 292
Comparator, IC, 189

Comparator, OTA, 123, 149
Compensation, op-amp, 73–91
Complementary inverter, 10
Complementary MOS logic, 333
Component arrays, 19, 189, 299
Connections ICs, 23
Constant-current source, op-amp, 64
Cooling, forced-air, 49
Core memory sense amplifier, 200
COS/MOS, 339
  bipolar interface, 372
  ECL interface, 376
  HTL interface, 375
  MOS interface, 378
  TTL/DTL interface, 363
Counter, frequency, 462
Current boosting, voltage regulator, 270
Current foldback, voltage regulator, 276
Current mirrors, 109
Current regulators, 271
Current requirements for ICs, 54

Data selector, IC, 211
Data synchronization, 315
Datasheets, logic elements, 354
DC motor control, 389
Decoder-demultiplexer, OTA, 139
Delay, logic element, 353
Demodulation, FM, 313
Design with ICs, 389
Desoldering tool tips, 35
Detector, AM, 297
Detector, FM and phase, 299
Detector, product, 296
Differential amplifier, IC, 202
Differential input, 57
Differential stage, 62, 63
Digital basics, 319
Digital control of power supplies, 403

Digital ICs, 19
Digital ICs, layout, 39
Digital recording with ICs, 415
Digital voltmeter, 442
Diode arrays, 301
Diode fabrication, 4
DIP (dual-in-line packaging), 2
DIP mounting, 29
Dissipation, logic element, 353
Doubler, 298
Drive, logic element, 355
DTL (diode-transistor logic), 320
Dual-ramp DVM, 442
DVM (digital voltmeter), 442

ECL (emitter-coupled logic), 328
  ECL-COS/MOS interface, 376
  interface, 384
  level translater, 381
Electrical characteristics, logic element, 355
Extended bandwidth compensation, 85

Fan-out, logic element, 354
Feedback, op-amp, 73, 74
FET (field-effect transistor), 14
Filters, op-amp, 163–71
Flat pack, 2
Flat pack mounting, 23–26
Floating inputs, op-amp, 66
FM amplifiers, IC, 228–49
FM demodulation, 313
FM detector, 299
Four-quadrant multipliers, IC, 249–66
Frequency counter, 462
Frequency response, op-amp, 73
Frequency synthesizer, 314
Frequency/gain characteristics, op-amp, 76
FSK (frequency shift keyer), 210

Gain control, OTA, 131
Gain, op-amp, 73
Gain/frequency characteristics, op-amp, 75
Gate, high-speed, 302
Gate, temperature-compensated, 212
Gate, TTL, 324
Ground currents, op-amp, 66
Ground loops, op-amp, 67
Gyrator, OTA, 130

Heatsinks, 46–49
    temporary, 33
High-frequency amplifier, IC, 205
HTL (high threshold logic), 321
    COS/MOS interface, 375
    interface, 382
    level translator, 381

IF amplifiers, IC, 228–49
Impedance, op-amp, 97
Induction motor control, 397
Industrial clock/timer, 451
Input bias, op-amp, 99
Input impedance modification, 79
Input offset, op-amp, 100–05
Interfacing logic elements, 356
Interfacing problems, summary, 388
Internal construction, IC, 3
Introduction, to ICs, 1
Inverting feedback, op-amp, 74
Inverting input, op-amp, 63

Layout of ICs, 39
Lead bending, 30
Level detector, IC, 191
Level shifter/translator, 373, 381
Limit detector, IC, 194
Line drivers, TTL, 325
Linear applications, Op-amp, 151

Linear component arrays, 19
Linear ICs, 17, 189
Linear ICs layout, 41
Load characteristics, logic element, 355
Lock-detection, PLL, 317
Logic forms, 319
Logic, TTL, 325

Maximum ratings, logic element, 355
McMOS, 334
MECL, 329
Metal can mounting, 27–29
Micropower op-amp, 70
Miller-effect, 81–91
Mixer, 297
Modulation, OTA, 131
Modulator
    amplitude, 295
    amplitude, IC, 213
    balanced, 292
    balanced, IC, 213
    pulse width, 219
MOS (metal oxide semiconductor)
    construction, 7
    COS/MOS interface, 378
    MOS interface, 358
    inverter, 12
    logic, 333
    protection, 38
    RTL interface, 378
MOS/LSI, 345
    interface, 358
MOSFET, 8
    versus two-junction size, 15
Multiplexer, OTA, 121
Multipliers, IC, 249–66
Multipliers, OTA, 133–139
Multivibrator, 195
    gated, IC, 217
    OTA, 147

Motor (DC) control, 389
Motor (induction) control, 397
Mounting ICs, 23
MTTL, 324
NMOS, 8

Noise voltage, op-amp, 105
Noise, logic, 352
Noninverting feedback, op-amp, 74
Noninverting input, op-amp, 63
Nonlinear applications, op-amp, 175

Offset
    null voltage, op-amp, 69
    offset, op-amp, 100–05
    problems, 57
Open-loop voltage gain (A_{VOL}, A_{OL}), 91
Op-amp, 18
    applications, 151, 175
    audio circuits, 171
    bandwidth, 73, 93
    basic, 61
    bias (input), 99
    characteristics, 91
    circuits, 62
    common-mode rejection, 99
    common-mode voltage swing, 98
    compensation, 73–90
    filters, 163–71
    frequency response, 73
    gain, 73
    micropower, 70
    noise voltage, 105
    offset null, 69
    offset problems, 100
    oscillators, 178
    output voltage, 96
    peak detector, 175
    phase shift, 93

Op-amp (continued)
    power dissipation, 106
    power supply sensitivity, 105
    ramp generator, 184
    setting time, 96
    slew rate, 96
    source follower, 158
    staircase generator, 184
    system design, 152
    temperature sensor, 182
    unity gain, 160
    universal, 68
    variable bias, 70
    voltage follower, 158
    voltage gain, 91
    zero offset suppression, 156
OTA (operational transconductance amplifier), 18, 107
    amplifier bias current, 107
    applications, 120
    characteristics, 113
    circuits, 108
    comparator, 123
    current output, 145
    decoder-demultiplexer, 139
    design, 114
    gain control, 131
    gyrator, 130
    micropower comparator, 149
    modulation, 131
    multiplexer, 121
    multiplier, 133–39
    multivibrator, 147
    output tilt, 127
    sample and hold circuits, 124
    slew rate, 128
    system, 115
    terms, 112
    threshold detector, 149
    transconductance, 107
Oscillator
    gated, IC, 218
    op-amp, 178

Output tilt, OTA, 127
Output, op-amp, 93
Overload protection, voltage regulator, 269

Package selection, 22
Packaging, IC, 1
PC (printed circuit) boards, 30
    replacement, 34
Peak detector, IC, 199
Peak detector, op-amp, 175
Phase comparator, 306
Phase compensation, op-amp, 73
Phase compensation, problems of, 78
Phase detector, 299
Phase-lag compensation, 81–91
Phase-lead compensation, 79–91
Phase shift, 93
    problems, 77
Planar construction, 3
PLL (Phase-locked loop), 305
PMOS, 8
Power amplifier, IC, 223–28
Power control with zero crossing switch, 430
Power dissipation, 43, 44
    op-amp, 106
    with heat sinks, 49
Power gates, TTL, 325
Power ICs, 44
Power source, logic element, 353
Power supply for ICs, 16, 51
Power supply sensitivity, op-amp, 105
Power supply with digital control, 403
Practical considerations for ICs, 21
Product detector, 296
Propagation delay, logic element, 353
Pull-down, pull-up resistors, 332
Pulse amplifier, IC, 221

Pulse generator, IC, 198
Pulse width modulation (PWM) for motor control, 389

Ramp generator, op-amp, 184
Recording digital information with ICs, 415
Regulator, shunt, 278
Regulator, switching, 282–92
Regulator, voltage, IC, 266
Resistor fabrication, IC, 5
Rolloff (early) compensation, 84
RTL (resistor-transistor logic), 320
    interface, 385
    MOS interface, 378

Sample and hold circuits, OTA, 124
SCR control with zero crossing switch, 437
Selecting logic ICs, 351
Sense amplifier, core memory, 200
Sensor, temperature, 182
Settling time, op-amp, 96
Shift register, MOS/LSI, 346
Short-circuit protection, voltage regulator, 269
Shunt regulator, 278
Shutdown, voltage regulator, 272
Single power supply for ICs, 56–60
Slewrate, op-amp, 93
Slewrate, OTA, 128
Socket mounting, IC, 23
Solder gobbler, 34
Soldering techniques, 36
Soldering tools, 32
Source follower, op-amp, 158
Speed control for motors, 397
Speed, logic element, 353
Split zener power supply, 56
Staircase generator, op-amp, 184
Strip soldering, 37

Supply voltages for ICs, 16
Switch, zero crossing, 430
Switching regulator, 282–292
Synchronization, data, 315
Synthesizer, frequency, 314

Temperature considerations, IC, 16
Temperature controller, 440
Temperature extremes, effects of, 50
Temperature sensor, op-amp, 182
Thermal resistance, 43–45
Thermal runaway, 45
Threshold detector, OTA, 149
TO-5 mounting, 27–29
TO-5 package, 2
Totem-pole output, 327
Touch-and-wipe desoldering, 36
Transconductance, OTA, 107
Transistor arrays, 303
Transistor fabrication, 4
Transistor package, 2
Transmission gate, 336
TTL (transistor-transistor logic), 322
TTL/DTL-COS/MOS interface, 363
Two-junction construction, 3

Unity gain op-amp, 160

Vacuum desolderers, 34
Variable bias op-amp, 70
Variable op-amp, 107
VCO (voltage-controlled oscillator), 309
Video amplifier, 205, 304
Video switch, 209
Voltage follower, op-amp, 158
Voltage regulators, IC, 266
Voltage regulators, specifications, 277
Voltages for ICs, 16

Wire-holding tools, 37
Working with ICs, 29

Zener (split) power supply, 56
Zero crossing pulse generator, IC, 198
Zero crossing switch for power control, 430
Zero offset, op-amp, 156